W9-BSA-963

The Selectins

Advances in Vascular Biology
A series of books bringing together important advances and reviewing all areas of vascular biology.
Edited by *Mathew A. Vadas, The Hanson Centre for Cancer Research, Adelaide, South Australia* and *John Harlan, Division of Hematology, University of Washington, Seattle, USA*.

Volume One
Vascular Control of Hemostasis
edited by *Victor W.M. van Hinsbergh*

Volume Two
Immune Functions of the Vessel Wall
edited by *Göran K. Hansson* and *Peter Libby*

Volume Three
The Selectins: Initiators of Leukocyte Endothelial Adhesion
edited by *D. Vestweber*

Volumes in Preparation

Platelets, Thrombosis and the Vessel Wall
edited by *M.C. Berndt*

Structure and Function of Endothelial Cell to Cell Junctions
edited by *E. Dejana*

The Role of Vascular Viral Infections in Atherosclerosis
edited by *S.M. Schwartz* and *D.P. Hajjar*

This book is part of a series. The publisher will accept continuation orders which may be cancelled at any time and which provide for automatic billing and shipping of each title in the series upon publication. Please write for details.

The Selectins
Initiators of Leukocyte Endothelial Adhesion

edited by

Dietmar Vestweber

Zentrum für Molekularbiologie der Entzündung
Münster
Germany

hoap

harwood academic publishers

Australia • Canada • China • France • Germany • India • Japan
Luxembourg • Malaysia • The Netherlands • Russia • Singapore
Switzerland • Thailand • United Kingdom

 Printed in Singapore.

Amsteldijk 166
1st Floor
1079 LH Amsterdam
The Netherlands

British Library Cataloguing in Publication Data

The selectins : initiators of leukocyte endothelial
adhesion. — (Advances in vascular biology ; 3)
1. Cardiovascular system
I. Vestweber, Dietmar
612.1′3

ISBN 9057020742

CONTENTS

SERIES PREFACE

It is our privilege to live at a time when scientific discoveries are providing insights into human biology at an unprecendented rate. It is also a time when the sheer quantity of information tends to obscure underlying principles, and when hypotheses or insights that simplify and unify may be relegated to the shadow of hard data.

The driving force for editing a series of books on Vascular Biology was to partially redress this balance. In inviting editors of excellence and experience, it is our aim to draw together important facts, in particular areas of vascular biology, and to allow the generation of hypotheses and principles that unite an area and define newer horizons. We also anticipate that, as is often the case in biology, the formulation and application of these principles will interrelate with other disciplines.

Vascular biology is a frontier that has been recognised since at least the time of Cohnheim and Metchikoff, but has really come into prominence over the last 10–15 years, once the molecules that mediate the essential functions of the blood vessel started to be defined. The boundaries of this discipline are, however, not clear. There are intersections, for example, with hypertension and atherogenesis that bring in, respectively, neuroendocrine control of vessel tone and lipid biochemistry which exist as separate bodies of knowledge. Moreover, it would be surprising if some regional vascular biology (for example, pulmonary, renal, etc.) were not to emerge as subgroups in the future. Our aims for the moment, however, are to concentrate on areas of vascular biology that have a wide impact.

It is our hope to publish two books each year for the next 3–4 years. Indeed the first five books have been commissioned and address areas primarily in endothelial biology (hemostasis and thrombosis), immunology, leukocyte adhesion molecules, platelet adhesion molecules, adhesion molecules that mediate cell–cell contact. Subsequent volumes will cover the physiology and pathology of other vascular cells as well as developmental vascular biology.

We thank the editors and contributors for their very hard work.

Mathew VADAS John HARLAN

LIST OF CONTRIBUTORS

Beaudet, Arthur L.
Dept. of Molecular & Human Genetics
Baylor College of Medicine
One Baylor Plaza
Houston, TX 77030-3491
USA

Bullard, Daniel C.
Dept. of Molecular & Human Genetics
Baylor College of Medicine
One Baylor Plaza
Houston, TX 77030-3491
USA

Etzioni, Amos
Faculty of Medicine, Technion
Rambam Medical Center, Haifa
Dept. of Pediatrics A,
B. Rappaport Medical School
Israel

Furie, Barbara C.
Tufts University, New England Medical Center
Center for Hemostasis & Thrombosis Research
750 Washington Street, NEMC #832
Boston, MA 02111
USA

Furie, Bruce
Tufts University, New England Medical Center
Center for Hemostasis & Thrombosis Research
750 Washington Street, NEMC #832
Boston, MA 02111
USA

Graves, Bradford J.
Dept. of Inflammation/Autoimmune Diseases
Hoffmann-La Rocke Inc.
340 Kingsland Street
Nutley, New Jersey 07110
USA

Hamann, Alf
Universitàts Kvanken Haus, Eppendorf
Uni. Hamburg, Med. Klinik
Abt. Immunologie
Martinistr. 52, D-20246 Hamburg
Germany

Harlan, John M.
Faculty of Medicine, Technion
Rambam Medical Center, Haifa
Dept. of Pediatrics A,
B. Rappaport Medical School
Israel

Huang, Kuo-Sen
Dept. of Inflammation/Autoimmune Diseases
Hoffmann-La Rocke Inc.
340 Kingsland Street
Nutley, New Jersey 07110
USA

Jonas, Petra
Universitàts Kvanken Haus, Eppendorf
Uni. Hamburg, Med. Klinik
Abt. Immunologie
Martinistr. 52, D-20246 Hamburg
Germany

Kei Kishimoto, Takashi
Dept. of Immunology
Boehringer Ingelheim Pharmaceuticals Inc.
900 Ridgebury Road, Box 369
Ridgefield, CT 06877
USA

Ley, Klaus
University of Virginia School of Medicine
Dept. of Biomedical Engineering
Health Sciences Center, Box 377
Charlottesville, VA 22908
USA

Lowe, John B.
Dept. of Pathology
The Howard Hughes Medical Institute
University of Michigan Medical School
Ann Arbor, 7148109
USA

McEver, Rodger P.
W.K. Warren Medical Research Inst.
University of Oklahoma Health Sciences Center
825 N.E. 13th Street
Oklahoma City, OK 73104
USA

Migaki, Grace I.
Dept. of Immunology
Boehringer Ingelheim Pharmaceuticals Inc.
900 Ridgebury Road, Box 369
Ridgefield, CT 06877
USA

Vestweber, Dietmar
Inst. of Cell Biology, ZMBE
University of Münster
Von-Esmarch-Str. 56
D-48149 Münster
Germany

Ward, Peter A.
Dept. of Pathology
The University of Michigan Medical School
M5240 Medical Science I, Box 0602
1301 Catherine Road
Ann Arbor, Michigan, 48109-0602
USA

Watson, Susan R.
805 Balra Drive
El Cerrito, CA 94530
USA

Wolitzky, Barry A.
Dept. of Inflammation/Autoimmune Diseases
Hoffmann-La Rocke Inc.
340 Kingsland Street
Nutley, New Jersey 07110
USA

INTRODUCTION

The migration of leukocytes from the vascular system to sites of pathogenic exposure is a key event in the process of inflammation. The inflammatory reaction enables the organism to defend itself against infectious microorganisms, but can also cause deleterious effects if the regulatory mechanisms of the inflammatory response are altered or if immune responses to residual microbial products, or to altered tissue components, trigger a persistent inflammatory reaction. In such instances, the defense attacks of the leukocytes can turn against the organism's own tissue and can lead to harmful destructions.

The entry of leukocytes into sites of injury or infection requires molecular mechanisms which enable the leukocyte to recognize such sites from within the vasculature and to form contact with the endothelium in order to exit and migrate through the blood vessel wall. Recognition as well as contact formation is mediated by several cell adhesion molecules which act in a sequential manner in concert with regulatory mediators such as the chemokines. The cell adhesion molecules which are involved in this process belong to three gene families which are: the selectins, the integrins and the immunoglobulin super gene family. The selectins mediate the initiation of the cell contact between leukocytes and endothelial cells. This selectin-mediated docking of leukocytes to the blood vessel wall in combination with the rapidly flowing blood stream leads to a rolling movement of the leukocytes on the endothelial cell surface. In contrast to the rapidly flowing cells in the blood stream, the rolling cells are able to sense signals from the endothelium which stimulate them to adhere more firmly to the endothelial cell surface. Such signals can be given by chemokines, which are presented and immobilized by proteoglycans on the endothelium, or by other mediators such as the phospholipid platelet activating factor (PAF). Their stimulatory effect causes activation of leukocyte integrins which bind to members of the Ig-superfamily on the endothelial cell surface. This leads to firm adhesion of the leukocytes to the endothelium and enables the leukocytes to actively migrate on the blood vessel wall along gradients of chemotactic factors and through the layer of the endothelium and the underlying basement membrane.

Capturing leukocytes from the rapidly flowing blood stream to the blood vessel wall is a very special example of cell contact formation which differs from most other cell adhesion phenomena, since considerable shear forces have to be overcome. The selectins, which mediate the very first steps in this process, seem to be specialized for this task. In contrast to the other gene families of cell adhesion molecules, the selectins are restricted to the leukocyte-vascular system. With only three members, the selectins form the smallest and most recently identified gene family among cell adhesion molecules. In contrast to the vast majority of all other adhesion molecules the selectins mediate cell contact via binding to carbohydrate ligand structures, thus they are lectins. L-selectin is found on most types of leukocytes, E-selectin is specific for endothelial cells and P-selectin is found on endothelium and platelets. The individual members of the group were designated by prefixes, which were chosen according to the cell type where the molecule was first identified.

L-selectin was first found as a 'lymphocyte homing' receptor, defined by the monoclonal antibody MEL 14 which blocked the binding of lymphocytes to lymph node high endothelial venules in lymph node tissue. Later L-selectin was also found on neutrophil granulocytes and monocytes and shown to be generally involved in leukocyte entry into sites of inflammation. E-selectin was also found by a monoclonal antibody approach when searching for cytokine inducible surface proteins on endothelial cells which would mediate the binding of neutrophils. P-selectin was originally found as a membrane protein of unknown function in platelet storage granules which later was also found in endothelial cells. Cloning and sequencing of all three selectins happened to occur at the same time and the results were published in 1989. Only then did it become apparent that a new gene family of cell adhesion molecules had been discovered.

Since the selectins were cloned, the field has rapidly progressed. This book will be an attempt to summarize the current knowledge about the physiological functions of the selectins and the molecular mechanisms by which the selectins mediate these functions. The importance of the selectins for the entry of neutrophils and lymphocytes into sites of inflammation has been documented *in vivo* in numerous inflammation models, based on the analysis of mice deficient for selectin genes and on studies with adhesion-blocking antibodies against the selectins. Analysis of the structural determinants of selectin function and the mechanisms which regulate the cell surface display of the selectins will provide necessary information to interfere with selectin-mediated leukocyte infiltration into tissue. Recent progress in the identification of selectin ligands will allow the generation of compounds which could serve as antagonists of the selectins. In addition, the identification of the selectin ligands allows the analysis of the question as to whether or not the selectins and their ligands also mediate signalling functions. As it is evolving from the functional analysis of all other families of cell adhesion molecules, evidence is starting to accumulate that the selectins and their ligands are also involved in signal transduction processes.

1 Functional Analysis of Selectin Structure

Kuo-Sen Huang*, Bradford J. Graves[+] and Barry A. Wolitzky*

Departments of Inflammation/Autoimmune Diseases and Physical Chemistry[+]*
Hoffmann-La Roche Inc., 340 Kingsland Street, Nutley, New Jersey 07110, USA

INTRODUCTION

Our current knowledge of the structure and function of selectin proteins has resulted from: 1) studies on the cloning and expression of selectin proteins, 2) three dimensional structural determination and molecular modeling of active domains, 3) mapping of neutralizing monoclonal antibody epitopes and 4) the analysis of site-directed mutations, domain deletions, and chimeric selectin molecules. This chapter will attempt to review these studies on the functional analysis of selectin structure and provide a comprehensive view of the nature of structural determinants as they relate to cell adhesion and the recognition of native carbohydrate ligands.

ORGANIZATION OF SELECTIN DOMAINS

The three members of the selectin family all share a common structural motif including a N-terminal C-type lectin domain (Drickamer, 1988), an epidermal growth factor (EGF)-like domain, a variable number of consensus repeats homologous to those in complement binding proteins (CR domains), a membrane spanning segment, and a short cytoplasmic region (Figure 1A). Human E-, P-, and L-selectin share an overall homology of approximately 40% and 60% at the nucleic acid and protein levels, respectively, with the highest degree of conservation within the lectin and EGF domains. Each of the selectins has now been cloned from several species. Figure 2 shows an alignment of the coding sequences within the lectin and EGF domains from multiple species. The sequence conservation (i.e. identity) is 60–80% between species within each family member. The "consensus selectin" described in Figure 2 and illustrated in Figure 1B demonstrates that ~42% of the amino acid residues are conserved within the lectin and EGF domains of all selectins from all species identified to date. This high degree of conservation supports the central role of these domains in the interactions with common carbohydrate determinants on ligands but

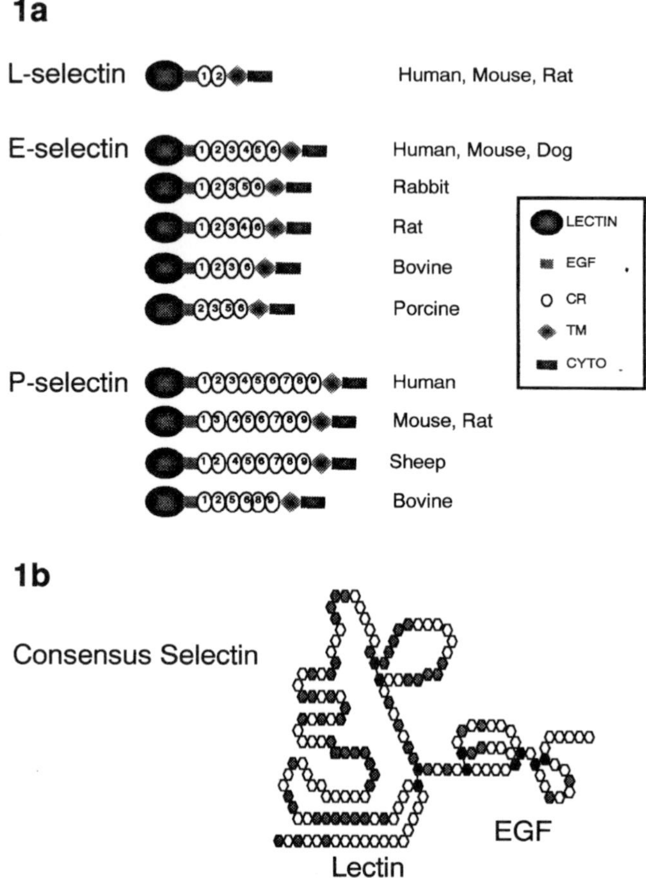

Figure 1 Domain composition of the selectin family of adhesion molecules and the consensus sequence of the lectin and EGF domains. **a**. L-selectin, E-selectin, and P-selectin have a common structural motif including: an amino terminal calcium-dependent lectin domain, a epidermal growth factor-like (EGF) domain, a variable number of repeat elements homologous to complement binding proteins (CR), a transmembrane region, and a short cytoplasmic sequence at the C-terminus. Sequence alignments of selectins from different species reveals that distinct CR repeats have been deleted during recent mammalian evolution. The CR domains are numbered with reference to the corresponding domain in the human selectin. **b.** A model of the consensus sequence for selectins within the lectin and EGF domain was generated by the sequence analysis shown in Figure 2. Amino acids are represented by diamonds with the black diamonds indicating cysteine residues, and the gray diamonds corresponding to residues that are identical in L-selectin (human, mouse, rat), E-selectin (human, mouse, rat, rabbit, dog, cow, and pig) and P-selectin (human, mouse, rat, sheep, and cow).

raises questions as to how the specificity of each selectin for its ligand is accomplished. Each CR domain is ~60 amino acids in length and contains three disulfide bonds. The number of CR domains varies between family members, with human L, E, and P-selectin containing two, six, and nine elements respectively. The cloning of E-selectin and P-selectin from numerous species has revealed that the number of CR domains is species specific. In comparison to the amino acid sequences of human E-selectin, rabbit E-selectin has lost CR domain 4, rat has deleted CR domain 5, while bovine and porcine contain only

```
                    Signal                                Lectin
                                                            +1
                                                            ▼
Human L-selectin    ......MIFP WKCQSTQRDL WNIFKLWGWT MLCCDFLAHH GTDCWTYHYS EKPMNWQRAR RFCRDNYTDL VAIQNKAEIE YLEKTLPFSR SYYWIGIRKI
Mouse L-selectin    ......MVFP WRCEGTYWGS RNILKLWVWT LLCCDFLIHH GTHCWTYHYS EKPMNWENAR KFCKQNYTDL VAIQNKREIE YLENTLPKSP YYYWIGIRKI
Rat L-selectin      ......MVFP WRCQSAQRGS WSFLKLWIRT LLCCDLLPHH GTHCWTYHYS ERSMNWENAR KFCKHNYTDL VAIQNKREIE YLEKTLPKNP TYYWIGIRKI

Consenesus L        ------M-FP W-C------- ----KLW--T -LCCD-L-HH GT-CWTYHYS E--MNW--AR -FC--NYTDL VAIQNK-EIE YLE-TLP--- -YYWIGIRKI

                                                            ▼
Human E-selectin    .......... .......... MIASQFLSAL TLVL.LI..K ESGAWSYNTS TEAMTYDEAS AYCQQRYTHL VAIQNKEEIE YLNSILSYSP SYYWIGIRKV
Mouse E-selectin    .......... .......... MNASRFLSAL VFV..LLA.G ESTAWYYNAS SELMTYDEAS AYCQRDYTHL VAIQNKEEIN YLNSNLKHSP SYYWIGIRKV
Rat E-selectin      .......... .......... MNASCFLSAL TFV..LLI.G KSIAWYYNAS SELMTYDEAS AYCQRDYTHL VAIQNKEEIN YLNSTLRYSP SYYWIGIRKV
Canine E-selectin   .......... .......... MITSQLLPAL TLVL.LLF.K EGGAWSYNAS TEAMTFDEAS TYCQQRYTHL VAIQNQEEIK YLNSMFTYTP TYYWIGIRKV
Rabbit E-selectin   .......... .......... MVASWLLSTL TFALVLLI.K ETSTWTYHFS AENMTYDEAS AYCQQNYTHL VAIQNKEEID YLNSILDYSP SYYWIGIRKV
Bovine E-selectin   .......... .......... MIVSQYLSAL TFVL.LLF.K ESRTWSYHAS TEMMTFEEAR DYCQKTYTAL VAIQNQEEIE YLNSTFSYSP SYYWIGIRKI
Porcine E-selectin  .......... .......... MIASQFLSAL PLVL.LLL.R ESGAWSYSAS TETMTFDDAS AYCQQRYTHL VAIQNHAEIE YLNSTFNYSA SYYWIGIRKI

Consensus E                               M--S--L--L -----LL--- ----W-Y--S -E-MT---A- -YCQ--YT-L VAIQN--EI- YLNS------ -YYWIGIRK-

                                                            ▼
Human P-selectin    MANCQIAILY QRFQ...RVV FGISQLLCFS ALISELTNQK EVAAWTYHYS TKAYSWNISR KYCQNRYTDL VAIQNKNEID YLNKVLPYYS SYYWIGIRKN
Mouse P-selectin    MAGCPKGSWT PRLR...SVI LGGAQLIWFS ALISELVNQK EVAAWTYNYS TKAYSWNNSR VFCRRHFTDL VAIQNKNEIA HLNDVIPFFN SYYWIGIRKI
Rat P-selectin      MAGCPKGSWK PRLR...SVV LGAAQLIWLS ALISELVNRK KVATWTYNYS TKAYSWNNSR AFCKRHFTDL VAIQNKNEIA HLNDVIPYVN SYYWIGIRKI
Bovine P-selectin   MASCPKAIWN WRFQ...RAV FRTVQLLCFS VLIFEVINQK EVSAWTYHYS NKTYSWNYSR AFCQKYYTDL VAIQNKNEIA YLNETIPYYN SYYWIGIRKI
Sheep P-selectin    MASCPKAIWS WRFQ...RVV FRSVQLLCFS ILIFELMTQK EVSAWTYHYS DKPYSWNYSR AFCQKYYTDL VAIQNKNEIA YLNETIPYYN SYYWIGIRKI

Consensus P         MA-C------ -R-------- ----QL---S -LI-E----K -V--WTY-YS -K-YSWN-SR --C----TDL VAIQNKNEI- -LN---P--- SYYWIGIRK-

Consensus selectin  ---------- ---------- ---------- ---------- ----W-Y--S ---------- --C----T-L VAIQN--EI- -L-------- -YYWIGIRK-

                                                                          EGF
                                                                   ▼                                ▼
Human L-selectin    GGIWTWVGTN KSLTEEAENW GDGEPNNKKN KEDCVEIYIK RNKDAGKWND DACHKLKAAL CYTASCQPWS CSGHGECVEI INNYTCNCDV GYYGPQCQFV
Mouse L-selectin    GKMWTWVGTN KTLTKEAENW GAGEPNNKKS KEDCVEIYIK RERDSGKWND DACHKRKAAL CYTASCQPGS CNGRGECVET INNHTCICDA GYYGPQCQYV
Rat L-selectin      GKTWTWVGTN KTLTKEAENW GTGEPNNKKS KEDCVEIYIK RERDSGKWND DACHKRKAAL CYTASCQPES CNRHGECVET INNNTCICDP GYYGPQCQYV

Consensus L         G--WTWVGTN K-LT-EAENW G-GEPNNKK- KEDCVEIYIK R--D-GKWND DACHK-KAAL CYTASCQP-S C---GECVE- INN-TC-CD- GYYGPQCQ-V   81%

                                                                   ▼                                ▼
Human E-selectin    NNVWVWVGTQ KPLTEEAKNW APGEPNNRQK DEDCVEIYIK REKDVGMWND ERCSKKKLAL CYTAACTNTS CSGHGECVET INNYTCKCDP GFSGLKCEQI
Mouse E-selectin    NNVWIWVGTG KPLTEEAQNW APGEPNNKQR NEDCVEIYIQ RTKDSGMWND ERCNKKKLAL CYTASCTNAS CSGHGECIET INSYTCKCHP GFLGPNCEQA
Rat E-selectin      NNVWIWVGTQ KPLTEEAKNW APGEPNNKQR NEDCVEIYIQ RPKDSGMWND ERCDKKKLAL CYTASCTNTS CSGHGECVET INSYTCKCHP GFLGPKCDQV
Canine E-selectin   NKKWTWIGTQ KLLTEEAKNW APGEPNNKQN DEDCVEIYIK RDKDSGKWND ERCDKKKLAL CYTASCTNTS CSGHGECVET INSYTCKCHP GFLGPKCDQV
Rabbit E-selectin   NNVWIWVGTH KPLTEGAKNW APGEPNNKQN NEDCVEIYIK RPKDTGMWND ERCSKKKLAL CYTAACTEAS CSGHGECIET INNYSCKCYP GFSGLKCEQV
Bovine E-selectin   NGTWTWIGTN KSLTKEATNW APGEPNNKQS DEDCVEIYIK REKDSGKWND EKCTKQKLAL CYKAACNPTP CGSHGECVET INNYTCQCHP GFKGLKCEQV
Porcine E-selectin  NGTWTWIGTK KALTPEATNW APGEPNNKQS NEDCVEIYIK RDKDSGKWND ERCSKKKLAL CYTAACTPTS CSGHGECIET INSSTCQCYP GFRGLQCEQ.
Consensus E         N--W-W-GT- K-LT--A-NW APGEPNN-Q- -EDCVEIYI- R-KD-G-WND E-C-K-KLAL CY-A-C---- C--HGEC-ET IN---C-C-P GF-G--C-Q-   61%

                                                                   ▼                                ▼
Human P-selectin    NKTWTWVGTK KALTNEAENW ADNEPNNKRN NEDCVEIYIK SPSAPGKWND EHCLKKKHAL CYTASCQDMS CSKQGECLET IGNYTCSCYP GFYGPECEYV
Mouse P-selectin    NNKWTWVGTN KTLTEEAENW ADNEPNNKKN NQDCVEIYIK SNSAPGKWND EPCFKRKRAL CYTASCQDMS CSNQGECIET IGSYTCSCYP GFYGPECEYV
Rat P-selectin      NNKWTWVGTN KTLTAEAENW ADNEPNNKRN NQDCVEIYIK SNSAPGKWND EPCFKRKRAL CYTASCQDMS CNSQGERIET IGSYTCSCYP GFYGPECEYV
Bovine P-selectin   NNKWTWVGTK KTLTEEAENW ADNEPNNKRN NQDCVEIYIK SLSAPGKWND EPCWKRKRAL CYRASCQDMS CSKQGECIET IGNYTCSCYP GFYGPECEYV
Sheep P-selectin    DNKWTWVGTK KTLTEEAENW ADNEPNNKKN NQDCVEIYIK SPSAPGKWND EPCGKRKRAL CYRASCQDMS CSKQGECIET IGNYTCSCYP GFYGPECEYV
Consensus P         ---WTWVGTK K-LT-EAENW ADNEPNNK-N N-DCVEIYIK S-SAPGKWND E-C-K-K-AL CY-ASCQDMS C--QGE--ET IG-YTCSCYP GFYGPECEYV   78%

Consensus selectin  ---W-W-GT- K-LT--A-NW ---EPNN--- --DCVEIYI- -----G-WND --C-K-K-AL CY-A-C---- C---GE--E- I----C-C-- G--G--C---   42%
```

Figure 2 Amino acid sequences of the lectin and EGF domains of L-selectin, E-selectin and P-selectin. The cloning and determination of the amino acid sequences of the selectins from multiple species has allowed for the alignment and analysis of homology within the region encoding the signal sequence, lectin, and EGF-like domains. The coding sequence begins at the residue marked +1, and the domain boundaries are indicated by the ▾'s. Consensus sequences for L-selectin, E-selectin, and P-selectin represent residues which are identical when compared between the known sequences identified for the respective family members with non-conserved residues indicated by a dash (–). The consensus selectin (also depicted in schematic form in Figure 1b) represents residues which are identical in all three selectins families from all known species. Sequences for the selectins were obtained from the following sources: human L-selectin (P14151) Siegelman and Weissman, 1989; murine L-selectin (P18337) Siegelman *et al.* 1989 and Lasky *et al.* 1989; rat L-selectin (P30836) Watanabe *et al.* 1992; human E-selectin (P16581) Bevilaqua *et al.* 1989; mouse E-sèlectin (Q00690) Becker-Andre *et al.* 1992; rat E-selectin (L25527); canine E-selectin (L23087); rabbit E-selectin (P27113) Larigan *et al.* 1992; bovine E-selectin (L12039) Nguyen *et al.* 1993; porcine E-selectin (L3906) Tsang *et al.* 1995; human P-selectin (P16109) Johnston *et al.* 1989; mouse P-selectin (Q01101) Weller *et al.* 1992; rat P-selectin (L23088) Auchampach *et al.* 1994; bovine P-selectin (JN0473) Strubel *et al.* 1993; sheep P-selectin (L34270). Accession numbers are given in parenthesis.

four CR domains and are missing CR domains 4 + 5 and 1 + 4 respectively. With respect to human P-selectin, mouse and rat have lost CR domain 2, sheep has deleted CR domain 3, and bovine is missing both CR domains 3 and 7. The variability in the number of CR domains and the fact that different CR domains have been lost during recent mammalian evolution suggest a less than critical role for these domains. However, as discussed below, a variety of functions have been attributed to the CR domains including; providing a structural scaffold for the proper presentation of the lectin and EGF domains at a defined distance from the plasma membrane, stabilizing the conformation of the lectin and EGF domains, or modulating receptor affinity by mediating oligomerization of selectins. While there is a moderate degree of conservation of amino acids within the transmembrane and cytoplasmic domains between species for a given selectin member, particularly for L- and P-selectin, the homology for the "consensus selectin" within these domains from all selectins is quite low.

An alternatively spliced form of P-selectin lacking the transmembrane domain has been described and may encode a soluble, secreted form of this selectin (Johnston *et al.*, 1990). Another group has reported that alternative splicing may generate a phosphotidylinositol linked form of human L-selectin (Camerini *et al.*, 1989). Soluble forms of selectins have been detected in a wide range of pathological settings including septic shock, rheumatoid arthritis, and asthma (Gearing and Newman, 1993). The levels of soluble selectins may be used as diagnostic markers or as an index of the efficacy of therapy in the treatment of inflammatory diseases. Soluble P-selectin is detected in normal individuals at a concentration (0.15 to 0.3 μg/ml), which could potentially be capable of modulating P-selectin dependent adhesion events (Dunlop *et al.*, 1992). Soluble E-selectin is a very poor inhibitor of adhesion *in vitro* and the potential significance of elevated circulating levels of this adhesion molecule are unclear. A low circulating level of L-selectin correlates with an increased risk of susceptibility, severity of symptoms, and mortality in patients with acute respiratory distress syndrome (Donnelly *et al.*, 1994).

The genes for the human and mouse selectins are closely linked on chromosome 1 (Watson *et al.*, 1990). The genes for other proteins which contain complement binding domains such as decay accelerating factor, C4 binding protein, and factor V are also located in this region of the genome. Analysis of the genes from human, mouse, and rabbit all revealed that the individual protein domains are encoded by separate exons and suggest that this family arose by exon shuffling and multiple gene duplication (Collins *et al.*, 1991; Dowbenko *et al.*, 1991; Larigan *et al.*, 1992).

THREE DIMENSIONAL STRUCTURE OF THE LECTIN AND EGF DOMAINS

E-selectin

As will be discussed in detail in subsequent sections of this chapter, the lectin and EGF domains alone are both necessary and sufficient for mediating cell adhesion and carbohydrate ligand recognition and thus represent the minimum functional domains of the selectins. In an effort to further characterize the structural determinants within this region, we have succeeded in crystallizing several forms of the lectin and EGF (lec-EGF)

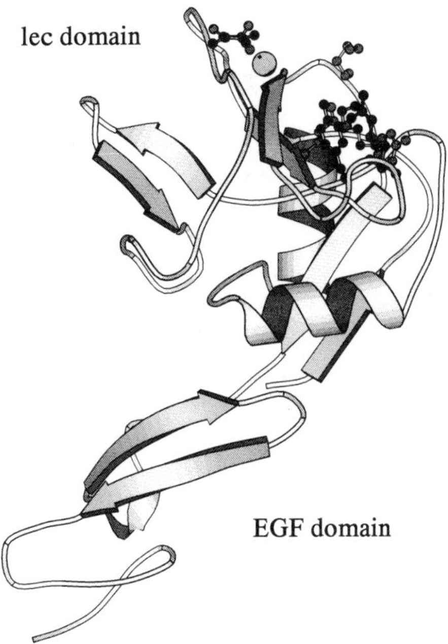

Figure 3 Ribbon diagram of lec-EGF E-selectin depicting the distinctness of the two domains and the remoteness of the EGF domain from the carbohydrate binding site. The diagram was generated with the program MOLSCRIPT (Kraulis, 1991) and displays β-strands as arrows, α-helices as coils and turns and unstructured regions as narrow tubes. Residues for which mutation renders the protein unable to sustain adhesion either completely or partially (Erbe *et al.*, 1992 and Graves *et al.*, 1994) are shown as black and dark gray balls. A close-up of this region from a more appropriate orientation is shown in Figure 5b. The lone calcium ion is depicted as the large light gray sphere.

domains from human E-selectin and have determined their three-dimensional structure (Graves *et al.*, 1994). Several forms of this protein were produced which helped expedite the progress of crystallization and structure determination. These include glycosylated forms from CHO cells and unglycosylated forms generated either by enzymatic deglycosylation or through expression of a form with the potential N-linked glycosylation sites mutated. All forms were eventually crystallized including some in the presence of a carbohydrate ligand, sLe^x (Table 1). These crystals exhibited diffraction to at least 2.1Å resolution. Potentially, these crystals would provide the structures of E-selectin in 1) its native state, 2) without calcium and 3) with the sLe^x ligand bound. However, as will be discussed, the carbohydrate ligand was never properly visualized.

The general structure of lec-EGF (Figure 3) is as expected in that the lectin domain looks very much like the mannose binding protein (MBP), the only other C-type lectin whose structure has been determined (Weis *et al.*, 1991; Weis *et al.*, 1992), and the EGF domain exhibits the same fold and disposition of disulfide bonds observed in a variety of structures with EGF-like domains (Hommel *et al.*, 1992: Padmanabhan *et al.*, 1993). However, the first real surprise of the E-selectin structure is the relative disposition of the lectin and EGF domains. The two domains are quite distinct with only a few limited interactions. In addition, the EGF domain is clearly remote from the region of the lectin domain that is thought to be important for binding carbohydrate (see below and Figure 3). These observations strongly suggest that the importance of the EGF domain is due to a direct role in binding the selectin ligand and not to any ability to act as a conformational modifier of the lectin domain. Another surprising feature of the structure is the nature of calcium binding.

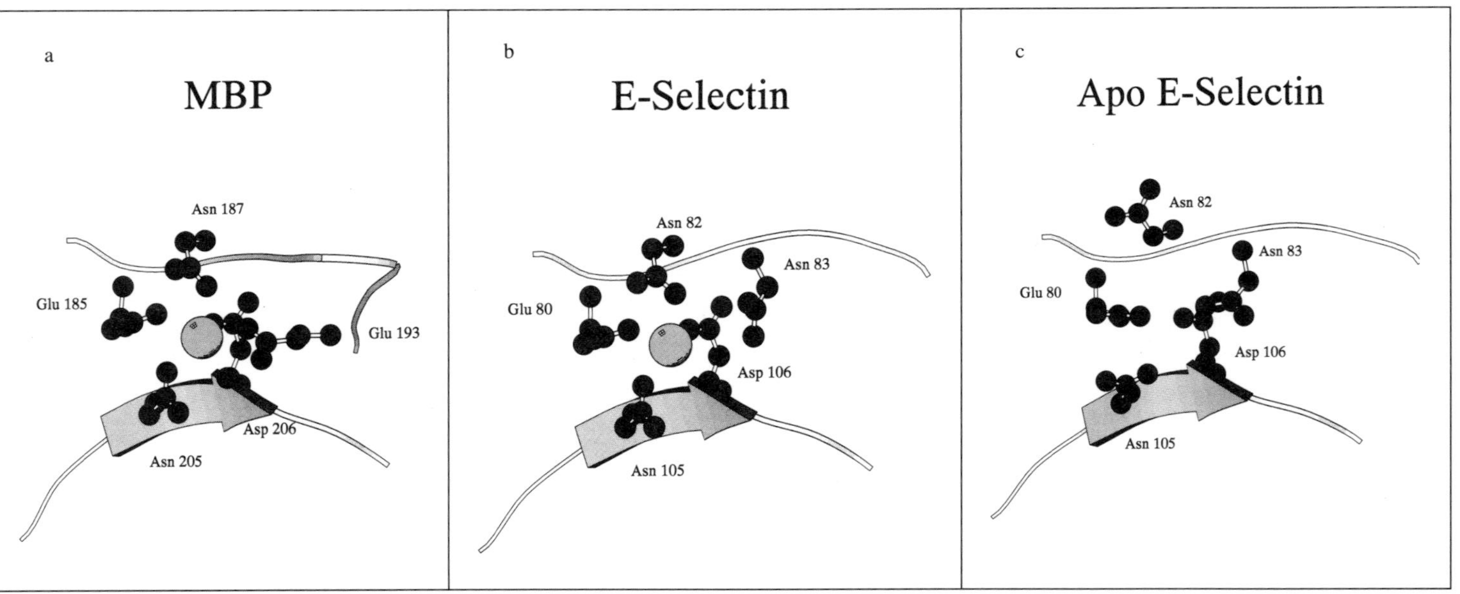

Figure 4 Comparison of calcium binding sites in C-type lectins. **a**, Close-up of the calcium binding site in the mannose binding protein that is conserved in the selectins. The coordinates of MBP from the complex with mannose (Weis *et al.*, 1992) were used with program MOLSCRIPT (Kraulis, 1991). The secondary structure is shown schematically for residues 184–194 and 202–209 and atoms (small, dark gray spheres) are shown explicitly for the side chains of residues which coordinate the calcium ion (large, light gray sphere). **b**, Same view for native E-selectin (Graves *et al.*, 1994) showing both the similarity (Glu 80, Asn 82, Asn 105 and Asp 106 are identical in structure and function to Glu 185, Asn 187, Asn 205 and Asp 206 in MBP) and the critical difference (Asn 83 coordinates calcium indirectly while Glu 193 of MBP binds directly to calcium) in the calcium coordination compared to MBP. **c**, Structural changes in the calcium binding site of E-selectin resulting from loss of calcium (Graves *et al.*, unpublished). The noticeable shifts include the adoption of a different χ_1 torsion angle for Asn 82 and a roughly 90° rotation of χ_2 in both Asn 83 and Asn 105.

Lec-EGF E-selectin has only one bound calcium ion and its pentagonal bipyramid coordination geometry (five protein ligands and two water molecules) is not unusual except when compared to the same site in the mannose binding protein. Figure 4 shows a comparison of the corresponding calcium binding sites in E-selectin (Figure 4b) and MBP (Figure 4a) and the remarkable observation is that even though Glu 203 of MBP is conserved in E-selectin (Glu 88) the glutamate residue in E-selectin is not utilized as a calcium ligand. This point is critically important for ligand binding since Glu 203 of MBP makes direct contact with bound mannose. Clearly, Glu 88 of E-selectin is not able to make the same contact and this calls into question whether, as has been predicted, sLe^x will bind with the fucose moiety adopting the same relative position as mannose does in its complex with MBP. Furthermore, though the mutation of Asn 82 to Asp (designed to retain calcium binding while eliminating hydrogen bond donating ability) in E-selectin abrogates carbohydrate binding in the same manner as Asn 187 to Asp does in MBP, a second-site mutation in E-selectin (Lys 86 to Ala) has been found which returns adhesion to wild-type levels (Burns *et al.*, unpublished data). The side chain of Lys 86 is well-removed and pointing away from the calcium ion and the reason for its ability to overcome the detrimental affects of the Asn 82 to Asp mutation is, as yet, unknown. But it is another indication that the binding of sLe^x to E-selectin may not parallel the interaction of mannose with MBP.

Somewhat surprisingly it was possible to crystallize the apo-form of lec-EGF (Table 1). The close proximity of acidic side chains in the calcium binding region and the increased sensitivity to proteases resulting from loss of bound calcium suggested that, at the very least, the loop involving residues Glu 80, Asn 82 and Asn 83 would likely alter its conformation. However, the apo-form of lec-EGF not only crystallized but its structure was solved by molecular replacement techniques using the native lec-EGF as the search model (eliminating residues 79 to 88) and the conformation of the protein backbone in the region where calcium normally is bound is virtually identical to that in the native protein (Figure 4c). This structural similarity is consistent with the observation that complete reconstitution is achieved extremely rapidly upon the addition of calcium to the apo protein. The interpretation of this must be tempered by the fact that the crystals were grown at a pH (4.5–5.0) where the acidic side chains might be partially protonated which would limit the electrostatic repulsion. However, the electron density maps clearly show the major population of side chain conformations follows that observed in the native protein. The major exception is the side chain of Asn 82 which adopts a different rotamer conformation at χ_1. Otherwise, there are only minor twists of the ends of the side chains for Glu 80, Asn 83 and Asn 105 which enable the formation of several new hydrogen bonds from Asn 105 to Glu 80 (Oε2), Tyr 94 (Oη), and Asp 106 (O) and from Asn 83 to Glu 80 (Oε2) and Asp 106 (O and Oδ1). This hydrogen bonding scheme must serve to dissipate any remaining charge repulsion remaining from the continued close proximity of Glu 80 and Asp 106.

Another interesting feature of the apo-structure is that the protein is glycosylated and the presence of the carbohydrate seems to have no impact on the overall structure. It has been shown that the loss of the three potential N-linked carbohydrate sites has no effect on cell adhesion. No significant density is observed at any of the three potential carbohydrate attachment sites. Finally, we do observe a twist of the EGF domain relative to the lec domain in the apo versus native structures. This approximately 5° rotation is presumably induced by crystal packing forces since even though the space groups are the same the cell constants and packing arrangements are different.

Table 1 Crystallization of LEC-EGF E-selectin

Protein	*Crystallization Conditions*	*Space Group*	*Cell Constants*	*Diffraction Maximum*
"Apo-E-Selectin"	2.5M PO_4 pH 4.5–5.0	$P2_12_12_1$	34.8Å 61.7 95.7	2.1Å
"Degly-E-Selectin"	25% PEG 4000 0.2M $CaCl_2$ 0.1M HEPES pH 7.5	$P2_12_12_1$	34.4Å 73.5 77.6	2.0Å
"Nongly-E-Selectin"	45% Sat'd. A.S. 2% PEG 400 0.1M HEPES pH 7.5	$P2_12_12_1$	33.8Å 73.9 77.4	2.0Å
"Nongly-E-Selectin" + SLe^x	45% Sat'd. A.S. 2% PEG 400 0.1M HEPES pH 7.5	$P2_12_12_1$	33.8Å 73.9 77.4	2.0Å

Proteins: All proteins were expressed in Chinese Hamster Ovary cells. A glycosylated form, which was used to grow the crystals of "apo-E-selectin", and the enzymatically deglycosylated protein ("degly-E-selectin") were generated and purified as indicated in Li *et al.* (1994). The unglycosylated form of the protein ("nongly-E-selectin") was generated by mutating the asparagine residues at the three potential N-linked carbohydrate sites to amino acids present in E-selectins from other species. Thus, Asn at positions 4 and 124 were changed to the corresponding rabbit E-selectin sequences (His and Glu, respectively) while Asn 139 was converted to Ser as it is in the mouse E-selectin. **Crystallization:** All crystallization experiments utilized the hanging drop vapor equilibration method and the crystallization conditions listed are the chemical conditions in the reservoir. Crystallization was initiated by mixing equal volumes of stock protein solutions (in all cases the protein was concentrated to about 20 mg/ml in 10 mM HEPES, pH 7.5, 2 mM $CaCl_2$) with the appropriate reservoir solution (typically 2 μl of each). Improved reproducibility for obtaining larger crystals with either the degly or nongly protein was achieved by macroseeding. For these experiments, the protein in the stock solutions was lowered to 12–15 mg/ml, mixed with an equal volume of reservoir containing 30% PEG 4000 and set over a final reservoir solution containing 16–18% PEG 4000. It should be noted that the nongly E-selectin yielded crystals under the same conditions listed for the degly protein. Attempts to form complexes with sLe^x or related molecules were performed both by cocrystallization and soaking preformed crystals in solution with the carbohydrate and utilized both the ammonium sulfate (A.S.) and polyethyleneglycol (PEG) conditions. **Structure determination:** The initial structure of native E-selectin with the degly protein was determined by the multiple heavy-atom isomorphous replacement method using gadolinium and platinum. Crystals would not grow using protein that already had the calcium substituted by gadolinium so it was necessary to slowly replace the calcium in the mother liquor containing preformed crystals. The platinum derivative was achieved by standard soaking procedures using 5 mM K_2PtCl_4. Program PHASES (Furey and Swaminathan; 1995) was employed to refine the heavy-atom positions. This was critical for the Gd derivative since it has a large anomalous scattering component and a small isomorphous difference (by virtue of its substituting for calcium) and the program utilizes the anomalous and isomorphous differences separately. This allowed the maximum phasing information to be extracted from the Gd derivative data. Data to 2.5Å resolution was used for phasing purposes. The phasing power of each component calculated by PHASES was 2.3 (Ptiso), 2.7 (Ptano), 1.4 (Gdiso), and 3.8 (Gdano). The combined mean figure of merit was 0.76. The resulting MIR map was of extremely high quality. All subsequent structures were solved by standard molecular replacement techniques as implemented in MERLOT (Fitzgerald, 1988).

P-selectin and L-selectin

Though no crystallographic structures have been published for either P- or L-selectin, several groups have made models (Erbe *et al.*, 1992; Hollenbaugh *et al.*, 1993; Chou, 1995). One limitation of most of these models is that they are based solely or predominantly on the structure of the rat mannose binding protein, which had been the only structure available of a C-type lectin domain. As noted previously, the lectin domains of selectins and mannose binding proteins are probably the least similar of all the members of the C-type lectin family (Bezouska *et al.*, 1991). Thus, any model of a selectin based on MBP is likely to have noticeable inaccuracies. In contrast, the lectin domains of the selectins are highly homologous (~65% identity) and, in fact, the hydrophobic cores are absolutely conserved — not only among the human versions of E-, P- and L-selectin but also among the different selectins from the several species that have been cloned (Figure 2). We have generated models of P- and L-selectin based on the crystallographic structure of E-selectin and find that these require only an occasional rotation of a surface side chain torsion angle in order to avoid a steric contact with another surface residue. Thus, we expect the structures of E-, P-, and L-selectin to be virtually identical. Of course, we can not eliminate the possibility that some of the surface loops may change conformation as a result of the sequence differences among the selectins but there is no obvious need to incorporate such drastic changes. Detailed analysis of the structure homology among the different selectins and the predicted sLe^x binding site will be discussed in the last section of this chapter.

FUNCTIONAL ROLE OF THE SELECTIN DOMAINS

The Lectin Domain

The lectin activity of selectins was first reported by Rosen and coworkers who observed that L-selectin function can be blocked by specific carbohydrates such as mannose-6-phosphate (M6P) and M6P-rich phosphomannan (PPME) (see reviews by Rosen, 1989 and 1990). Later, this was supported by the cDNA cloning of L-selectin which showed it to contain a lectin domain homologous to that of the C-type, Ca^{+2}-dependent animal lectins (Drickamer, 1988). To date, much data exist to demonstrate that the lectin domain of selectins plays a major role in ligand recognition. The evidence is based on epitope mapping of blocking antibodies, blocking peptides, and site-specific mutagenesis of the lectin domain.

Epitopes for blocking monoclonal antibodies

Many anti-selectin blocking antibodies have been shown to recognize the lectin domain of selectins. For example, we generated three anti-human E-selectin monoclonal antibodies (7H5, 8E4, and 3B7) that block E-selectin-mediated HL-60 cell adhesion (Erbe *et al.*, 1992). By evaluating the effects of mutations in the E-selectin lectin domain on binding to these antibodies, we showed that all three antibodies recognize epitopes in the lectin domain (Figure 5a). Monoclonal antibody (mAb) 8E4 recognizes aspects of the NH_2-terminal 9 amino acids and mAb 3B7 recognizes a loop including Val 101 and Glu 98. On the other hand, mAb 7H5 seems to recognize spatially distinct epitopes located in the NH_2 (Thr 7, Ala 9, Ser 47) and COOH terminal (Lys 113) region of the lectin domain. Clearly, these epitopes are in the vicinity of the bound calcium ion. We also mapped the epitopes of two commercially available anti-E-selectin blocking antibodies, mAb BBA2 and mAb ENA-1, to the NH_2-terminal region of the lectin domain. By using E-selectin domain deletion constructs as probes, Pigott *et al.* (1991) also showed that a blocking antibody directed against the lectin domain inhibits cell adhesion.

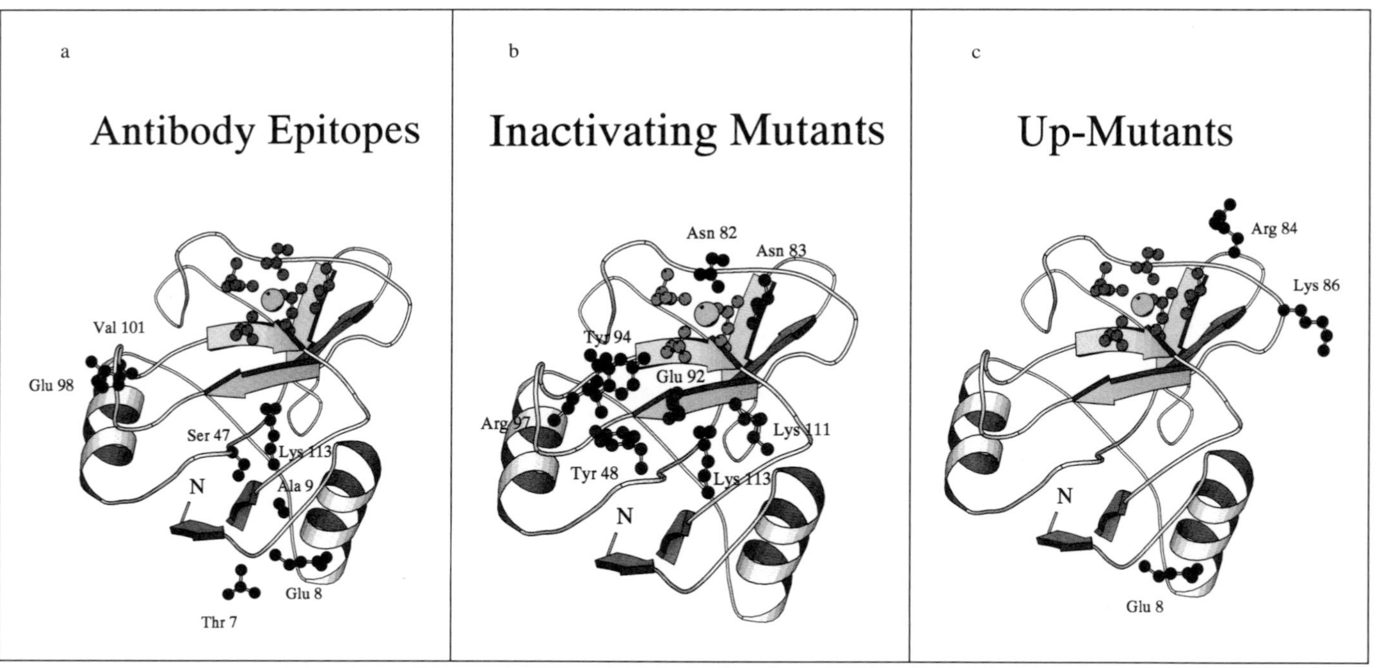

Figure 5 Structure-function data for E-selectin. Program MOLSCRIPT (Kraulis, 1991) has been used to depict the structure of the lectin domain of E-selectin in schematic fashion and explicit side chain atoms are shown for residues which are either (**a**) part of the epitopes (black spheres) for neutralizing antibodies (Erbe *et al.*, 1992), (**b**) sites of mutation which render the protein either completely (black spheres) or partially (dark gray spheres) deficient for adhesion (Erbe *et al.*, 1992 and Graves *et al.*, 1994), or (**c**) sites of mutation (black spheres) which increase the strength of adhesion (Erbe *et al.*, 1992 and Graves *et al.*, 1994). In all three panels, the important residues are labelled along with the N-terminus (**N**). In addition, the calcium ion (large, light gray sphere) and the residues which coordinate to it (medium gray spheres) are shown for reference. The carbohydrate binding site is crudely defined by the antibody epitopes but is more precisely delineated by the mutagenesis results. The role of the residues denoted as up-mutants is uncertain for the wild-type protein.

Bowen *et al.* (1990) demonstrated that an adhesion-blocking anti-L-selectin antibody, Mel 14, recognizes a conformational determinant in the NH_2-terminal 53 amino acids of the L-selectin lectin domain. Kansas *et al.* (1991) reported on an anti-L-selectin blocking mAb LAMI-3 recognizing the COOH-terminal 67 amino acid residues of the lectin domain. At least two blocking anti-P-selectin monoclonal antibodies (G3 and AK-6) have been shown to recognize the lectin domain of P-selectin (Geng *et al.*, 1990; Erbe *et al.*, 1993). Geng *et al.* (1992) showed that antibody G3 recognizes residues 19–34.

Berg *et al.* (1995) recently reported the isolation of three monoclonal antibodies cross-reactive with E- and P-selectin. These antibodies recognize the same or overlapping epitopes within the lectin domains of E- and P-selectin and block their functions. Determination of the exact epitopes recognized by these antibodies may identify ligand recognition regions that are commonly used by P- and E-selectins. These dual-reactive E- / P-selectin-specific antibodies may also provide a more effective and broadly useful reagent for treating inflammatory diseases.

Identification of critical amino acids

Erbe *et al.* (1992) evaluated the effects of various mutations in the E-selectin lectin domain on sLe^x binding. These mutants were expressed as dimers of an IgG chimera comprising the lectin, EGF and 2 CR domains followed by the hinge regions (CH2 and CH3) of human IgG_1. These E-selectin-IgG alanine mutants were then examined for their ability to bind sLe^x glycolipid that had been immobilized on plastic microtiter wells. The structural integrity of these mutants was assessed by their reactivities to a panel of blocking and non-blocking antibodies. The results allowed Erbe *et al.* (1992) to define a small region of the E-selectin lectin domain that is critical for carbohydrate recognition. Based on the crystal structure of the lectin and EGF domains of E-selectin, we also constructed a set of site-specific E-selectin mutants consisting of the entire extracellular domain anchored to membranes by glycosyl phosphatidylinositol linkages (Graves *et al.*, 1994). These mutants were expressed in COS cells and evaluated for their ability to mediate neutrophil adhesion. These results were in general agreement with those of Erbe *et al.* (1992) and have been summarized below and depicted in Figure 5b.

Mutations at Asn 82, Tyr 94, Arg 97 and Lys 113 completely abolish neutrophil adhesion. As previously discussed, Asn 82 is a calcium ligand (see Figure 4b), and is equivalent in position to Asn 187 in rat mannose-binding protein (MBP). Substitution of Asn 82 $\rightarrow$ Asp in E-selectin suggests that the role of Asn 82 in binding sLe^x may be similar to that of Asn 187 in binding mannose in the MBP. Interestingly, mutation of Lys 86 to Ala is able to rescue the null mutation at position 82 (Asn 82 to Asp) but has no effect as a single mutation. Again, the mechanism for this recovery of adhesion is not known but is not believed to result from a nonspecific, compensatory mechanism since even the double mutation at positions 84 and 86 is not able to rescue either the Tyr 94 to Phe or the Lys 113 to Ala mutation.

Tyr 94 and Arg 97, which lie in the loop connecting two β-strands ($\beta 4$ and $\beta 5$) seem essential for neutrophil adhesion. In addition, Lys 99 may also play a role since Erbe *et al.* (1992) observed that substitution of Lys 99 with Ala almost abolished ligand binding. Arg 97 was particularly sensitive to substitution. Even the conservative substitution with a lysine residue completely diminished cell adhesion. Arg 97 forms a strong hydrogen bond to the side chain of Asp 100, which may play a critical role in ligand binding. It is striking that a mutation at Tyr 94 with Phe results in the loss of ligand binding. The ring of Tyr 94 is largely buried under Arg 97 and blocked on the side by the chains of Tyr 48 and Asp 100. Since the hydroxyl group of Tyr 94 only forms two hydrogen bonds to water molecules and makes no contact with any other part of the protein, it is possible that upon ligand binding the hydroxyl group of Tyr 94 forms hydrogen bonds to the carbohydrate ligand. Lys 113 is an interesting residue in that the side chain amino group forms an ion pair with Glu 92 while the alkyl portion of the side chain forms a hydrophobic contact with Trp 50. Thus, a mutation of Lys 113 to Ala may abolish these interactions and, consequently, ligand binding activity. However, Kogan *et al.* (1995) recently showed that changing Lys 113 to Gln or Glu does not affect HL-60 cell adhesion activity, implying that the ion pair formation between the side chains of Lys 113 and Glu 92 may not be critical for ligand binding. It also argues against the participation of Lys 113 in directly interacting with the sLe^x.

We also identified several mutations that partially abolished neutrophil adhesion. These changes were Tyr 48 $\rightarrow$ Phe, Asn 83 $\rightarrow$ Ala, Glu 92 $\rightarrow$ Ala, and Lys 111 $\rightarrow$ Ala. These residues are located in the vicinity of the critical residues described above, with their side chains extending in the direction of the top face of the lectin domain. Although the role of these residues in ligand recognition is not fully understood, they interact with amino acids that are critical to ligand binding. For example, the aromatic rings of Tyr 48 and Tyr 94 form an edge to face and Glu 92 forms an ion pair with Lys 113. In addition, Glu 92 also forms a hydrogen bond with the amide group of Asn 105, which is a calcium ligand. Lys 111 forms an ion pair with Glu 107, which is adjacent to a calcium ligand, Asp 106. Asn 83 is interesting because the side chain carbonyl oxygen forms a hydrogen bond with a water molecule, which is coordinating with the Ca^{+2}. The involvement of Asn 83 in Ca^{+2} binding is surprising because the equivalent residue in MBP coordinates with another Ca^{+2}, one that does not exist in E-selectin. Mutation of Asn 83 to Ala would abolish hydrogen bonding with the water molecule and consequently affect Ca^{+2} and ligand binding.

Erbe *et al.* (1992) observed that mutants Ser 45 $\rightarrow$ Ala and Ser 47 $\rightarrow$ Ala exhibited diminished sLe^x binding, suggesting that the side chain hydroxyl groups may be involved in ligand binding. However, close examination of the serine side chains in the E-selectin structure suggests that these substitutions might alter the local packing environment causing a slight structural perturbation that could account for the decreased binding. The crystal structure of lec-EGF reveals that the Ser 45-Pro 46-Ser 47 segment forms a tight Type I reverse turn (Pro being the i + 1 residue) with side chains of the two serines forming a strong (2.96Å) hydrogen bond. Otherwise, these two side chains project into a hydrophobic pocket formed by Ala 9, Leu 114, Lys 113 and Tyr 49. Ser 45 makes van der Waals contacts with Leu 114 so this residue probably can not tolerate much additional bulk without adopting a different torsion angle. This probably occurs in P-selectin in which Ser 45 is replaced by an aromatic residue. In contrast, Ser 47 has more space around it and may accommodate larger sidechains. Though Ser 47 is conserved in most of the selectins, murine L-selectin utilizes a Tyr at this position. It is possible that larger side chains at position 47 might restore the tight turn that smaller substitutions and the loss of the hydrogen bond disrupt. To test this hypothesis, a double mutant Ser45Ala/Ser47Val was constructed and found to bind to neutrophils at levels equivalent to wild type. Thus, it seems clear that neither Ser 45 nor Ser 47 are involved in sLe^x binding.

A number of mutants with increased levels of ligand binding activity were observed (Figure 5c). Erbe *et al.* (1992) found that the Glu 8 → Ala mutant (IgG chimera) exhibited 4–5 fold higher binding activity for sLe^x glycolipid than the wild type. However, in a HL-60 cell adhesion assay, it showed the same activity as the wild type. In addition, we identified an Arg 84 → Ala mutation which resulted in five fold higher neutrophil adhesion and sLe^x binding. Since all other members of the selectin family have a Lys at this position, we made a Arg 84 → Lys mutation in E-selectin and found it enhanced neutrophil adhesion as well. This result suggests that the increased activity in neutrophil adhesion with substitutions at position 84 can not be readily explained by simply removing the large, bulky sidechain. Since Arg 84 and Glu 8 are located outside the putative sLe^x binding region, it is possible that these residues interact with non-sLe^x components of native ligand to enhance the binding avidity.

Since the lectin domains of selectins are highly homologous, it is of interest to determine whether or not common sites are used for carbohydrate recognition. Based on the crystal structure of MBP, Erbe *et al.* (1993) generated a three-dimensional model of the lectin domain for P- and E-selectin. Comparison of the two models revealed that of the residues which appeared important for E-selectin binding to sLe^x, five are conserved in P-selectin: Tyr 48, Asn 82, Tyr 94, Lys 111 and Lys 113. Since mutation of these residues in E-selectin affected ligand binding activity, Erbe *et al.* (1993) examined similar substitutions in P-selectin (Y48F, K111A and K113A). In sLe^x glycolipid binding and HL-60 cell adhesion assays, the activity of Y48F and K113A was completely abolished whereas K111A was mostly abolished. These results were the same as those obtained using E-selectin mutants. However, when they were tested for 2′6 sLe^x ($\alpha 2 \rightarrow 6$ siaylated derivative of Le^x) binding, both K111A and K113A ablated activity while Y48F had no effect. In sulfatide binding assays, only the K113A substitution eliminated binding. These results suggest that P- and E-selectin use common amino acids for 2′3 sLe^x and HL-60 cell recognition. However, an overlapping, but not identical set of amino acids may be used for P-selectin binding to 2′6 sLe^x and sulfatides.

Based on the crystal structure of MBP, Hollenbaugh *et al.* (1993) also constructed a three dimensional model of the lectin domain of P-selectin. Examination of the model and its solvent accessible surface indicated a shallow depression in the vicinity of the calcium site. This region is mainly formed by polar and charged residues including Lys 113, Tyr 48, and Tyr 94. To investigate the importance of residues in this groove for HL-60 cell adhesion, they generated site-specific P-selectin mutants comprising the extracellular domain of P-selectin fused to the hinge regions of human IgG_1. The single substitutions K113A, Y48A, Y48F, Y94A, and Y94F all abolished P-selectin binding to HL-60 cells, suggesting the critical role of Lys 113, Tyr 48 and Tyr 94 in ligand binding. In addition, mutations E92A, N107A and K111A also partially diminished HL-60 cell adhesion. Interestingly, a mutation at Asn 105 with Asp resulted in impaired HL-60 cell adhesion. Since Asn 105 is a proposed Ca^{+2} ligand, the mutation of Asn 105 to Asp is likely to preserve Ca^{+2} binding. The impaired ligand binding activity exhibited by N105D implied that Asn 105 not only serves as a Ca^{+2} ligand but also interacts with ligand. Taken together, these results were in agreement with those obtained by Erbe *et al.* (1993) and suggested that residues such as Tyr 48, Tyr 94, Glu 92, Lys 111 and Lys 113 play a common functional role in P- and E-selectin binding sites.

To further investigate the role of the side chain of Tyr 48 and Lys 113 in P-selectin ligand binding, Bajorath *et al.* (1994) generated Y48S and K113R mutants and found that neither bound HL-60 cells, indicating that a conservation of charged and hydrogen-bonding side chains is not sufficient to maintain P-selectin function. In a related paper, Hollenbaugh *et al.* (1995) investigated the side chain functions of Lys 111 and Lys 113 in P-selectin ligand binding. They generated K111C and K113C mutants, and as expected, both mutants exhibited reduced binding activity compared to the wild-type. These mutants as well as the wild-type P-selectin were then treated with thio group modifying reagents such as aziridine (AZ) and nipsylcysteamine (NC) to generate side chain amino groups that are 0.5–2.0Å longer than that of lysine. The resulting chemically modified K111C mutants displayed significant increases in HL-60 cell binding, suggesting that the side chain length at position 111 is not critical. However, this may not be the case at the 113 position because the chemically modified K113C did not lead to any detectable increase in binding, supporting the previous observation that the exact conformation of the side chain of Lys 113 is very critical for P-selectin ligand binding.

Characterization of blocking peptides

Synthetic peptides corresponding to residues 23–30, 54–63 and 70–79 of the NH_2-terminal lectin domain of P-selectin completely inhibit leukocyte adhesion to P-selectin (Geng *et al.*, 1992; Heavner *et al.*, 1993). Related peptides corresponding to the homologous 23–30 and 54–63 regions of E-selectin and L-selectin also prevent cell adhesion to P-selectin. Immobilized albumin conjugates of these P-selectin peptides support HL-60 cell adhesion. Geng *et al.* (1992) further showed that calcium interacts with the 23–30 and 54–63 peptides of all three selectins. However, the crystal structure of the E-selectin lectin domain reveals that calcium does not bind to these two regions (Graves *et al.*, 1994). Murphy and McGregor (1994) also showed that two peptides (residues 19–34 and 51–61) derived from the P-selectin lectin domain inhibit the adhesion of peripheral blood mononuclear cells and monocytes to thrombin-activated endothelial cells. However, site-specific mutagenesis of several residues in these regions of E- and P-selectin did not affect ligand binding and cell adhesion activity (Erbe *et al.*, 1992 and Hollenbaugh *et al.*, 1993). The crystal structure of the lectin and EGF domains of E-selectin also reveals that these regions are located outside the putative sLe^x binding region (Figure 6a). The role of these regions in P-selectin dependent cell adhesion remains unclear.

Ligand binding specificity of selectins

Although it has been shown that all three selectins can recognize sialylated, fucosylated lactosaminoglycans (such as sLe^x) (Polley *et al.*, 1991; Lasky, 1992; Berg *et al.*, 1992; Foxall *et al.*, 1992; Bevilacqua and Nelson, 1993; Mebius and Watson, 1993; Varki, 1994), many studies also indicate the clear differences in carbohydrate recognition by different selectins (see review by Varki, 1994). For example, in contrast to E-selectin, both L- and P-selectin can bind sulfatides (e.g. galactose-4-sulfate ceramide). In fact, a

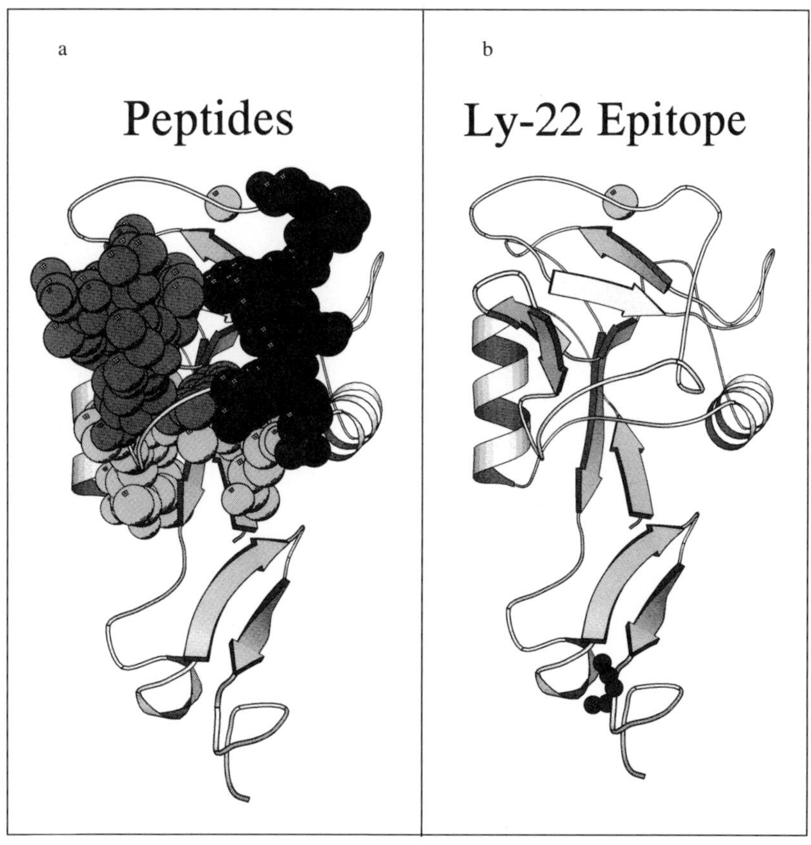

Figure 6 Some structure-function data which may implicate the involvement of the 'back side' of the selectins in ligand binding. **a,** The same schematic model with explicit atoms shown for peptides derived from P-selectin which have been shown to inhibit adhesion (Heavner *et al.*, 1993). The first peptide (light gray) consists of residues 23–30 and stretches along the underside of the lectin domain and forms much of the boundary between the lectin and EGF domains. The second peptide (medium gray) consists of residues 54–63 and is highly exposed on the back side of the protein. Four of these residues are conserved among the known sequences of E- and P-selectin. The third peptide (dark gray) is also highly exposed and consists of residues 70–79. Five of the residues are conserved **b**, MOLSCRIPT (Kraulis, 1991) ribbon diagram of lec-EGF E-selectin showing the site in the EGF domain (residue 146, dark gray spheres) that is part of the epitope for the Ly-22 neutralizing antibody directed against L-selectin (Siegelman *et al.*, 1990). This site is at the bottom of the EGF domain near the C-terminus and is quite remote from the carbohydrate binding site but close to the first CR element. This view is similar in orientation to Figure 7c.

recent report indicates that the major capping group on the carbohydrates of GlyCAM-1, a specific ligand protein for L-selectin, is 6′-sulfated sLe^x (Hemmerich and Rosen, 1994). Larsen and colleagues (1992) also reported that although the P- and E-selectin ligand share the same components (sialic acid and Le^x), they are structurally distinct. A protein component containing 2′3 sLe^x in proximity to sialyl-2,6βGal structures on the P-selectin ligand may contribute to its specificity for P-selectin. In fact, P-selectin is the only selectin that binds to the 2′6 sLe^x (Foxall *et al.*, 1992). Patel *et al.* (1994) reported that tetraantennary N-linked structures with a sialyl di-Le^x (a carbohydrate structure containing sLe^x plus Le^x) moiety isolated from HL-60 cells bind to E-selectin with much higher affinity than sLe^x, suggesting that sialyl di-Le^x is responsible for the specificity of E-selectin-dependent adhesion. Recent progress on the isolation of specific ligand proteins for each selectin should help to characterize the exact carbohydrate structures recognized by individual selectins.

To determine if the ligand binding differences of E- and P-selectin are attributable to differences in their lectin domains, Erbe *et al.* (1993) generated a chimera (PE-1) consisting of E-selectin-IgG with the E-selectin lectin domain replaced with the lectin domain from P-selectin. When PE-1 was tested for binding to various glycolipids, it appeared to closely mimic P-selectin-IgG in binding to 2′3 sLe^x, 2′6 sLe^x and sulfatides, suggesting that the ligand binding specificity is dictated by the lectin domain. Similar experiments were also carried out by Kansas *et al.* (1991) who generated chimeric selectins, in which the lectin and EGF-like domains of L-selectin were substituted into P-selectin. These chimeric selectins were found to bind to PPME and fucoidin, soluble complex carbohydrates that specifically define the lectin activity of L-selectin. Hence, the lectin domain of L-selectin determined carbohydrate binding specificity.

The role of the Ca^{+2} binding site

By equilibrium dialysis, we determined that E-selectin contains a high affinity Ca^{+2} site with a Kd of ≈3.5 μM (Anostario and Huang, 1995). However, at very high Ca^{+2} concentrations, we observed that more than one Ca^{+2} was able bind to E-selectin. Scatchard plot analysis indicated that the affinity for the second site was much lower. These results were in agreement with observations about the crystal structure of lec-EGF (Graves *et al.*, 1994). When lec-EGF was crystallized under high $CaCl_2$ concentrations, three sites were observed. However, only one showed high affinity coordination, and the other two were adventitious resulting from the crystallization conditions. When crystals were grown under low Ca^{+2} concentrations, only one binding site was observed. Interestingly, Geng *et al.* (1991) reported that P-selectin contains two indistinguishable high affinity Ca^{+2} sites (Kd = 22 μM by equilibrium dialysis and 4.8 μM by fluorescence emission intensity). In addition, P-selectin also contains distinct Mg^{+2} binding sites with 10 times lower affinities than the Ca^{+2} sites. Although the presence of Mg^{+2} alone is not sufficient to support neutrophil adhesion to P-selectin, addition of Mg^{+2} allows lower concentrations of Ca^{+2} to mediate cell adhesion.

Geng *et al.* (1991) showed that binding of Ca^{+2} or Mg^{+2} to P-selectin apparently induces a conformational change as indicated by a reduction in intrinsic fluorescence emission intensity and emission wavelength. However, when similar experiments were carried out on E-selectin, we observed only a very small difference in fluorescence intensity between apo- and the Ca^{+2}-bound form (Anostario and Huang, 1995). Furthermore, we did not detect any significant emission wavelength shift. We then used limited proteolysis as a probe to monitor E-selectin conformation. Apo-E-selectin was sensitive to limited proteolysis by Glu-C endoproteinase. Upon Ca^{+2} binding, the protein was protected from proteolysis. This Ca^{+2}-dependent protection was further augmented upon sLe^x ligand binding. These results implied that Ca^{+2} binding to E-selectin induces a conformational change and perhaps facilitates ligand binding. The sLe^x-bound complex in turn stabilizes Ca^{+2} binding.

When Glu-C-digested fragments of E-selectin were analyzed by N-terminal sequencing analysis, the results indicated that the major cleavage site is at Glu 98, which is in the loop (residues 94–103) adjacent to the Ca^{+2} binding region of the lectin domain. Although residues in this loop do not directly interact with Ca^{+2} (18Å away), they adopt an interesting conformation that tilts the loop towards Ca^{+2}. Removal of Ca^{+2} from E-selectin renders Asn 105 and Asp 106 (two Ca^{+2} ligands) more conformationally flexible which consequently influences the adjacent loop conformation such that it is susceptible to proteolysis. Since Tyr 94 and Arg 97 in this loop are critical for mediating cell adhesion, it is possible that the unusual conformation of this loop in the Ca^{+2} form of the protein facilitates its interaction with carbohydrate ligand. We also observed that although the proteolytic fragments were associated with each other through disulfide bonds, they did not retain Ca^{+2} or ligand binding activity. Thus, maintaining the rigid conformation of this loop is critical for E-selectin function. In addition to the major cleavage site at Glu 98, we also observed two minor cleavage sites at Glu 92 and Glu 107. Since these two residues are close to each other and adjacent to the Ca^{+2} site (≈ 6Å), it is possible that removal of Ca^{+2} from E-selectin also induces a conformational change in this region.

Asa *et al.* (1992) showed that Sr^{+2} is capable of replacing Ca^{+2} for restoring selectin function (all three selectins). In addition, Cu^{+2} and Mg^{+2} can also substitute for Ca^{+2} to restore L- and P-selectin function, respectively. By competitive equilibrium dialysis and proteolysis analysis, we demonstrated that Ba^{+2} bound to apo-E-selectin 5-fold tighter than Ca^{+2}. Interestingly, Ba^{+2} regenerated E-selectin did not show significant activity in binding to either HL-60 cells or an sLe^x carrying protein, carcinoembryonic antigen (CEA) (Anostario *et al.*, 1994; Anostario and Huang, 1995). Thus, Ba^{+2} is a potent antagonist. Sr^{+2} also bound to apo-E-selectin tighter than Ca^{+2}. However, Sr^{+2}-regenerated E-selectin exhibited 50% ligand binding activity. Mg^{+2} bound to apo-E-selectin with much weaker affinity than Ca^{+2} and did not show any activity. In summary, our results seem to suggest that Ca^{+2} binding to E-selectin induces a minor, yet critical, conformational change. Perturbations in the conformation of the Ca^{+2} binding region by either limited proteolysis or substitutions with other metal ions can completely abolish its function.

The EGF-like Domain

Unlike the lectin domain, the functional role of the EGF-like domain is less understood. By studying domain-deletion constructs of E-selectin, we (Li *et al.*, 1994) and Pigott *et al.* (1991) demonstrated that the minimum functional unit of E-selectin consists of both the lectin and EGF-like domains. These results suggested that the EGF-like domain is also involved in cell adhesion, either in stabilizing the conformation of the lectin domain or in directly interacting with the ligand.

Kansas *et al.* (1994) created a panel of chimeric selectins by exchange of domains between L- and P-selectin and examined their ligand binding specificities. Exchange of only the lectin domains between L- and P-selectin conferred the adhesion properties of the lectin domain of the parent molecule. However, chimeric selectins containing both the lectin domain of L-selectin and the EGF-like domain of P-selectin exhibited the adhesive properties of both L- and P-selectin. These chimeric proteins supported adhesion both to HL-60 cells and to high endothelial venules (HEV) of lymph nodes and mesenteric venules. These results suggested that the P-selectin EGF-like domain may participate directly in ligand recognition via protein-protein interactions. For example, the L-selectin lectin domain may naturally bind sLe^x or other carbohydrates on HL-60 cells with low affinity. This binding alone does not support HL-60 cell adhesion, but is enhanced by the presence of an adjoining P-selectin EGF-like domain. The adhesive function of the EGF-like domain of P-selectin seems unique to this selectin because the EGF-like domain of L-selectin does not appear to be required for maximal adhesion. This may explain why the amino acids of the P-selectin EGF-like domain are more highly conserved when compared with residues in either the E- or L-selectin EGF-like domains (Figure 2). We have shown that the EGF-like domain from human clotting Factor IX, can substitute for the endogenous EGF-like domain in human E-selectin and maintains the ability to mediate neutrophil adhesion which is indistinguishable from controls (Graves *et al.* 1994). Since the Factor IX EGF-like domain is only 40% homologous to the native domain, it may be likely that this region only plays a structural role in E-selectin and does not contribute specifically in ligand recognition.

It has also been shown that antibodies against the EGF domain of L-selectin can inhibit L-selectin-mediated cell adhesion (Siegelman *et al.*, 1990; Spertini *et al.*, 1991; Kansas *et al.*, 1994). The Ala at position 146 in the EGF domain of mouse L-selectin is part of the epitope for the neutralizing antibody Ly-22. As indicated in Figure 6b, residue 146 is near the C-terminus of the EGF domain. Clearly, even a neutralizing antibody which recognizes this region is not likely to impact directly on the carbohydrate binding site. A peptide derived from the EGF-like domain (residues 127–139) effectively inhibited monocyte or U937 cell adhesion to activated endothelial cells (Murphy and McGregor, 1994). However, results by Freedman *et al.* (1993) showed that the isolated EGF-like domain of P-selectin failed to inhibit HL-60 cell adhesion to CHO cells expressing P-selectin, which may argue against the direct involvement of the EGF-domain in adhesion.

A critical issue when studying requirements for selectin-mediated cell adhesion is the selectin density under the assay conditions. In early studies, Larkin *et al.* (1992) showed that the nature of the carbohydrate component of the ligand recognized by CHO cells expressing E-selectin varies with the density of E-selectin on the cell surface. Cells expressing E-selectin were able to recognize carbohydrates without terminal sialic acid only at very high densities. More recently, Gibson *et al.* (1995) incorporated purified P-selectin (membrane bound-form) at varying concentrations into phospholipid bilayers and studied density-dependent binding to HL-60 cells. The data suggest that the optimal density for P-selectin-mediated adhesion is in the range of 100–200 molecules/μm^2, which is comparable to the P-selectin density found on stimulated platelets. Chimeras of P- and L-selectin were then stably expressed in CHO cells, and clones that expressed the chimeras at the estimated physiologic density were isolated and examined in a HL-60 cell adhesion assay. The authors concluded that the lectin domain alone did support leukocyte binding, but was insufficient for maximal binding. The lectin and EGF-like domains together are required for optimal recognition.

The CR Domains

Although much evidence indicates that the CR domains of selectins are not required for ligand recognition, they may have a functional role in enhancing the ligand binding affinity. Watson *et al.* (1991) reported that the Mel 14 mAb, an adhesion blocking antibody which recognizes a conformational determinant in the lectin domain of L-selectin, showed very weak binding to a construct lacking the CR domains. This construct also showed a profound decrease in lectin-specific interactions with the carbohydrate PPME and a lack of interaction with the peripheral lymph node endothelium, suggesting that the CR domains may be involved in induction of lectin domain conformation and enhancing its activity.

We generated soluble E-selectin constructs containing the lectin and EGF domains plus different numbers of CR domains and examined their ability to block neutrophil and HL-60 cell adhesion to either immobilized E-selectin or cytokine-stimulated HUVEC monolayers (Li *et al.*, 1994). Only the construct containing all six CR domains could effectively block cell adhesion, suggesting that the CR domains of E-selectin enhance ligand binding. Although the detailed mechanism for this interaction is not clear, it is probably not due to CR-mediated oligomerization of the protein because all the constructs exist as monomers in solution. It is more likely that the CR domains interact with the non-sLe^x region of the ligand molecule to enhance overall avidity. Alternatively, the CR domains may be required for maintaining a proper molecular conformation for high affinity ligand binding. However, Hensley *et al.* (1994) reported that the soluble form of E-selectin is an asymmetric monomer with dimensions of approximately 25Å by 270Å as determined by velocity ultracentrifugation. By electron microscopy, Ushiyama *et al.* (1993) also showed that soluble P-selectin is an extended rod-like molecule. These results argue against strong interactions between the CR domains and the rest of molecule.

The functional importance of the CR domains is also supported by the isolation of a mAb, EL-246 that recognizes both L- and E-selectin (Jutila *et al.*, 1992). Domain mapping studies localized the EL-246 epitope in a common region of the CR domains of these two selectins. In *in vitro* cell adhesion experiments under either static or shear conditions, EL-246 effectively blocked the function of both E- and L-selectin. Furthermore, in an *in vivo* homing experiment, pretreatment of bovine lymphocytes with EL-246 blocked their ability to home to mouse peripheral lymph nodes by >65% (Bargatze *et al.*, 1994). These results demonstrate that a conserved epitope in the CR domains of L- and E-selectin is crucial for leukocyte-endothelial interactions.

The Cytoplasmic Domain

The cytoplasmic domain of selectins consists of only 20–35 amino acids. There is no sequence homology between the cytoplasmic domains of the different selectins. However, each of the selectin cytoplasmic tails is well conserved between different species (Bevilacqua and Nelson, 1993), suggesting that this region has distinct functions. Crovello *et al.* (1993) reported that P-selectin is rapidly phosphorylated in platelets upon thrombin-stimulation, with maximum phosphorylation observed at 15–30s. Approximately, a 0.5 molar ratio of phosphate to P-selectin was observed. Phosphoamino acid analysis of the phosphorylated P-selectin revealed that phosphorylation occurred on serine, threonine as well as tyrosine residues. Interestingly, phosphotyrosine and phosphothreonine disappeared within 5 min of platelet activation while the level of phosphoserine remained stable. A similar observation was also reported by Fujimoto and McEver (1993). They further demonstrated that Ser 788 in the cytoplasmic domain was the principal site of phosphorylation. More recently, Crovello *et al.* (1995) observed that P-selectin was also rapidly phosphorylated and dephosphorylated on histidine residues upon thrombin or collagen stimulation. Although the functional significance of P-selectin phosphorylation remains unknown, it may be involved in intracellular trafficking or activation-dependent signal transduction. For example, phosphorylation may be required for the secretion and fusion of the granule to the plasma membrane. Alternatively, phosphorylation may be required for P-selectin activation for interacting with its ligand, or it may facilitate the initiation of signalling events at the platelet plasma membrane.

In addition to phosphorylation, Fujimoto *et al.* (1993) also showed that the cytoplasmic domain of P-selectin is acylated with palmitic acid and stearic acid at Cys 766 through a thioester bond. In platelets, acylation does not appear to be dependent on thrombin stimulation, and does not seem to promote binding of P-selectin to the cytoskeleton. It will be interesting to find out if acylation regulates sorting, endocytosis, or other trafficking functions.

The cytoplasmic domain of P-selectin has been shown to control sorting of newly synthesized P-selectin to α-granules and Weibel-Palade bodies (Disdier *et al.*, 1992). It has also been shown to contain a sorting determinant that mediates rapid degradation in lysosomes (Green *et al.*, 1994). Deletion of 10 amino acids from the cytoplasmic domain of P-selectin slows down the internalization process. This sorting event may represent a means for P-selectin to down regulate expression in activated endothelial cells.

Although E-selectin is not targeted to secretory granules, it is also phosphorylated on one or more serine residues on the cell surface. Following cell surface expression, E-selectin is rapidly degraded in cultured endothelial cells (Smeets *et al.*, 1993). However, it is not clear whether phosphorylation is related to internalization.

L-selectin expression is rapidly down-regulated in response to leukocyte activation (Jung *et al.*, 1988; Jung and Dailey, 1990; Kishimoto *et al.*, 1990). Based on the rapid kinetics of downregulation, it has been postulated that L-selectin is cleaved by an activated membrane protease (Kishimoto *et al.*, 1989; Berg and James, 1990; June and Dailey, 1990). A 6-kDa L-selectin transmembrane and cytoplasmic domain peptide has recently been identified on activated leukocytes and on L-selectin transfectants (Kahn *et al.*, 1994). Appearance of the 6-kDa fragment on activated leukocytes correlated with the disappearance of the intact membrane-bound L-selectin and with the appearance of a soluble form of L-selectin. Amino acid sequencing analysis of the 6-kDa fragment revealed that the cleavage occurs between Lys 321 and Ser 322 in a short region (membrane-proximal cleavage domain consisting of 15 amino acids) between the second CR domain and the transmembrane domain. Interestingly, this short membrane proximal region is highly divergent among selectins. This is consistent with the observation that only L-selectin is rapidly downregulated in this manner.

To further understand the sequence specificity surrounding the cleavage site, Migaki *et al.* (1995) replaced the membrane-proximal cleavage domain of L-selectin with the corresponding region of E-selectin and found that the resulting chimera prevents L-selectin from shedding. Deletions of four or five amino acids in the L-selectin cleavage domain also inhibit L-selectin downregulation. However, point mutations of the cleavage site, as well as mutations of multiple conserved amino acids within the cleavage domain, do not significantly affect L-selectin shedding. These results imply that the proteolytic processing of L-selectin may depend on the length or secondary structure of the cleavage domain.

Although the cytoplasmic domain of selectins may not be directly involved in carbohydrate recognition, Kansas *et al.* (1993) showed that deletion of the COOH-terminal 11 amino acids from the cytoplasmic domain of L-selectin eliminated binding of lymphocytes to HEV in an *in vitro* frozen section assay, and also abolished leukocyte rolling *in vivo* in exteriorized rat mesenteric venules. More recently, Pavalko *et al.* (1995) demonstrated that the cytoplasmic domain of L-selectin interacts directly with cytoskeletal α-actinin and forms a complex with vinculin and possibly talin. The interaction is mediated by a region within the carboxy terminal 11 amino acids of the cytoplasmic domain. Interestingly, this 11 carboxy terminal sequence is not required for proper localization of L-selectin on the cell surface. These results imply that the cytoplasmic domain of L-selectin regulates leukocyte adhesion to endothelium by controlling cytoskeletal interactions and binding avidity. A recent publication suggests that the presentation of L-selectin on the microvillus of leukocytes is critical for the efficient interaction with ligands under conditions of flow (von Andrian *et al.*, 1995). The disruption of interactions with cytoskeletal elements would be expected to alter the cellular localization within the microvillus and may be responsible for modulating L-selectin dependent adhesion.

SELECTIN HOMOLOGY

Based on the E-selectin structure and models of P- and L-selectin, we analyzed the nature and extent of selectin homology. Figure 7 depicts the identical residues among several sequences of E- and P-selectins. The results are only mildly different if L-selectin sequences are also included. What can not be discerned from these space-filling renditions is that the entire hydrophobic core is completely conserved. As mentioned above, this makes it highly likely that all of the selectins have very similar three-dimensional structures. While internal conservation is expected for related proteins, surface conservation is generally not observed except in those regions which have functional importance (or unless the proteins diverged in the evolutionary tree relatively recently). Figure 7a focuses on the predicted binding site for sLe^x and shows an area of conservation that includes not only the residues involved in calcium coordination but also most of the residues identified by mutagenesis as being important for sugar binding. The conservation in this region is expected since all of the selectins bind calcium and sLe^x. Other surface regions generally do not exhibit a high level of conservation (Figure 7b) but there is an extended patch on the "back side" of the protein that is highly conserved (Figure 7c). This region is contiguous with the conserved region around the calcium and sLe^x binding sites and extends downward to include residues in the EGF domain (not shown). Since this level of conservation is not observed for most other surface regions, an intriguing possibility is that this may represent the binding site for common elements of the protein:protein interaction between the selectins and their ligands. The only bit of evidence that would argue against this proposal is the finding that Lys 74, which is not conserved but is on the edge of this region, appears to be part of the epitope for several non-neutralizing monoclonal antibodies (Erbe *et al.*, 1992). However, it should be mentioned that Lys 74 would appear to be an unlikely recognition point for an antibody since it is marginally exposed (the aliphatic portion interacts with the side chains of Leu 69 and Trp 76 and it is covered on the other side by Glu 71) and thus there is a good chance that

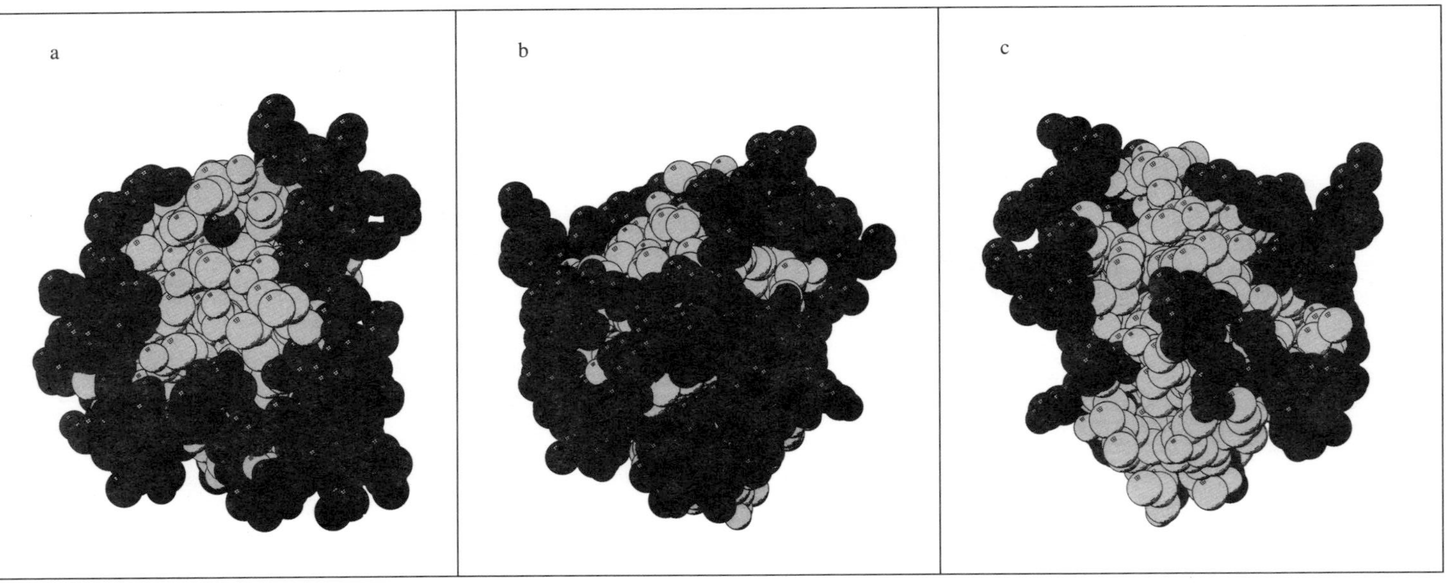

Figure 7 Selectin homology. The lectin domain of E-selectin is depicted as a CPK model in different views using the program MOLSCRIPT (Kraulis, 1991). The color coding is as follows: residues which are identical in all known sequences of E- and P-selectin (see Figure 2) are shown as light gray spheres, residues for which there is any divergence are shown as dark gray spheres and the calcium ion is black. **a**, Top view of the molecule showing the conservation of residues in the region around the calcium and carbohydrate binding sites. The orientation of this view is the same as in Figure 5. **b**, A side of the lectin domain which shows the typical lack of conservation expected for non-functional surface regions. The calcium ion can barely be seen in the top-middle portion of the figure and much of the conservation which is observed in this view is part of the site shown in **a**. **c**, The 'back side' of the lectin domain showing extended regions of absolute conservation. The conserved surface region continues to include residues in the EGF domain as well (not shown). The orientation of this view is such that Asn 82 and Asn 83 are at the top-middle part of the figure (also the top-middle in **a**). Thus, this region is contiguous with the conserved region around the calcium and carbohydrate binding sites.

mutation to alanine might cause a structural rearrangement that could affect the antibody binding at a distance.

A recent report from a group at Affymax (Martens *et al.*, 1995) details the discovery of peptides which bind to E-selectin and block neutrophil adhesion. These peptides were derived from random 12-mers, and bind specifically to E-selectin. These peptides do not depend on calcium for binding and are not competitive with sLe^x. Furthermore, within the 12-mers there are four residues which do not tolerate substitution and four more which maintain character. Of these eight positions, six are hydrophobic and if one analyzes the surface of lec-EGF E-selectin there are only two areas where there is a concentration of hydrophobic residues and both are on the "back side" of the molecule. The larger grouping is in the lectin domain and includes residues V56, V59, W60, V61, V63, P68, L69, W76, P78, P81, V101, and M103. This patch overlaps the region of E- and P-selectin identities mentioned above (Figure 7c). The other grouping is in the EGF domain and is comprised of V134, P146, F148, I156 and V157. As mentioned above, residue 146 is part of the epitope for the Ly-22 antibody and F148 is a residue commonly found to be important for protein:protein interactions with EGF domains. Thus, though we can not discount the possibility that these peptides bind to one or more of the CR domains, it seems that there are two reasonable interpretations of this data. Either the peptides bind to one or both of the hydrophobic patches on E-selectin and block binding of the protein component of the E-selectin ligand directly or the peptides bind to the crevice between the lectin and EGF domains and alter the spatial relationship between critical aspects of the lectin and EGF domains, thereby blocking binding indirectly. It would be exciting to test the blocking activity of these peptides against the lec-EGF form of E-selectin and, assuming they are still active, to pursue the structure of the complex. The selectivity and specificity of members of the selectin family for distinct ligands can not be explained entirely by the putative carbohydrate/Ca^{+2} binding site alone. The role of additional regions remains to be explored.

MODELS OF BOUND SLe^x

Unfortunately, all attempts to generate crystals with bound carbohydrate seemed to fail in that no significant electron density was ever observed that could be interpreted as sLe^x (or any of the numerous derivatives that were tested). Since the likely binding site was completely open to a solvent channel in the crystal, the possible interpretations are 1) sLe^x binds elsewhere on the surface of the protein, 2) the carbohydrate is there but adopts so many different conformations and/or orientations that the resulting density is too smeared out to observe, or 3) we were unable to achieve sufficient concentrations of the carbohydrate to shift the equilibrium toward the bound state. While explanation #2 seems the most likely, we have no direct evidence to confirm this. Thus, anything we might learn about the nature of bound carbohydrate will have to be surmised from the available structure-activity relationships and modeling efforts.

It has already been shown that the inactivating mutants map out a binding site for sLe^x which is consistent with the size and structure of this carbohydrate molecule. The two major, unresolved issues are the stoichiometry of sugar binding and, if only one molecule is bound, which end of the tetrasaccharide is interacting with the calcium ion. The initial assumption had been that the fucose moiety would coordinate to the calcium in a manner similar to mannose binding to MBP. This has been questioned on the following grounds: 1) given a choice, it probably makes chemical sense for the carboxylate of the sialic acid to coordinate calcium, 2) the fact that Glu 88 is not a calcium ligand (unlike the analogous Glu 193 in MBP) removes one of the hydrogen bonding interactions from the fucose hydroxyl, and 3) the observation that a second-site mutation can rescue the binding for the Asn 82 to Asp mutation. Thus, there had been some thought that the sialic acid moiety might be closest to the calcium. However, recently Kogan and colleagues (1995) have argued on the basis of additional mutagenesis and modeling results that indeed it is the fucose group that coordinates to the calcium ion. In addition, they speculate that the carboxylate of the sialic acid group interacts with the side chain of Arg 97 and that Lys 113 is not involved in sLe^x recognition. The non-involvement of Lys 113 is reasonable since it was argued (Graves *et al.*, 1994) that this residue seemed to have primarily a structural role. While the Kogan model of sLe^x binding may be correct, it is based on several assumptions that might argue for a more cautious interpretation. First and foremost, much of the argument is based on the ability to generate a mannose binding site from the sLe^x binding site on E-selectin. Given the structural similarity of the proteins in this region, it is not surprising that one could make E-selectin look and function like MBP. However, what this tells us about the binding of sLe^x is uncertain. The only direct bit of evidence is that the A77K mutation which confers mannose binding also abrogates sLe^x binding. The suggestion that lysine at position 77 causes a steric clash with the proposed bound conformation of sLe^x seems unlikely since the lysine side chain could readily rotate to avoid the interaction (see Figure 5 of Kogan *et al.*, 1995). Furthermore, as these authors have themselves cautioned, one must be careful in interpreting mutagenesis results. This is particularly true for A77K since if the lysine adopts a conformation identical to that observed in the mannose complex of MBP, the side chain will bump into Asp 100 which would require either one of the side chains to move (but Asp 100 has no room to rotate) or a rearrangement of the backbone. In either event, the $N\zeta$ of Lys 77 is not likely to be located at the same place as the equivalent atom in the MBP complex and the effect on sLe^x binding could result indirectly from a mild alteration of the structure. Second, the model of bound sLe^x assumes that it adopts the conformation as determined by Cooke *et al.* (1994), but the results of Hensley *et al.* (1994) and our own unpublished results (D. Fry *et al.*) indicate that sLe^x does not change its conformation significantly upon binding to E-selectin. Third, the modeling was done with the C2 and C3 hydroxyls coordinating the calcium ion but it is also possible that sLe^x could employ the C3 and C4 hydroxyls for this purpose (Cooke *et al.*, 1994). Thus, it seems clear that we still do not know enough to create a definitive model. However, the model proposed by Kogan *et al.* (1995) may provide a basis for formulating hypotheses about the important interactions between sLe^x and E-selectin.

SUMMARY

The structure of lec-EGF E-selectin has provided a sound basis for beginning to understand the molecular aspects of selectin-based adhesion. First, it is clear that the selectins will all have very similar structures and correlating the mutagenesis results with the structure has defined a reasonable binding site for the carbohydrate component of the selectin ligands. Secondly, the availability of the structure has given rise to further hypotheses concerning the involvement of other areas of the selectins in binding to their ligands and these hypotheses are immediately testable. Progress in the identification and characterization of native ligands may provide an opportunity for generating a crystal structure of a selectin-ligand complex. This structural information will further define the binding specificity among the different selectins and may prove to be critical for driving the rationale design of high affinity small molecule selectin antagonists. The pivotal role played by the selectin family of adhesion molecules in initiating the process of extravasation of leukocytes provides a strong rationale for developing selectin antagonists for the treatment of inflammatory and autoimmune diseases. The challenge for the future will be our ability to convert our understanding of leukocyte-endothelial cell interactions into working therapies for the treatment diseases.

Acknowledgements

We would like to acknowledge the following colleagues for their scientific contributions to the research described in this chapter: Michael Anastario, Dan Burns, Chitra Candran, Robert Crowther, Philip Familletti, David Fry, Shirley Li, Chris Norton, David Presky, Rob Ramos and John Rumberger.

References

Anostario, M., Jr., Li, S. and Huang, K-S. (1994). A ligand binding assay for E-selectin. *Anal. Biochem.*, **221**:317–322.

Anostario, M., Jr. and Huang, K-S. (1995). Modulation of E-selectin structure/function by metal ions. Studies on limited proteolysis and metal ion regeneration. *J. Biol. Chem*,. **270**:8138–8144.

Asa, D., Gant, T., Oda, Y. and Brandley, B.K. (1992). Evidence for two classes of carbohydrate binding sites on selectins. *Glycobiology*, **2**:395–400.

Auchampach, J.A., Oliver, M.G., Anderson, D.C. and Manning, A.M. (1994). Cloning, sequence comparison and *in vivo* expression of the gene encoding rat P-selectin. *Gene*, **145**:251–255.

Bajorath, J., Hollenbaugh, D., King, G., Harte, W., Eustice, D.C., Darveau, R.P. and Aruffo, A. (1994). CD 62/p-selectin binding sites for myeloid cells and sulfatides are overlapping. *Biochemistry*, **33**: 1332–1339.

Bargatze, R.F., Kurk, S., Watts, G., Kishimoto, T.K., Speer, C.A. and Jutila, M.A. (1994). *In vivo* and *in vitro* functional examination of a conserved epitope of L- and E-selectin crucial for leukocyte-endothelial cell interactions. *J. Immunol.*, **152**:5814–5825.

Becker-Andre, M., Van Huijsduijnen, R.H., Losberger, C., Whelan, J. and Delamarter, J.F. (1992). Murine endothelial leukocyte adhesion molecule-1 is a close structural and functional homologue of the human protein. *Eur. J. Biochem.*, **206**:401–411.

Berg, M. and James, S.P. (1990). Human neutrophils release the Leu-8 lymph node homing receptor during cell activation. *Blood*, **76**:2381–2388.

Berg, E.L., Magnani, J., Warnock, R.A., Robinson, M.K. and Butcher, E.C. (1992). Comparison of L-selectin and E-selectin ligand specificities: the L-selectin can bind the E-selectin ligands sialyl Le^x and sialyl Le^a. *Biochem. Biophys. Res. Commun.*, **184**:1048–1055.

Berg, E.L., Fromm, C., Melrose, J. and Tsurushita, N. (1995). Antibodies cross-reactive with E- and P-selectin block both E- and P-selectin functions. *Blood*, **85**:31–37.

Bevilacqua, M.P., Stengelin, S., Gimbrone, M.A., Jr. and Seed, B. (1989). Endothelial leukocyte adhesion molecule-1: an inducible receptor for neutrophils related to complement regulatory proteins and lectins. *Science*, **243**:1160–1165.

Bevilacqua, M.P. and Nelson, R.M. (1993). Selectins. *J. Clin. Invest.*, **91**: 379–387.

Bezouska, K., Crichlow, G.V., Rose, J.M., Taylor, M.E. and Drickamer, K. (1991). Evolutionary conservation of intron position in a subfamily of genes encoding carbohydrate-recognition domains. *J. Biol. Chem.*, **266**:11604–11609.

Bowen, B., Fennie, C., and Lasky, L.A. (1990). The Mel 14 antibody binds to the lectin domain of the murine peripheral lymph node homing receptor. *J. Cell. Biol.*, **110**:147–153.

Camerini, D., James, S.P., Stamenkovic, I. and Seed, B. (1989). Leu-8/TQ 1 is the human equivalent of the Mel-14 lymph node homing receptor. *Nature*, **342**:78–82.

Chou, K.C. (1995). The convergence-divergence duality in lectin domains of selectin family and its implications *FEBS Lett.*, **363**:123–126.

Collins, T., Williams, A., Johnston, G.I., Kim, J., Eddy, R., Shows, T., Gimbrone, M.A. and Bevilaqua, M.P. (1991). Structure and chromosomal location of the gene for endothelial-leukocyte adhension molecule 1. *J. Biol. Chem.*, **266**:2466–2473.

Cooke, R.M., Hale, R.S., Lister, S.G. and Weir, M.P. (1994). The conformation of the sialyl Lewis x ligand changes upon binding to E-selectin. *Biochemistry*, **33**:10591–10596.

Crovello, C.S., Furie, B. and Furie, B. (1993). Rapid phosphorylation and selective dephosphorylation of P-selectin accompanies platelet activation. *J. Biol. Chem.*, **268**:14590–14593.

Crovello, C.S., Furie, B. and Furie, B. (1995). Histidine phosphorylation of P-selectin upon stimulation of human platelets: a novel pathway for activation-dependent signal transduction. *Cell*, **82**:279–286.

Disdier, M., Morrissey, J.H., Fugate, R.D., Bainton, D.F. and McEver, R.P. (1992). Cytoplasmic domain of P-selectin (CD62) contains the signal for sorting into the regulated secretory pathways. *Molecular Biology of the Cell*, **3**:309–321.

Donnelly, S.C., Haslett, C., Dransfield, I., Robertson, C.E., Carter, D.C., Ross, J.A., Grant, I.S. and Tedder, T.F. (1994). Role of selectins in development of adult respiratory distress syndrome. *Lancet*, **344**:215–219.

Dowbenko, D.J., Diep, A., Taylor, B.A., Lusis, A.J. and Lasky, L.A. (1991). Characterization of the murine homing receptor gene reveals correspondence between protein domains and coding exons. *Genomics*, **9**:270–277.

Drickamer, K. (1988). Two distinct classes of carbohydrate-recognition domains in animal lectins. *J. Biol. Chem.*, **263**:9557–9560.

Dunlop, L.C., Skinner, M.P., Bendall, L.J., Favaloro, E.J., Castaldi, P.A., Gorman, J.J., Gamble, J.R., Vadas, M.A. and Berndt, M.C. (1992). Characterization of GMP-140 (P-selectin) as a circulating plasma protein. *J. Exp. Med.*, **175**:1147–1150.

Erbe, D.V., Wolitzky, B.A., Presta, L.G., Norton, C.R., Ramos, R.J., Burns, D.K., Rumberger, J.M., Rao, N., Foxall, C., Brandley, B.K. and Lasky, L.A. (1992). Identification of an E-selectin region critical for carbohydrate recognition and cell adhesion. *J. Cell Biol.*, **119**:215–227.

Erbe, D.V., Watson, S.R., Presta, L.G., Wolitzky, B.A., Foxall, C., Brandley, B.K. and Lasky, L.A. (1993). P- and E-selectin use common sites for carbohydrate ligand recognition and cell adhesion. *J. Cell Biol.*, **120**:1227–1235.

Fitzgerald, P.M.D. (1988). MERLOT: An integrated package of computer programs for the determination of crystal structures by molecular replacement. *J. Appl. Crystallogr.*, **21**:273.

Foxall, C., Watson, S.R., Dowbenko, D., Fennie, C., Lasky, L.A., Kiso, M., Hasegawa, A., Asa, D. and Brandley, B.K. (1992). The three members of the selectin receptor family recognize a common carbohydrate epitope, the sialyl Lewis^x oligosaccharide. *J. Cell Biol.*, **117**:895–902.

Freedman, S.J., Furie, B. and Furie, B.C. (1993). The lectin domain but not the EGF domain is a potent inhibitor of P-selectin-mediated cellular adhesion. *Blood*, **82**:341a (suppl. abstr.)

Fujimoto, T. and McEver, R.P. (1993). The cytoplasmic domain of P-selectin is phosphorylated on serine and threonine residues. *Blood*, **82**:1758–1766.

Fujimoto, T., Stroud, E., Whatley, R.E., Prescott, S.M., Muszbek, L., Laposata, M. and McEver, R.P. (1993). P-selectin is acylated with palmitic acid and stearic acid at cysteine 766 through a thioester linkage. *J. Biol. Chem.*, **268**:11394–11400.

Furey, W. and Swaminathan, S. (1995). PHASES–95: A program for the processing and analysis of diffraction data from macromolecules. Methods in Enzymology (C. Carter & R. Sweet, eds., Academic Press), in press.

Gearing, A.J.H. and Newman, W. (1993). Circulating adhesion molecules in disease. *Immunol. Today*, **14**:506–512.

Geng, J.G., Bevilacqua, M.P., Moore, K.L., McIntyre, T.M., Prescott, S.M., Kim, J.M., Bliss, G.A., Zimmerman, G.A. and McEver, R.P. (1990). Rapid neutrophil adhesion to activated endothelium mediated by GMP140. *Nature*, **343**:757–760.

Geng, J.G., Moore, K.L., Johnson, A.E. and McEver, R.P. (1991). Neutrophil recognition requires a Ca^{+2}-induced conformational change in the lectin domain of GMP-140. *J. Biol. Chem.*, **266**: 22313–22318.

Geng, J.G., Heavner, G.A. and McEver, R.P. (1992). Lectin domain peptides from selectins interact with both cell surface ligands and Ca^{+2} ions. *J. Biol. Chem.*, **267**:19846–19853.

Gibson, R.M., Kansas, G.S., Tedder, T.F., Furie, B. and Furie, B.C. (1995). Lectin and epidermal growth factor domains of P-selectin at physiologic density are the recognition unit for leukocyte binding. *Blood*, **85**:151–158.

Graves, B.J., Crowther, R., Chandran, C., Rumberger, J.M., Li, S., Huang, K.-S., Presky, D.H., Familletti, P.C., Wolitzky, B.A. and Burns, D.K. (1994). Insight into E-selectin/ligand interaction from the crystal structure and mutagenesis of the lec/EGF domains. *Nature*, **367**:532–538.

Green, S.A., Setiadi, H., McEver, R.P. and Kelly, R.B. (1994). The cytoplasmic domain of P-selectin contains a sorting determinant that mediates rapid degradation in lysosomes. *J. Cell Biol.*, **124**:435–448.

Heavner, G.A., Falcone, M., Kruszynski, M., Epps, L., Mervic, M., Riexinger, D. and McEver, R.P. (1993). Peptides from multiple regions of the lectin domain of P-selectin inhibiting neutrophil adhesion. *Int. J. Pept. Protein Res.*, **42**:484–489.

Hemmerich, S. and Rosen, S.D. (1994). 6′-Sulfated sialyl Lewis x is a major capping group of GlyCAM-1. *Biochemistry*, **33**:4830–4835.

Hensley, P., McDevitt, P.J., Brooks, I., Trill, J.J., Feild, J.A., McNulty, D.E., Conor, J.R., Griswold, D.E., Kumar, N.V., Kopple, K.D., Carr, S.A., Dalton, B.J. and Johanson, K. (1994). The soluble form of E-selectin is an asymmetric monomer. Expression, purification and characterization of the recombinant protein. *J. Biol. Chem.*, **269**:23949–23958.

Hollenbaugh, D., Bajorath, J., Stenkamp, R. and Aruffo, A. (1993). Interaction of P-selectin (CD62) and its cellular ligand: analysis of critical residues. *Biochemistry*, **32**:2960–2966.

Hollenbaugh, D., Aruffo, A. and Senter, P.D. (1995). Effects of chemical modification on the binding activities of P-selectin mutants. *Biochemistry*, **34**:5678–5684.

Hommel, U., Harvey, T.S., Driscoll, P.C. and Campbell, I.D. (1992). Human epidermal growth factor — high resolution solution structure and comparison with human transforming growth factor a. *J. Molec. Biol.*, **227**:271–282.

Johnston, G.I, Cook, R.G. and McEver, R.P. (1989). Cloning of GMP-140, a granule membrane protein of platelets and endothelium: sequence similarity to proteins involved in cell adhesion and inflammation. *Cell*, **56**:1033–1044.

Johnston, G.I., Bliss, G.A., Newman, P.J. and McEver, R.P. (1990). Structure of the human gene encoding granule membrane protein-140, a member of the selectin family of adhesion receptors for leukocytes. *J. Biol. Chem.*, **265**:21381–21385.

Jung, T.M., Gallatin, W.M., Weissman, I.L. and Dailey, M.O. (1988). Downregulation of homing receptors after T-cell activation. *J. Immunol.*, **141**:4110–4117.

Jung, T.M. and Dailey, M.O. (1990). Rapid modulation of homing receptors (gp90 Mel-14) induced by activators of protein kinase C. *J. Immunol.*, **144**:3130–3136.

Jutila, M.A., Watts, G., Walcheck, B. and Kansas, G.S. (1992). Characterization of a functionally important and evolutionarily well-conserved epitope mapped to the short consensus repeats of E-selectin and L-selectin. *J. Exp. Med.*, **175**:1565–1573.

Kahn, J., Ingraham, R.H., Shirley, F., Migaki, G.I. and Kishimoto, T. (1994). Membrane proximal cleavage of L-selectin: identification of the cleavage site and a 6-kD transmembrane peptide fragment of L-selectin. *J. Cell Biol.*, **125**:461–470.

Kansas, G.S., Spertini, O., Stoolman, L.M. and Tedder, T.F. (1991). Molecular mapping of functional domains of the leukocyte receptor for endothelium, LAM-1. *J. Cell Biol.*, **114**:351–358.

Kansas, G.S., Ley, K., Munro, J.M. and Tedder, T.F. (1993). Regulation of leukocyte rolling and adhesion to high endothelial venules through the cytoplasmic domain of L-selectin. *J. Exp. Med.*, **177**:833–838.

Kansas, G.S., Saunders, K.B., Ley, K., Zakrzewicz, A., Gibson, R.M., Furie, B.C. and Tedder, T.F. (1994). A role for the epidermal growth factor-like domain of P-selectin in ligand recognition and cell adhesion. *J. Cell Biol.*, **124**:609–618.

Kishimoto, T.K., Jutila, M.A., Berg, E.L. and Butcher, E.C. (1989). Neutrophil Mac-1 and MEL-14 adhesion proteins inversely regulated by chemotactic factors. *Science*, **245**:1238–1241.

Kishimoto, T.K., Jutila, M.A. and Butcher, E.C. (1990). Identification of a human peripheral lympho node homing receptor: a rapidly downregulated adhesion molecule. *Proc. Natl. Acad. Sci. USA*, **87**:2244–2248.

Kogan, T.P., Revelle, B.M., Tapp, S., Scott, D. and Beck, P.J. (1995). A single amino acid residue can determine the ligand specificity of E-selectin. *J. Biol. Chem.*, **270**:14047–14055.

Kraulis, P.J. (1991). MOLSCRIPT: A program to produce both detailed and schematic plots of protein structures. *J. Appl. Crystallogr.*, **24**:946–950.

Larigan, J.D., Tsang, T.C., Rumberger, J.M. and Burns, D.K. (1992). Characterization of cDNA and genomic sequences encoding rabbit ELAM-1: conservation of structure and functional interactions with leukocytes. *DNA and Cell Biol.*, **11**:149–162.

Larkin, M., Ahern, T.J., Stoll, M.S., Shaffer, M., Sako, D., O'Brien, J., Yuen, C.-T., Lawson, A.M., Childs, R.A., Baron, K.M., Langer-Safer, P.R., Hasegawa, A., Kiso, M., Larsen, G.R. and Feizi, T. (1992). Spectrum of sialylated and nonsialylated fuco-oligosaccharides bound by the endothelial-leukocyte adhesion molecule E-selectin. *J. Biol. Chem.*, **267**:13661–13668.

Larsen, G.R., Sako, D., Ahern, T.J., Shaffer, M., Erban, J., Sajer, S.A., Gibson, R.M., Wagner, D.D., Furie, B.C. and Furie, B. (1992). P-selectin and E-selectin: distinct but overlapping leukocyte ligand specificities. *J. Biol. Chem.*, **267**:11104–11110.

Lasky, L.A., Singer, M.S., Yednock, T.A., Dowbenko, D., Fennie, C., Rodriguez, H., Nguyen, T., Stachel, S. and Rosen, S.D. (1989). Cloning of a lymphocyte homing receptor reveals a lectin domain. *Cell*, **56**:1045–1055.

Lasky, L.A. (1992). Selectins: Interpreters of cell-specific carbohydrate Information during inflammation. *Science*, **258**:964–969.

Li, S.H., Burns, D.K., Rumberger, J.M., Presky, D.H., Wilkinson, V.L., Anostario, M. Jr., Wolitzky, B.A., Norton, C.R., Familletti, P.C., Kim, K.J., Goldstein, A.L., Cox, D.C. and Huang, K-S. (1994). Consensus repeat domains of E-selectin enhance ligand binding. *J. Biol. Chem.*, **269**:4431–4437.

Martens, C.L., Cwirla, S.E., Lee, R.Y., Whitehorn, E., Chen, E.Y., Bakker, A., Martin, E.L., Wagstrom, C., Gopalan, P. and Smith, C.W. (1995). Peptides which bind to E-selectin and block neutrophil adhesion. *J. Biol. Chem.*, **270**:21129–21136.

Mebius, R.E. and Watson, S.R. (1993). L- and E-selectin can recognize the same naturally occuring ligands on high endothelial venules. *J. Immunol.*, **151**:3252–3260.

Migaki, G.I., Kahn, J. and Kishimoto, T.K. (1995). Mutational analysis of the membrane-proximal cleavage site of L-selectin: relaxed sequence specificity surrounding the cleavage site. *J. Exp. Med.*, **182**:549–557.

Murphy, J.F. and McGregor, J.L. (1994). Two sites on P-selectin (the lectin and epidermal growth factor-like domains) are involved in the adhesion of monocytes to thrombin-activated endothelial cells. *Biochem. J.*, **303**:619–624.

Nguyen, M., Strubel, N.A. and Bischoff, J. (1993). A role for sialyl Lewis-X/A glycoconjugates in capillary morphogenesis. *Nature*, **16**:267–269.

Padmanabhan, K., Padmanabhan, K.P., Tulinsky, A., Park, C.H., Bode, W., Huber, R., Blankenship, D.T., Cardin, A.D. and Kisiel, W. (1993). Structure of Human Des(1–45) Factor Xa at 2.2Å Resolution. *J. Molec. Biol.*, **232**:947–966.

Patel, T.P., Goelz, S.E., Lobb, R.R. and Parekh, R.B. (1994). Isolation and characterization of natural protein-associated carbohydrate ligands for E-selectin. *Biochemistry*, **33**:14815–14824.

Pavalko, F.M., Walker, D.M., Graham, L., Goheen, M., Doerschuk, C.M. and Kansas, G.S. (1995). The cytoplasmic domain of L-selectin interacts with cytoskeletal protein via α-actinin: Receptor positioning in microvilli does not require interaction with α-actinin. *J. Cell Biol.*, **129**:1155–1164.

Pigott, R., Needham, L.A., Edwards, R.M., Walker, C. and Power, C. (1991). Structural and functional studies of the endothelial activation antigen endothelial leukocyte adhesion molecule-1 using a panel of monoclonal antibodies. *J. Immunol.*, **147**:130–135.

Polly, M.J., Phillips, M.L., Wayner, E., Nudelman, E., Finghal, A.K., Hackamore, S. and Paulson, J.C. (1991). CD62 and endothelial cell-leukocyte adhesion molecule 1 (ELAM-1) recognize the same carbohydrate ligand, sialyl-Lewis x. *Proc. Natl. Acad. Sci. USA*, **88**:6224–6228.

Rosen, S.D. (1989). Lymphocyte homing: progress and prospects. *Curr. Opin. Cell Biol.*, **1**:913–919.

Rosen, S.D. (1990). The LEC-CAMs: an emerging family of cell adhesion receptors based upon carbohydrate recognition. *Am. J. Respir. Cell Mol. Biol.*, **3**:397–402.

Siegelman, M. and Weissman, I. (1989). Homolog of mouse lymph node homing receptor: evolutionary conservation at tandem cell interaction domains. *Proc. Natl. Acad. Sci. USA*, **86**:5562–5566.

Siegelman, M., van de Rijn, M. and Weissman, I. (1989). Mouse lymph node homing receptor cDNA clone encodes a glycoprotein revealing tandem interaction domains. *Science*, **243**:1165–1172

Siegelman, M.H., Cheng, I.C., Weissman, I.L. and Wakeland, E.K. (1990). The mouse lymph node homing receptor is identical with the lymphocyte cell surface marker Ly-22: Role of EGF domain in endothelial binding. *Cell*, **61**:611–622.

Smeets, E.F., de Vries, T., Leeuwenberg, J.F.M., van den Eijnden, D., Buurman, W.A. and Neefjes, J.J. (1993). Phosphorylation of surface E-selectin and the effect of soluble ligand (sialyl Lewis^x) on the half-life of E-selectin. *Eur. J. Immunol.*, **23**:147–151.

Spertini, O., Kansas, G.S., Reinmann, K.A., Mackay, C.R. and Tedder, T.F. (1991). Functional and evolutionary conservation of distinct epitopes on the leukocyte adhesion molecule (LAM-1) that regulate leukocyte migration. *J. Immunol.*, **147**:942–949.

Strubel, N., Nguyen, M., Kansas, G. S., Tedder, T. F. and Bischoff, J. (1993). Isolation and characterization of a bovine endothelial cDNA encoding a functional homolog of human P-selectin. *Biochem. Biophys. Res. Commun.*, **192**:338–344.

Tsang, Y.T.M., Stephens, P.E., Licence, S.T., Haskard, D.O., Binns, R.M. and Robinson, M.K. (1995). Porcine E-selectin: cloning and functional characterization. *Immunology*, **85**:140–145.

Ushiyama, S., Laue, T.M., Moore, K.L., Erickson, H.P. and McEver, R.P. (1993). Structural and functional characterization of monomeric soluble P-selectin and comparison with membrane P-selectin. *J. Biol. Chem.*, **268**:15229–15237.

Varki, A. (1994). Selectin ligands. *Proc. Natl. Acad. Sci. USA*, **91**:7390–7397.

von Andrian, U.H., Hasslen, S.R., Nelson, R.D., Eriandsen, S.L. and Butcher, E.C. (1995). A central role for microvillons receptor presentation in leukocyte adhesion under flow. *Cell*, **82**:989–999.

Watanabe, T., Song, Y., Hirayama, Y., Tamatani, T., Kuida, K. and Miyasaka, M. (1992). Sequence and expression of a rat cDNA for LECAM-1. *Biochim. Biophys. Acta.*, **15**:321–324.

Watson, M.L., Kingsmore, S.F., Johnston, G.I., Siegelman, M.H., Le Beau, M.M., Lemons, R.S., Bora, N.S., Howard, T.A., Weissman, I.L., McEver, R.P. and Seldin, M.F. (1990). Genomic organization of the selectin family of leukocyte adhesion molecules on human and mouse chromosome 1. *J. Exp. Med.*, **172**:263–272.

Watson, S.R., Imai, Y., Fennie, C., Geoffrey, J., Singer, M., Rosen, R. and Lasky, L.A. (1991). The complement binding-like domains of the murine homing receptor facilitate lectin activity. *J. Cell Biol.*, **115**:235–243.

Weller, A., Isenmann, S. and Vestweber, D. (1992). Cloning of the mouse endothelial selectins.Expression of both E- and P-selectin is inducible by tumor necrosis factor alpha. *J. Biol. Chem.*, **25**:5176–5183.

Weis, W.I., Kahn, R., Fourme, R., Drickamer, K. and Hendrickson, W.A. (1991). Structure of the calcium-dependent lectin domain from a rat mannose-binding protein determined be MAD phasing. *Science*, **254**:1608–1615.

Weis, W.I., Drickamer, K. and Hendrickson, W.A. (1992). Structure of a C-type mannose-binding protein complexed with an oligosaccharide. *Nature*, **360**:127–134.

2 Regulation of Expression of E-selectin and P-selectin

Rodger P. McEver

The W.K. Warren Medical Research Institute, Departments of Medicine and Biochemistry and Molecular Biology, University of Oklahoma Health Sciences Center, and the Cardiovascular Biology Research Program, Oklahoma Medical Research Foundation, Oklahoma City, OK 73104, USA

Address correspondence to: Rodger P. McEver, M.D., W.K. Warren Medical Research Institute, University of Oklahoma Health Sciences Center, 825 N. E. 13th Street, Oklahoma City, OK 73104, USA

E- and P-selectin are synthesized by endothelial cells, and P-selectin is also synthesized by megakaryocytes, where it is incorporated into platelets. The surface expression of both molecules is regulated at the level of synthesis, subcellular trafficking, and degradation. E- and P-selectin function as adhesion receptors for neutrophils, monocytes, eosinophils, basophils, and subsets of lymphocytes. Physiologic regulation of the surface expression of these glycoproteins helps control the degree and duration of the inflammatory response. Conversely, dysregulated expression of E- and P-selectin may contribute to a variety of inflammatory and thrombotic disorders. This chapter describes our current understanding of the regulation of expression of E-selectin and P-selectin.

REGULATION OF EXPRESSION OF E-SELECTIN

Inducible Synthesis and Subcellular Trafficking of E-selectin

E-selectin was first identified as a cytokine-inducible adhesion receptor for myeloid cells on cultured human umbilical vein endothelial cells (HUVEC) (Bevilacqua *et al.*, 1987). Unstimulated HUVEC synthesize little or no E-selectin. However, addition of interleukin-1 (IL-1), tumor necrosis factor-α (TNF-α), or liposaccharide (LPS) causes transient expression of E-selectin mRNA and protein, with peak levels reached after 4–6 hours (Figure 1). Levels of both mRNA and protein then decline to basal levels over the next 12–24 hours, even in the continuous presence of agonist (Bevilacqua *et al.*, 1989). Nuclear run-on assays indicate that cytokines directly induce transcription of mRNA encoding E-selectin (Montgomery *et al.*, 1991; Whelan *et al.*, 1991). Upon cessation of transcription, steady-state mRNA levels rapidly decline, presumably because of AUUUA destabilizing sequences present in the 3′ untranslated region (Bevilacqua *et al.*, 1989). Newly synthesized E-selectin protein enters the endoplasmic reticulum, where several high mannose N-linked oligosaccharides are attached. These oligosaccharides are modified to complex forms as the protein passes through the Golgi complex before reaching

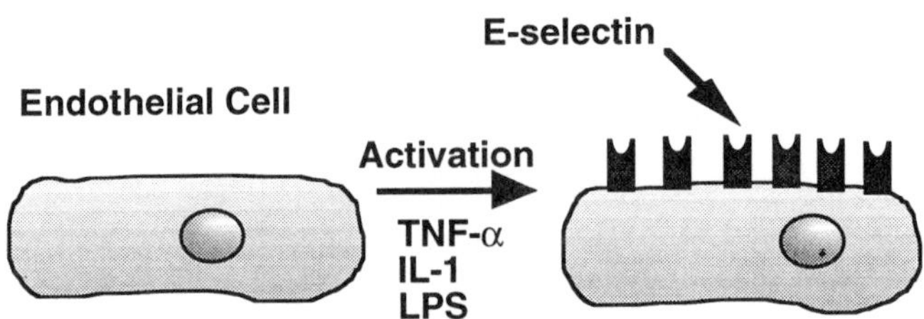

Figure 1 Inducible expression of E-selectin in endothelial cells. Inflammatory mediators such as TNF-α, IL-1, or LPS induce the transient synthesis of E-selectin, which is then transported to the cell surface.

the cell surface (Bevilacqua *et al.*, 1989). E-selectin has a short half-life (Bevilacqua *et al.*, 1989), because it is internalized from the cell surface and then delivered much more rapidly than most membrane proteins to lysosomes, where it is degraded (Von Asmuth *et al.*, 1992; Smeets *et al.*, 1993; Kuijpers *et al.*, 1994). Thus, in the absence of continued mRNA synthesis, E-selectin disappears from the cell surface in only a few hours. The cytoplasmic domain of E-selectin probably contains signals for internalization into coated pits and for rapid movement from endosomes to lysosomes. This possibility has not been studied extensively as it has for P-selectin (see below). However, deletion of the cytoplasmic domain of E-selectin does prevent endocytosis (H. Setiadi and R.P. McEver, unpublished observations). The cytoplasmic tail of E-selectin is phosphorylated on serine residues, but this modification does not obviously affect subcellular trafficking (Smeets *et al.*, 1993).

Soluble E-selectin antigen is found in the medium of cultured cytokine-activated endothelial cells (Pigott *et al.*, 1992), and in human plasma and synovial fluids (Gearing and Newman, 1993). The soluble antigen is most likely derived from proteolytic cleavage or by shedding of small membrane vesicles. The levels of antigen are extremely low and unlikely to be physiologically significant. Antigen levels are elevated in some inflammatory disorders, making them potential markers of disease activity (Carson *et al.*, 1993; Carson *et al.*, 1994).

In vivo, E-selectin is not constitutively synthesized by human endothelial cells or other tissues. However, cytokines or LPS induce the expression of E-selectin at sites of inflammation, as detected by immunocytochemistry (Cotran *et al.*, 1986). Expression of E-selectin is limited to endothelial cells, particularly in postcapillary venules. In some human inflammatory tissues, particularly skin, the expression of E-selectin is prolonged rather than transient as in cultured HUVEC (Cotran *et al.*, 1986). This persistent expression may reflect use of an alternative polyadenylation site in the E-selectin transcript that deletes the AUUUA instability elements in the 3′ untranslated region. *In vitro*, such transcripts have much longer half-lives (Chu *et al.*, 1994).

IL-1, TNF-α, and LPS also induce transient expression of E-selectin mRNA and protein in cultured murine and bovine endothelial cells (Weller *et al.*, 1992; Hahne *et al.*, 1993) and in murine endothelial cells *in vivo* (Weller *et al.*, 1992). Thus, at least some mechanisms for inducible expression are conserved across species. Phorbol esters and oxygen radicals have similar stimulatory effects on E-selectin expression (Montgomery *et al.*, 1991). Interferon γ, while not directly inducing synthesis of E-selectin, prolongs the half-life of E-selectin transcripts induced by IL-1 or TNF-α (Leeuwenberg *et al.*, 1990). In contrast, transforming growth factor-β (Gamble *et al.*, 1993), IL-4 (Thornhill and Haskard, 1990), corticosteroids (Cronstein *et al.*, 1992), and elevation of cAMP (Parhami *et al.*, 1993; Pober *et al.*, 1993; Ghersa *et al.*, 1994) inhibit E-selectin mRNA induction by IL-1, TNF-α, or LPS. These observations indicate that the inducible expression of E-selectin can be modulated, perhaps through different mechanisms.

Transcription Regulation of E-selectin

The human gene encoding E-selectin is divided into exons that encode structurally distinct domains (Collins *et al.*, 1991). The murine E-selectin gene has a similar architecture (Becker-André *et al.*, 1992). The 5′ flanking regions of both the human and murine genes have been cloned and sequenced (Collins *et al.*, 1991; Becker-André *et al.*, 1992), and the regulatory elements in the human sequence have been studied intensively. The human gene has a canonical TATA box that initiates transcription from a single site. The first 160 base pairs upstream from the transcriptional start site have sufficient information to direct cytokine-inducible expression of a reporter gene in transfected endothelial cells. However, this sequence does not confer tissue-specific expression *in vitro*, as the reporter gene is also inducibly expressed in non-endothelial cells (Whelan *et al.*, 1991). Little is known about the mechanisms that limit expression of E-selectin to the endothelium, except that CpG methylation of the promoter may repress E-selectin transcription in some non-endothelial cells (Smith *et al.*, 1993). In contrast, considerable information has accumulated in regard to the mechanisms for transcriptional induction by inflammatory mediators.

Four positive regulatory elements have been identified in the human E-selectin promoter. Three of these elements bind members of the NF-κB/Rel family of transcription factors. The DNA-binding forms of these factors are homodimers or heterodimers that bind to elements in many genes involved in inflammation, acute phase responses, cell proliferation, and differentiation (Siebenlist *et al.*, 1995). The proteins share a Rel homology domain and can be divided into two classes (Figure 2). One class includes Rel A (p65), Rel (c-Rel), Rel B, v-Rel, and the *Drosophila* proteins Dorsal and Dif. These proteins share an acidic transactivation domain. The other class includes the precursor proteins p105 (NF-κB1) and p100 (NF-κB2), which are proteolytically cleaved to the mature p50 and p52 proteins, respectively. Dimers containing p65 or c-Rel are retained in the cytoplasm by complex formation with IκBα and related proteins (Beg and Baldwin, 1993), or with the p105 or p100 precursor (Figures 2 and 3). These complexes mask nuclear localization signals in p65 or c-Rel. Cellular stimulation by IL-1, TNF-α, LPS, oxygen radicals, and other mediators induces phosphorylation of IκBα, p105, and p100 (Siebenlist *et al.*, 1995), targeting them for degradation through the proteasome proteolytic pathway (Palombella *et al.*, 1994; Traencker *et al.*, 1994). This unmasks nuclear localization signals in p65 or c-Rel. The dimers, including the prototypical p50/p65 NF-κB protein, then migrate to the nucleus, bind to κB elements in specific genes, and activate gene expression in conjunction with other transcription factors. Significantly, the promoter of the IκBα gene contains a κB element; when occupied by p50/p65, transcription of IκBα is dramatically enhanced (Siebenlist *et al.*, 1995). The newly synthesized IκBα protein binds to p65 and c-Rel in the cytoplasm, inhibiting nuclear translocation (Figure 3). Thus, NF-κB and IκBα interact in an autoregulatory manner, limiting the inflammatory response.

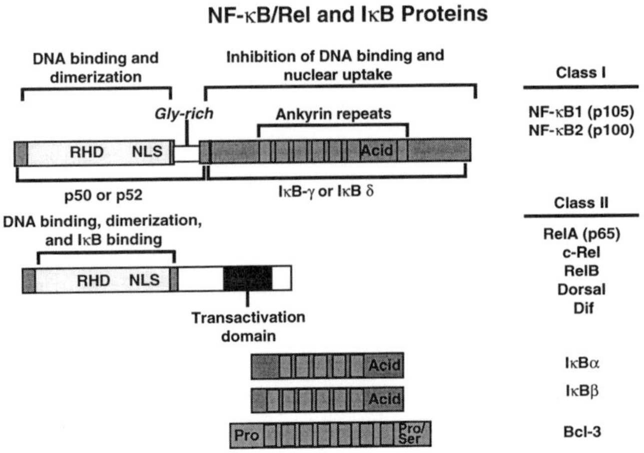

Figure 2 NF-κB/Rel and IκB proteins. The NF-κB/Rel proteins are divided into two classes. Both classes have a Rel homology domain (RHD) and a nuclear localization signal (NLS). Class I proteins have a C-terminal domain with a series of ankyrin repeats and a region with many acidic residues (Acid); this domain masks the nuclear localization signal. Class II proteins have a C-terminal transactivation domain. The IκB proteins also have tandem ankyrin repeats. IκBα and IκBβ bind to class II proteins in the cytoplasm, masking their nuclear localization signals. In contrast, Bcl-3 binds to class I proteins in the nucleus.

All three κB elements in the E-selectin promoter bind p50/p65 heterodimers and perhaps other heterodimers (Montgomery *et al.*, 1991; Whelan *et al.*, 1991; Whitney *et al.*, 1994; Lewis *et al.*, 1994; Schindler and Baichwal, 1994). Endothelial cells express transcripts and protein for a variety of Rel family members, including p65, p105 (p50), and IκBα (Read *et al.*, 1994). Therefore, cytokines and other inflammatory mediators appear to induce E-selectin transcription, in part, by mobilization of p50/p65 to the nucleus, where it binds to the κB elements in the E-selectin promoter (Figure 3). Transcription is then terminated by newly synthesized IκBα, which again sequesters p50/p65 in the cytoplasm (Read *et al.*, 1994). IκBα also dissociates p50/p65 from κB elements in the E-selectin promoter (Read *et al.*, 1995a). The importance of the NF-κB pathway is emphasized by the observation that addition of proteasome inhibitors to HUVEC decreases nuclear accumulation of NF-κB and blocks TNF-α-induced expression of E-selectin (Read *et al.*, 1995b).

The κB elements are not sufficient to mediate cytokine-induced transcription of E-selectin (Whelan *et al.*, 1991). An element at –104 to –100 complexes with unknown nuclear proteins; mutation of this element inhibits cytokine induction of the E-selectin promoter (Van Huijsduijnen *et al.*, 1992). A more extensively characterized element at –154 to –147 interacts with several members of the ATF family of transcription factors (Whitley *et al.*, 1994; Van Huijsduijnen *et al.*, 1992; Kaszubska *et al.*, 1993; De Luca *et al.*, 1994). ATF proteins have a leucine zipper domain and heterodimerize with c-fos and c-jun proteins (Lamb and McKnight, 1991). Mutation of the ATF element inhibits cytokine induction of the E-selectin promoter (Whitley *et al.*, 1994; Kaszubska *et al.*, 1993). Changes in E-selectin expression are also correlated with differences in the composition of the proteins binding to the ATF element. This suggests that transcription can be augmented

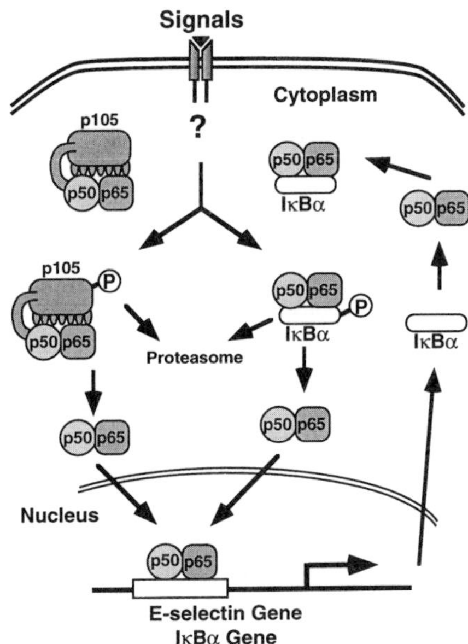

Figure 3 Transient expression of E-selectin mediated through the NF-κB/IκBα autoregulatory system. Inactive NF-κB (p50/p65) is maintained in the cytoplasm either by association with the C-terminal half of the p50 precursor (p105) or by association with IκBα. Inflammatory mediators generate intracellular signals that phosphorylate p105 and IκBα, targeting them for degradation by the proteasome. The nuclear localization signals in p65 are then unmasked, allowing p50/p65 heterodimers to activate the expression of many genes, including those for E-selectin and IκBα. Newly synthesized IκBα enters the cytoplasm, where it sequesters p50/p65, downregulating the inflammatory response. Adapted from Collins *et al.*, 1995, with permission.

or repressed, depending on the specific ATF proteins that bind to the ATF element. Members of the ATF family physically interact with NF-κB proteins, further suggesting the importance of the ATF element (Kaszubska *et al.*, 1993). ATF-2 and c-jun are among the proteins that bind to the ATF element. Both factors must be phosphorylated by the c-jun NH_2-terminal protein kinase for transcriptional activity, and this kinase is activated by inflammatory cytokines (Gupta *et al.*, 1995). These observations suggest that inflammatory signals induce E-selectin expression through both the NF-κB and ATF pathways.

Recent studies suggest that binding of proteins to the E-selectin promoter/enhancer bends the DNA, facilitating a higher-order complex of multiple DNA-protein and protein-protein interactions (Whitley *et al.*, 1994; Lewis *et al.*, 1994) (Figure 4). The organization of this complex is remarkably similar to that of the viral-responsive enhancer in the human interferon β gene (Thanos and Maniatis, 1992). Similar higher-order complexes may assemble on the enhancers of other cytokine-induced genes (Collins *et al.*, 1995). HMG I(Y) proteins bind to AT-rich sequences in the minor groove of DNA. These interactions bend the DNA and increase the affinity of NF-κB and ATF proteins for DNA and for each other. This higher-order structure requires precise spacing of the DNA elements that bind protein. Consistent with this prediction, moving one of the elements 5 base pairs (the distance of a half-helix) reduces E-selectin promoter activity, whereas moving it 10 base pairs (a full helix turn) has little effect (Meacock *et al.*, 1994).

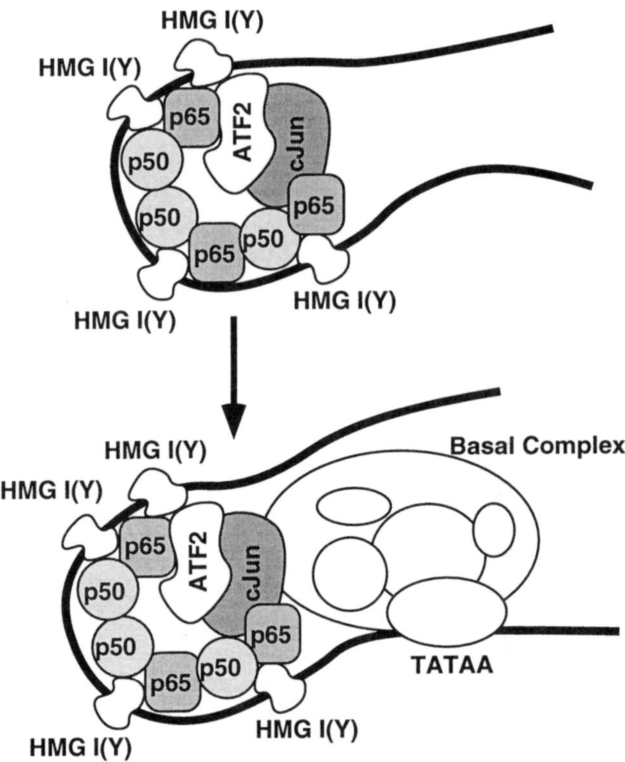

Figure 4 Model of the cytokine-induced E-selectin enhancer. Top, after induction by cytokine, p50/p65 heterodimers of NF-κB and an ATF-2 homodimer or an ATF-2/c-Jun heterodimer bind to the promoter. The binding of HMG I(Y) at multiple sites increases the binding of NF-κB and ATF-2 for their respective sites and bends DNA to facilitate the formation of a higher order complex necessary for transcriptional activation. This complex may bind to additional co-activator proteins. Bottom, the complex interacts as a unit with the basal transcriptional apparatus. Adapted from Collins *et al.*, 1995, with permission.

It is likely that the organization of these proteins in the complex facilitates interactions with the basal transcription machinery as well as with unknown coactivators.

REGULATION OF EXPRESSION OF P-SELECTIN

Constitutive Synthesis and Subcellular Trafficking of P-selectin

Unlike E-selectin, which is only synthesized when endothelial cells are exposed to inflammatory mediators, P-selectin is constitutively synthesized by endothelial cells, primarily in postcapillary venules (McEver *et al.*, 1989; Johnston *et al.*, 1989a). P-selectin is also constitutively synthesized by megakaryocytes, where it is incorporated into circulating platelets (Hsu-Lin *et al.*, 1984; McEver and Martin, 1984). In both cell types, core high mannose N-linked oligosaccharides are attached to the newly synthesized protein as it enters the endoplasmic reticulum. These glycans are modified into complex forms as the protein traverses the Golgi complex (Johnston *et al.*, 1989b; McEver *et al.*, 1989).

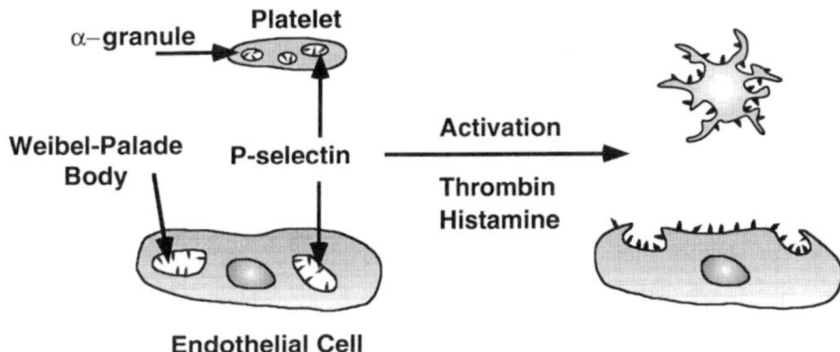

Figure 5 Inducible expression of P-selectin in platelets and endothelial cells. P-selectin is constitutively synthesized by megakaryocytes and endothelial cells, where it is sorted into the membrane of α-granules of platelets and the Weibel-Palade bodies of endothelial cells. Upon cellular activation by mediators such as thrombin or histamine, P-selectin is rapidly redistributed to the plasma membrane.

Unlike E-selectin, which passes directly to the cell surface, P-selectin is sorted into secretory storage granules: the α granules of megakaryocytes/platelets (Stenberg *et al.*, 1985; Berman *et al.*, 1986) and the Weibel-Palade bodies of endothelial cells (McEver *et al.*, 1989; Hattori *et al.*, 1989a; Bonfanti *et al.*, 1989). Sorting probably occurs in the trans-Golgi network (TGN), where proteins are diverted to other subcellular destinations such as the lysosome (Pelham and Munro, 1993). The sorting of P-selectin into secretory granules provides a pool of molecules that can be rapidly mobilized to the cell surface by agonists that induce the fusion of secretory granule membranes with the plasma membrane (Figure 5). Physiologic agonists of this type include thrombin and histamine (Hattori *et al.*, 1989a). Complement components and oxygen radicals also induce degranulation (Hattori *et al.*, 1989b; Patel *et al.*, 1991; Foreman *et al.*, 1994), as do pharmacologic mediators such as phorbol esters and calcium ionophores (Hattori *et al.*, 1989a). Basal levels of nitric oxide may serve to dampen the response of endothelial cells to secretagogues (Davenpeck *et al.*, 1994).

Thrombin or histamine induces surface expression of P-selectin on platelets within seconds (McEver and Martin, 1984; Hsu-Lin *et al.*, 1984) and on endothelial cells within minutes (McEver *et al.*, 1989; Hattori *et al.*, 1989a). *In vitro*, P-selectin remains on the surface of platelets for at least 1 hour after activation, probably because these short-lived cell fragments lack all the components required for efficient endocytosis (George *et al.*, 1986). In contrast, surface levels of P-selectin on cultured endothelial cells reach maximal levels 5–10 minutes after activation and decline to basal levels after 30–60 minutes (Hattori *et al.*, 1989a; Lorant *et al.*, 1991). This transient surface expression results from rapid endocytosis of the protein (Hattori *et al.*, 1989a), which occurs in clathrin-coated pits (Setiadi *et al.*, 1995). Like many internalized membrane proteins such as the transferrin receptor, P-selectin appears to recycle between endosomes and the plasma membrane (Hattori *et al.*, 1989a) (Figure 6). Some of the P-selectin molecules then move from endosomes to the TGN, where they are resorted into newly formed Weibel-Palade bodies (Subramaniam *et al.*, 1993). However, the efficiency of this process is unclear, because P-selectin is also delivered rapidly from endosomes to lysosomes, where it is degraded (Green *et al.*, 1994). This probably explains why P-selectin, after translocation to the surface of venular endothelial cells by mild dermal tissue trauma, disappears for 12–24 hours before it is again detected by immunoperoxidase staining of tissue sections. The lag in detection probably

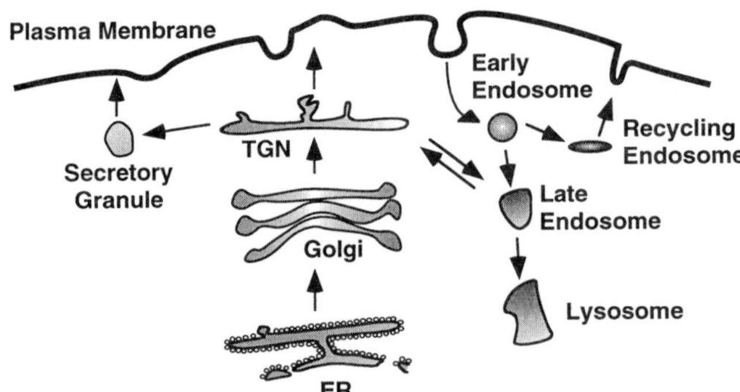

Figure 6 Subcellular trafficking of E-selectin and P-selectin. Following translation, both proteins are glycosylated as they pass from the ER through the Golgi network to the TGN. E-selectin then proceeds directly to the cell surface. P-selectin is sorted into secretory granules, where it can be rapidly redistributed to the cell surface in response to secretagogues. Both proteins are internalized, probably in clathrin-coated pits, and then recycle to the plasma membrane. In addition, both E- and P-selectin are directed more efficiently than most membrane proteins from endosomes to lysosomes, where they are degraded. An unknown fraction of E- and P-selectin may return to the TGN, where P-selectin may again be sorted into secretory granules. The steady-state distribution of the proteins reflects the balance between the rates of synthesis, the sorting efficiencies into various compartments, and the rates of degradation.

reflects the time required for newly synthesized P-selectin to accumulate in Weibel-Palade bodies (Silber *et al.*, 1994).

The 35-residue cytoplasmic domain of P-selectin is highly conserved across species, suggesting that it has important conserved functions (Johnston *et al.*, 1989a; Weller *et al.*, 1992; Sanders *et al.*, 1992; Auchampach *et al.*, 1994; Strubel *et al.*, 1993). Consistent with this prediction, the cytoplasmic domain contains at least three different cellular trafficking signals: for sorting of P-selectin into newly formed secretory granules (Disdier *et al.*, 1992), for internalization of the protein in clathrin-coated pits (Setiadi *et al.*, 1995; Subramaniam *et al.*, 1993), and for movement of P-selectin from endosomes to lysosomes (Green *et al.*, 1994). The cytoplasmic domain presumably interacts with sorting molecules that mediate delivery of vesicles carrying P-selectin to its appropriate destination. These molecules are unknown, except for the cytoplasmic adaptins that are constituents of the endocytic machinery of clathrin-coated pits (Robinson, 1994). The structural features in the cytoplasmic domain of P-selectin that constitute each sorting signal require further definition. A considerable literature describes short cytoplasmic-domain sorting signals that are centered around a tyrosine or a dileucine motif (Trowbridge *et al.*, 1993). However, the cytoplasmic domain of P-selectin has no dileucine sequence, and a putative tyrosine-based motif can be deleted without significantly affecting the internalization rate (Setiadi *et al.*, 1995). Indeed, extensive mutational analysis suggests that residues throughout the cytoplasmic domain contribute to the internalization efficiency of P-selectin (Setiadi *et al.*, 1995). An 11-residue internal deletion in the cytoplasmic domain prolongs the half-life of P-selectin in transfected cells without affecting the internalization rate; this deletion presumably disrupts the endosomal sorting signal (Green *et al.*, 1994). The cytoplasmic domain may require a specific three-dimensional structure to present all its sorting signals. The cytoplasmic domain is acylated on a single cysteine (Fujimoto *et al.*, 1993), and is transiently phosphorylated on serine, threonine, histidine, and tyrosine following platelet activation (Fujimoto and McEver, 1993; Crovello *et al.*, 1993; Crovello *et al.*, 1995). To date, however, mutational analysis has not documented a function for these modifications in subcellular trafficking (Green *et al.*, 1994). The precise sites in the cytoplasmic domain of P-selectin, or of other membrane proteins, that dock to sorting molecules remain unknown.

The sorting signals in the cytoplasmic domain allow tight regulation of the surface expression of P-selectin. Because sorting systems within cells are saturable, the subcellular distribution of P-selectin is likely to reflect a balance between its rate of synthesis and its rate of delivery to various organelles (Figure 6). Under basal conditions in endothelial cells, most of the constitutively synthesized P-selectin is probably sorted from the TGN into secretory granules. Low levels of protein reaching the cell surface will be rapidly internalized and either delivered from endosomes to lysosomes for degradation, or returned to the TGN for another opportunity to be sorted into secretory granules. Upon agonist-induced degranulation, a large cohort of P-selectin molecules rapidly appears on the cell surface. This cohort of proteins distributes within minutes between the plasma membrane and recycling endosomes. However, the efficient removal of P-selectin from endosomes to lysosomes clears P-selectin from the cell surface within a relatively short period of 30–60 minutes.

Pathologic inflammatory mediators may cause dysregulated expression of P-selectin on the cell surface. Oxygen radicals, which are generated at sites of ischemia-reperfusion injury and other inflammatory disorders, induce degranulation and surface expression of P-selectin. Significantly, these radicals also prolong the surface expression of P-selectin for many hours, probably because they impair endocytosis in clathrin-coated pits (Patel *et al.*, 1991). Viral infection of endothelial cells also prolongs the surface expression of P-selectin (Etingin *et al.*, 1992).

Soluble P-selectin has been found in plasma (Dunlop *et al.*, 1992; Katayama *et al.*, 1993; Ushiyama *et al.*, 1993). Unlike soluble E-selectin, soluble P-selectin appears to be derived from an alternatively spliced transcript that lacks the exon encoding the transmembrane domain (Johnston *et al.*, 1990; Johnston *et al.*, 1989a; Ishiwata *et al.*, 1994). Soluble P-selectin has no sequence facing the cytoplasm and is transported entirely within the lumen of transport vesicles. Therefore, it is unable to interact with cytoplasmic sorting molecules and is expected to be transported by default from the TGN to the cell surface, where it is secreted. Alternatively spliced P-selectin is secreted in monomeric form from transfected cells (Ushiyama *et al.*, 1993). Plasma levels are low and a biologic function for circulating P-selectin has not been defined. However, modestly elevated levels of plasma P-selectin have been identified in some inflammatory disorders (Chong *et al.*, 1994; Katayama *et al.*, 1993; Ikeda *et al.*, 1994; Takeda *et al.*, 1994; Sakamaki *et al.*, 1995).

Inducible Synthesis of P-selectin

The initial studies of P-selectin focused on its constitutive synthesis and its inducible redistribution from secretory granule stores to the plasma membrane, where it could serve as a rapidly mobilizable adhesion molecule during acute inflammation. As mentioned above, however, the surface expression of P-selectin should reflect the balance between the rate of synthesis and the efficiency of sorting. Thus, a high synthetic rate might saturate the pathway for sorting of P-selectin from the TGN to secretory granules, resulting in increased constitutive delivery to the plasma membrane. Rapid internalization and endosomal sorting tend to clear the cell surface of P-selectin. If the synthetic rate is higher, however, steady-state levels of P-selectin on the cell surface might be sufficient to mediate leukocyte adhesion during chronic inflammation, without a requirement for degranulation. Furthermore, a higher synthetic rate could deliver more newly synthesized protein to Weibel-Palade bodies, providing a larger pool of P-selectin that could be mobilized by thrombin or other secretogogues.

In vitro, TNF-α and LPS cause 2–4-fold increases in steady-state levels of P-selectin mRNA in cultured murine, rat, and bovine endothelial cells (Weller *et al.*, 1992; Hahne *et al.*, 1993; Bischoff and Brasel, 1995). The elevated transcripts are associated with an increase in P-selectin protein; some of this protein is immediately displayed on the cell surface, and a larger pool is also mobilized by stimulation with histamine (Weller *et al.*, 1992; Hahne *et al.*, 1993). The kinetics of induction of P-selectin transcripts are similar to those for E-selectin, with levels of mRNA reaching maximum 4–6 hours after addition of cytokine and returning to basal levels after 12–24 hours (Weller *et al.*, 1992; Hahne *et al.*, 1993; Bischoff and Brasel, 1995). *In vivo*, TNF-α and LPS induce even larger increases in P-selectin transcripts, as measured by Northern blot analysis. The increases are particularly evident in vascular organs such as lung, liver, heart, and spleen (Sanders *et al.*, 1992; Weller *et al.*, 1992; Auchampach *et al.*, 1994; Mayadas *et al.*, 1993). The kinetics of induction are similar to those observed *in vitro*, with mRNA levels peaking after 4–6 hours and declining to basal levels after 12–24 hours. Immunocytochemical analysis indicates that the TNF-α induced expression of P-selectin is restricted to endothelial cells (Gotsch *et al.*, 1994).

The induction of P-selectin mRNA and protein by TNF-α and LPS has some similarities to the induction of E-selectin mRNA and protein. However, several observations suggest that there are important differences in the mechanisms for regulating the transcripts encoding these two proteins. First, P-selectin is also constitutively synthesized by endothelial cells as well as by megakaryocytes. Second, TNF-α, IL-1, and LPS do not increase P-selectin mRNA levels in cultured HUVEC (L. Yao, J. Pan, and R.P. McEver, unpublished observations). Third, murine P-selectin mRNA levels rapidly decline after cytokine induction *in vitro* or *in vivo*. However, the half-life of P-selectin mRNA in cultured HUVEC is at least 12 hours (L. Yao, J. Pan, and R.P. McEver, unpublished observations). Fourth, P-selectin, but not E-selectin, is displayed *in vivo* on the apical surface of human endothelial cells at sites of allergic inflammation (Symon *et al.*, 1994), in synovial tissues from patients with rheumatoid arthritis (Grober *et al.*, 1993), and overlying atherosclerotic plaques (Johnson-Tidey *et al.*, 1994). These observations could indicate that different species use distinct mechanisms for regulating the synthesis and stability of transcripts for P- or E-selectin, and/or that the same species differentially regulates the transcription of P- and E-selectin.

Transcriptional Regulation of P-selectin

The human gene for P-selectin is also divided into exons that encode structurally distinct domains (Johnston *et al.*, 1990). The 5′ flanking region of the human gene has been cloned, and a preliminary analysis of its properties has been conducted (Pan and McEver, 1993). Significantly, the sequence of the P-selectin promoter has little resemblance to that of the E-selectin promoter. There is no canonical TATA box, and transcription is initiated from multiple sites ranging from –95 to –25 nucleotides upstream of the ATG codon where protein-coding sequence begins. The first 249 base pairs upstream of the translational start site are sufficient to confer constitutive expression of a reporter gene in cultured endothelial cells, but not in COS-7, 293, or Hela cells. Thus, unlike the E-selectin promoter, the P-selectin promoter confers constitutive expression in a tissue-specific manner, at least *in vitro*. Serial deletions of the P-selectin promoter reveal at least three positive regulatory regions. A GATA element in the region from –197 to –147 is functional, as mutation of the element reduces constitutive expression of the reporter gene in endothelial cells by ≈80%. The region from –249 to –197 contains putative elements for ETS proteins and an element for NK-κB/Rel proteins.

The κB element in the P-selectin promoter is novel, in that it binds homodimers of p50 and p52, but not homodimers or heterodimers containing p65 (Pan and McEver, 1995). Most studies indicate that homodimers containing p50 or p52 do not activate gene expression (Siebenlist *et al.*, 1995). Homodimers of p50 are present constitutively in the nucleus of some cells, where they may prevent binding of inducible dimers containing p65 (Kang *et al.*, 1992). Bcl-3, a protein structurally related to IκBα (Figure 2), dissociates bound p50 homodimers from DNA, allowing heterodimers containing p65 to bind DNA and activate gene expression (Franzoso *et al.*, 1993). In contrast, Bcl-3 forms a ternary complex with p52 homodimers, resulting in transactivation (Bours *et al.*, 1993; Fujita *et al.*, 1993). Co-transfection experiments indicate that interactions of Bcl-3 and p52 homodimers with the P-selectin κB element augment expression of a reporter gene in cultured endothelial cells. In contrast, interactions of p50 homodimers with the κB element repress expression; however, repression is prevented by co-expression of Bcl-3 (Pan and McEver, 1995). These data suggest that differential interactions of Bcl-3 with p50 and p52 homodimers regulate the constitutive and perhaps the inducible expression of P-selectin, and raise the possibility that other genes may have κB elements with similar specificity (Figure 7). The results also confirm that there are important differences in the transcriptional regulation of the human genes for E-selectin and P-selectin. The E-selectin promoter can be organized into a higher-order, cytokine-inducible structure in which p50/p65 heterodimers of NF-κB are essential components. In contrast, the P-selectin promoter has a GATA element and a κB element that binds only p50 or p52 homodimers. It is likely that the P-selectin promoter is also organized into a higher order structure through interactions with several transcription factors that mediate constitutive or inducible gene expression. However, less is known about these components than those associated with the E-selectin promoter.

SUMMARY

Neither E- nor P-selectin is normally present on the surface of endothelial cells. However, a variety of inflammatory agents induce the surface display of these proteins, which mediate the initial rolling adhesion of leukocytes on the vessel wall. E-selectin does not reach the cell surface until 1–2 hours after endothelial cell activation, because it must first be synthesized. P-selectin can be mobilized to the cell surface within minutes, because it is constitutively synthesized and stored in secretory granules. P-selectin is probably the earliest adhesion receptor expressed on the endothelial cell surface in response to tissue injury or

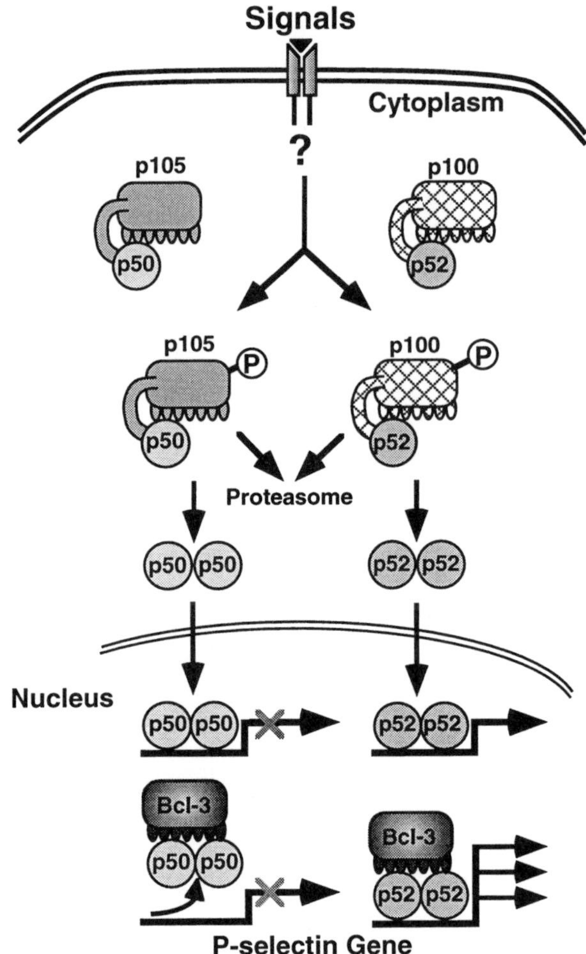

Figure 7 Model for NF-κ-dependent expression of P-selectin. Unknown signals cause phosphorylation of the C-terminal domains of p105 and p100, targeting them for degradation by the proteasome. The liberated p50 and p52 proteins form homodimers, which migrate to the nucleus and bind to some but not all κB elements. The κB element in the human P-selectin promoter binds these homodimers, but does not bind dimers containing p65 or c-Rel. Bound p50 homodimers do not activate gene expression, whereas bound p52 homodimers may weakly activate expression. Other signals regulate the expression and function, perhaps by phosphorylation, of the cofactor Bcl-3. Bcl-3 dissociates p50 homodimers from DNA. In contrast, ternary complexes of Bcl-3, p52 homodimers, and DNA augment expression of P-selectin.

infection. However, inflammatory mediators also augment its synthesis so that P-selectin, like E-selectin, may function in chronic as well as acute inflammation. Much more needs to be learned about the mechanisms for regulating the tissue-specific, constitutive, and/or inducible transcription of E- and P-selectin. There are already clearly defined differences in the transcriptional regulation of the two genes. To date, the most striking differences is the use of DNA elements with distinct specificities for dimeric members of the NF-κB/Rel family. The mechanisms for regulating mRNA half-life may also differ. The half-life of

E-selectin transcripts is usually short, whereas the half-life of P-selectin transcripts is usually long; however, there appear to be exceptions for each transcript.

After E- and P-selectin reach the endothelial cell surface, they are internalized. They are then transported from the pool of recycling endosomes to lysosomes much more efficiently than most membrane proteins. The net effect is to quickly remove both selectins from the cell surface, unless they are replaced with newly synthesized proteins. The cytoplasmic domain of P-selectin clearly mediates both endocytosis and lysosomal targeting, and the cytoplasmic domain of E-selectin probably has similar functions. The cytoplasmic domain of P-selectin has a third unique role, the sorting of P-selectin into newly forming secretory granules. This sorting determinant provides a pool of P-selectin that is protected from lysosomal degradation and can be rapidly translocated to the cell surface in response to secretagogues. When P-selectin is expressed in heterologous cells that lack regulated secretory granules, it is immediately transported from the TGN to the cell surface, like E-selectin.

Thus, endothelial cells use multiple mechanisms to control the synthesis and degradation of mRNA for E- and P-selectin, and to control the surface expression of the proteins after they are synthesized. These mechanisms are used to regulate the density of E- and P-selectin on the activated endothelial cell surface, and the time that they remain on the surface before they are removed. In physiologic situations, these mechanisms serve to limit the duration of the inflammatory response. In pathologic disorders, however, dysregulated expression of E- and P-selectin may cause the endothelial cell surface to remain adhesive for leukocytes for prolonged periods. The activation of adherent leukocytes may then result in tissue destruction within and without the vasculature. Understanding the physiologic mechanisms that control the expression of E- and P-selectin may facilitate the development of drugs that block their pathologic expression.

References

Auchampach, J.A., Oliver, M.G., Anderson, D.C. and Manning, A.M. (1994). Cloning, sequence comparison and *in vivo* expression of the gene encoding rat P-selectin. *Gene*, **145**:251–255.

Becker-André, M., Van Huijsduijnen, R.H., Losberger, C., Whelan, J. and Delamarter, J.F. (1992). Murine endothelial leukocyte-adhesion molecule 1 is a close structural and functional homologue of the human protein. *Eur. J. Biochem.*, **206**:401–411.

Beg, A.A. and Baldwin, A.S., Jr. (1993). The IκB proteins: multifunctional regulators of Rel/NF-κB transcription factors. *Genes & Dev.*, **7**:2064–2070.

Berman, C.L., Yeo, E.L., Wencel-Drake, J.D., Furie, B.C., Ginsberg, M.H. and Furie, B. (1986). A platelet alpha granule membrane protein that is associated with the plasma membrane after activation. *J. Clin. Invest.*, **78**:130–137.

Bevilacqua, M.P., Pober, J.S., Mendrick, D.L., Cotran, R.S. and Gimbrone, M.A., Jr. (1987). Identification of an inducible endothelial-leukocyte adhesion molecule. *Proc. Natl. Acad. Sci. USA*, **84**:9238–9242.

Bevilacqua, M.P., Stengelin, S., Gimbrone, M.A., Jr. and Seed, B. (1989). Endothelial leukocyte adhesion molecule 1: an inducible receptor for neutrophils related to complement regulatory proteins and leuctins. *Science*, **243**:1160–1165.

Bischoff, J. and Brasel, C. (1995). Regulation of P-selectin by tumor necrosis factor-α. *Biochem. Biophys. Res. Commun.*, **210**:174–180.

Bonfanti, R., Furie, B.C., Furie, B. and Wagner, D.D. (1989). PADGEM (GMP 140) is a component of Weibel-Palade bodies of human endothelial cells. *Blood*, **73**:1109–1112.

Bours, V., Granzoso, G., Azarenko, V., Park, S., Kanno, T., Brown, K. and Siebenlist, U. (1993). The oncoprotein Bcl-3 directly transactivates through κB motifs via association with DNA-binding p50B homodimers. *Cell*, **72**:729–739.

Carson, C.W., Beall, L.D., Hunder, G.G., Johnson, C.M. and Newman, W. (1993). Serum ELAM-1 is increased in vasculitis, scleroderma, and systemic lupus erythematosus. *J. Rheumatol.*, **20**:809–814.

Carson, C.W., Beall, L.D., Hunder, G.G., Johnson, C.M. and Newman, W. (1994). Soluble E-selectin is increased in inflammatory synovial fluid. *J. Rheumatol.*, **21**:605–611.

Chong, B.H., Murray, B., Berndt, M.C., Dunlop, L.C., Brighton, T. and Chesterman, C.N. (1994). Plasma P-selectin is increased in thrombotic consumptive platelet disorders. *Blood*, **83**:1535–1541.

Chu, W., Presky, D.H., Swerlick, R.A. and Burns, D.K. (1994). Alternatively processed human E-selectin transcripts linked to chronic expression of E-selectin *in vivo*. *J. Immunol.*, **153**:4179–4189.

Collins, T., Williams, A., Johnston, G.I., Kim, J., Eddy, R., Shows, T., Gimbrone, M.A. Jr. and Bevilacqua, M.P. (1991). Structure and chromosomal location of the gene for endothelial-leukocyte adhesion molecule 1. *J. Biol. Chem.*, **266**:2466–2473.

Collins, T., Read, M.A., Neish, A.S., Whitley, M.Z., Thanos, D. and Maniatis, T. (1995). Transcriptional regulation of endothelial cell adhesion molecules: NF-κB and cytokine-inducible enhancers. *FASEB J.*, **9**:899–909.

Cotran, R.S., Gimbrone, M.A. Jr., Bevilacqua, M.P., Mendrick, D.L. and Pober, J.S. (1986). Induction and detection of a human endothelial activation antigen *in vivo*. *J. Exp. Med.*, **164**:661–666.

Cronstein, B.N., Kimmel, S.C., Levin, R.I. and Martiniuk, F. (1992). A mechanism for the antiinflammatory effects of corticosteroids: the glucocorticoid receptor regulates leukocyte adhesion to endothelial cells and expression of endothelial leukocyte adhesion molecule 1 and intercellular adhesion molecule 1. *Proc. Natl. Acad. Sci. USA*, **89**:9991–9995.

Crovello, C.S., Furie, B.C. and Furie, B. (1993). Rapid phosphorylation and selective dephosphorylation of P-selectin accompanies platelet activation. *J. Biol. Chem.*, **268**:14590–14593.

Crovello, C.S., Furie, B.C. and Furie, B. (1995). Histidine phosphorylation of P-selectin upon stimulation of human platelets: A novel pathway for activation-dependent signal transduction. *Cell*, **82**:279–286.

Davenpeck, K.L., Gauthier, T.W. and Lefer, A.M. (1994). Inhibition of endothelial-derived nitric oxide promotes P-selectin expression and actions in the rat microcirculation. *Gastroenterology*, **107**:1050–1058.

De Luca, L.G., Johnson, D.R., Whitley, M.Z., Collins, T. and Pober, J.S. (1994). cAMP and tumor necrosis factor competitively regulate transcriptional activation through and nuclear factor binding to the cAMP-responsive element/activating transcription factor element of the endothelial leukocyte adhesion molecule-1 (E-selectin) promoter, *J. Biol. Chem.*, **269**:19193–19196.

Disdier, M., Morrissey, J.H., Fugate, R.D., Bainton, D.F. and McEver, R.P. (1992). Cytoplasmic domain of P-selectin (CD62) contains the signal for sorting into the regulated secretory pathway. *Mol. Biol. Cell*, **3**:309–321.

Dunlop, L.C., Skinner, M.P., Bendall, L.J., Favaloro, E.J., Castaldi, P.A., Gorman, J.J., Gamble, J.R., Vadas, M.A. and Berndt, M.C. (1992). Characterization of GMP-140 (P-selectin) as a circulating plasma protein. *J. Exp. Med.*, **175**:1147–1150.

Etingin, O.R., Silverstein, R.L. and Hajjar, D.P. (1992). Identification of a monocyte receptor on herpesvirus-infected endothelial cells. *Proc. Natl. Acad. Sci. USA*, **88**:7200–7203.

Foreman, K.E., Vaporciyan, A.A., Bonish, B.K., Jones, M.L., Johnson, K.J., Glovsky, M.M., Eddy, S.M. and Ward, P.A. (1994). C5a-induced expression of P-selectin in endothelial cells. *J. Clin. Invest.*, **94**:1147–1155.

Franzoso, G., Bours, V., Azarenko, V., Park, S., Tomita-Yamaguchi, M., Kanno, T., Brown, K. and Siebenlist, U. (1993). The oncoprotein Bcl-3 can facilitate NF-κB-mediated transactivation by removing inhibiting p50 homodimers from select κB sites. *EMBO J.*, **12**:3893–3901.

Fujimoto, T. and McEver, R.P. (1993). The cytoplasmic domain of P-selectin is phosphorylated on serine and threonine residues. *Blood*, **82**:1758–1766.

Fujimoto, T., Stroud, E., Whatley, R.E., Prescott, S.M., Muszbek, L., Laposata, M. and McEver, R.P. (1993). P-selectin is acylated with palmitic acid and stearic acid at cysteine 766 through a thioester linkage. *J. Biol. Chem.*, **268**:11394–11400.

Fujita, T., Nolan, G.P., Liou, H-C., Scott, M.L. and Baltimore, D. (1993). The candidate proto-oncogene *bcl*-3 encodes a transcriptional coactivator that activates through NF-κB p50 homodimers. *Genes & Dev.*, **7**:1354–1363.

Gamble, J.R., Khew-Goodall, Y. and Vadas, M.A. (1993). Transforming growth factor-β inhibits E-selectin expression on human endothelial cells. *J. Immunol.*, **150**:4494–4503.

Gearing, A.J.H. and Newman, W. (1993). Circulating adhesion molecules in disease. *Immunol. Today*, **14**:506–512.

George, J.N., Pickett, E.B., Saucerman, S., McEver, R.P., Kunicki, T.J., Kieffer, N. and Newman, P.J. (1986). Platelet surface glycoproteins. Studies on resting and activated platelets and platelet membrane microparticles in normal subjects, and observations in patients during adult respiratory distress syndrome and cardiac surgery. *J. Clin. Invest.*, **78**:340–348.

Ghersa, P., Hooft van Huijsduijnen, R., Whelan, J., Cambet, Y., Pescini, R. and Delamarter, J.F. (1994). Inhibition of E-selectin gene transcription through a cAMP-dependent protein kinase pathway. *J. Biol. Chem.*, **269**:29129–29137.

Gotsch, U., Jager, U., Dominis, M. and Vestweber, D. (1994). Expression of P-selectin on endothelial cells is upregulated by LPS and TNF-α *in vivo*. *Cell Adhes. Commun.*, **2**:7–14.

Green, S.A., Setiadi, H., McEver, R.P. and Kelly, R.B. (1994). The cytoplasmic domain of P-selectin contains a sorting determinant that mediates rapid degradation in lysosomes. *J. Cell Biol.*, **124**:435–448.

Grober, J.S., Bowen, B.L., Ebling, H., Athey, B., Thompson, C.B., Fox, D.A. and Stoolman, L.M. (1993). Monocyte-endothelial adhesion in chronic rheumatoid arthritis: in situ detection of selectin and integrin-dependent interactions. *J. Clin. Invest.*, **91**:2609–2619.

Gupta, S., Campbell, D., Derijard, B. and Davis, R.J. (1995). Transcription factor ATF2 regulation by the JNK signal transduction pathway. *Science*, **267**:389–393.

Hahne, M., Jäger, U., Isenmann, S., Hallmann, R. and Vestweber, D. (1993). Five tumor necrosis factor-inducible cell adhesion mechanisms on the surface of mouse endothelioma cells mediate the binding of leukocytes. *J. Cell Biol.*, **121**:655–664.

Hattori, R., Hamilton, K.K., Fugate, R.D., McEver, R.P. and Sims, P.J. (1989a). Stimulated secretion of endothelial von Willebrand factor is accompanied by rapid redistribution to the cell surface of the intracellular granule membrane protein GMP-140. *J. Biol. Chem.*, **264**:7768–7771.

Hattori, R., Hamilton, K.K., McEver, R.P. and Sims, P.J. (1989b). Complement proteins C5b-9 induce secretion of high molecular weight multimers of endothelial von Willebrand factor and translocation of granule membrane protein GMP-140 to the cell surface. *J. Biol. Chem.*, **264**:9053–9060.

Hsu-Lin, S-C., Berman, C.L., Furie, B.C., August, D. and Furie, B. (1984). A platelet membrane protein expressed during platelet activation and secretion. Stuides using a monoclonal antibody specific for thrombin-activated platelets. *J. Biol. Chem.*, **259**:9121–9126.

Ikeda, H., Nakayama, H., Oda, T., Kuwano, K., Muraishi, A., Sugi, K., Koga, Y. and Toshima, H. (1994). Soluble form of P-selectin in patients with acute myocardial infarction. *Coronary Artery Dis.*, **5**:515–518.

Ishiwata, N., Takio, K., Katayama, M., Watanabe, K., Titani, K., Ikeda, Y. and Handa, M. (1994). Alternatively spliced isoform of P-selectin is present *in vivo* as a soluble molecule. *J. Biol. Chem.*, **269**:23708–23715.

Johnson-Tidey, R.R., McGregor, J.L., Taylor, P.R. and Poston, R.N. (1994). Increase in the adhesion molecule P-selectin in endothelium overlying atherosclerotic plaques. Coexpression with intercellular adhesion molecule-1. *Am. J. Pathol.*, **144**:952–961.

Johnston, G.I., Cook, R.G. and McEver, R.P. (1989a). Cloning of GMP-140, a granule membrane protein of platelets and endothelium: sequence similarity to proteins involved in cell adhesion and inflammation. *Cell*, **56**:1033–1044.

Johnston, G.I., Kurosky, A. and McEver, R.P. (1989b). Structural and biosynthetic studies of the granule membrane protein, GMP-140, from human platelets and endothelial cells. *J. Biol. Chem.*, **264**:1816–1823.

Johnston, G.I., Bliss, G.A. Newman, P.J. and McEver, R.P. (1990). Structure of the human gene encoding granule membrane protein-140, a member of the selectin family of adhesion receptors for leukocytes. *J. Biol. Chem.*, **265**:21381–21385.

Kang, S-M., Tran, A-C., Grilli, M. and Lenardo, M.J. (1992). NF-κB subunit regulation in nontransformed CD4$^+$T lymphocytes. *Science*, **256**:1452–1456.

Kaszubska, W., Hooft van Huijsduijnen, R., Ghersa, P., DeRaemy-Schenk, A-M., Chen, B.P.C., Hai, T., Delamarter, J.F. and Whelan, J. (1993). Cyclic AMP-independent ATF family members interact with NF-κB and function in the activation of the E-selectin promoter in response to cytokines. *Mol. Cell. Biol.*, **13**:7180–7190.

Katayama, M., Handa, M., Araki, Y., Ambo, H., Kawai, Y., Watanabe, K. and Ikeda, Y. (1993). Soluble P-selectin is present in normal circulation and its plasma level is elevated in patients with thrombotic thrombocytopenic purpura and haemolytic uraemic syndrome. *Br. J. Haematol.*, **84**:702–710.

Kuijpers, T.W., Raleigh, M., Kavanagh, T., Janssen, H., Calafat, J., Roos, D. and Harlan, J.M. (1994). Cytokine-activated endothelial cells internalize E-selectin into a lysosomal compartment of vesiculotubular shape: A tubulin-driven process. *J. Immunol.*, **152**:5060–5069.

Lamb, P. and McKnight, S.L. (1991). Diversity and specificity in trascriptional regulation: the benefits of heterotypic dimerization. *Trends Biochem. Sci.*, **16**:417–422.

Leeuwenberg, J.F.M., Von Asmuth, E.J.U., Jeunhomme, T.M.A.A. and Buurman, W.A. (1990). IFN-γ regulates the expression of the adhesion molecule ELAM-1 and IL-6 production by human endothelial cells *in vitro*. *J. Immunol.*, **145**:2110–2114.

Lewis, H., Kaszubska, W., Delamarter, J.F. and Whelan, J. (1994). Cooperativity between two NF-κB complexes, mediated by high-mobility-group protein I(Y), is essential for cytokine-induced expression of the E-selectin promoter. *Mol. Cell. Biol.*, **14**:5701–5709.

Lorant, D.E., Patel, K.D., McIntyre, T.M., McEver, R.P., Prescott, S.M. and Zimmerman, G.A. (1991). Coexpression of GMP-140 and PAF by endothelium stimulated by histamine or thrombin: A juxtacrine system for adhesion and activation of neutrophils. *J. Cell Biol.*, **115**:223–234.

Mayadas, T.N., Johnson, R.C., Rayburn, H., Hynes, R.O. and Wagner, D.D. (1993). Leukocyte rolling and extravasation are severely compromised in P selectin-deficient mice. *Cell*, **74**:541–554.

McEver, R.P. and Martin, M.N. (1984). A monoclonal antibody to a membrane glycoprotein binds only to activated platelets. *J. Biol. Chem.*, **259**:9799–9804.

McEver, R.P., Beckstead, J.H., Moore, K.L., Marshall-Carlson, L. and Bainton, D.F. (1989). GMP-140, a platelet alpha-granule membrane protein, is also synthesized by vascular endothelial cells and is localized in Weibel-Palade bodies. *J. Clin. Invest.*, **84**:92–99.

Maecock, S., Pescini-Gobert, R., Delamarter, J.F. and Van Huijsduijnen, R.H. (1994). Trasncription factor-induced, phased bending of the E-selectin promoter. *J. Biol. Chem.*, **269**:31756–31762.

Montgomery, K.F., Osborn, L., Hession, C., Tizard, R., Goff, D., Vassallo, C., Tarr, P.I., Bomsztyk, K., Lobb, R., Harlan, J.M. and Pohlman, T.H. (1991). Activation of endothelial-leukocyte adhesion molecule 1 (ELAM-1) gene transcription. *Proc. Natl. Acad. Sci. USA*, **88**:6523–6527.

Palombella, V.J., Rando, O.J., Goldberg, A.L. and Maniatis, T. (1994). The ubiquitin-proteasome pathway is required for processing the NF-κB1 precursor and the activation of NF-κB. *Cell*, **78**:773–785.

Pan, J. and McEver, R.P. (1993). Characterization of the promoter for the human P-selectin gene. *J. Biol. Chem.*, **268**:22600–22608.

Pan, J. and McEver, R.P. (1995). Regulation of the human P-selectin promoter by Bcl-3 and specific homodimeric members of the NF-κB/Rel family. *J. Biol. Chem.*, **270**:23077–23083.

Parhami, F., Fang, Z.T., Fogelman, A.M., Andalibi, A., Territo, M.C. and Berliner, J.A. (1993). Minimally modified low density lipoprotein-induced inflammatory responses in endothelial cells are mediated by cyclic adenosine monophosphate. *J. Clin. Invest.*, **92**:471–478.

Patel, K.D., Zimmerman, G.A., Prescott, S.M., McEver, R.P. and McIntyre, T.M. (1991). Oxygen radicals induce human endothelial cells to express GMP-140 and bind neutrophils. *J. Cell Biol.*, **112**:749–759.

Pelham, H.R.B. and Munro, S. (1993). Sorting of membrane proteins in the secretory pathway. *Cell*, **75**:603–605.

Pigott, R., Dillon, L.P., Hemingway, I.H. and Gearing, A.J.H. (1992). Soluble forms of E-selectin, ICAM-1 and VCAM-1 are present in the supernatants of cytokine activated cultured endothelial cells. *Biochem. Biophys. Res. Commun.*, **187**:584–589.

Pober, J.S., Slowik, M.R., De Luca, L.G. and Ritchie, A.J. (1993). Elevated cyclic AMP inhibits endothelial cell synthesis and expression of TNF-induced endothelial leukocyte adhesion molecule-1, and vascular cell adhesion molecule-1, but not intercellular adhesion molecule-1. *J. Immunol.*, **150**:5114–5123.

Read, M.A., Whitley, M.Z., Williams, A.J. and Collins, T. (1994). NF-κB and IκBα: An inducible regulatory system in endothelial activation. *J. Exp. Med.*, **179**:503–512.

Read, M.A., Neish, A.S., Gerritsen, M.E. and Collins, T. (1995a). Nuclear IκB-α and the post-induction transcriptional repression of E-selectin and VCAM-1. *J. Biol. Chem.*, submitted.

Read, M.A., Neish, A.S., Luscinskas, F.W., Palombella, V.J., Maniatis, T. and Collins, T. (1995b). The proteasome pathway is required for cytokine-induced endothelial-leukocyte adhesion molecule expression. *Immunity*, **2**:493–506.

Robinson, M.S. (1994). The role of clathrin, adaptors and dynamin in endocytosis. *Curr. Opin. Cell Biol.*, **6**:538–544.

Sakamaki, F., Ishizaka, A., Handa, M., Fujishima, S., Urano, T., Sayama, K., Nakamura, H., Kanazawa, M., Kawashiro, T., Katayama, M. and Ikeda, Y. (1995). Soluble form of P-selectin in plasma is elevated in acute lung injury. *Am. J. Respir. Crit. Care Med.*, **151**:1821–1826.

Sanders, W.E., Wilson, R.W., Ballantyne, C.M. and Beaudet, A.L. (1992). Molecular cloning and analysis of *in vivo* expression of murine P-selectin. *Blood*, **80**:795–800.

Schindler, U. and Baichwal, V.R. (1994). Three NF-κB binding sites in the human E-selectin gene required for maximal tumor necrosis factor α-induced expression. *Mol. Cell. Biol.*, **14**:5820–5831.

Setiadi, H., Disdier, M., Green, S.A., Canfield, W.M. and McEver, R.P. (1995). Residues throughout the cytoplasmic domain affect the internalization efficiency of P-selectin. *J. Biol. Chem.*, in press.

Siebenlist, U., Franzoso, G. and Brown, K. (1995). Structure, regulation and function of NF-κB. *Annu. Rev. Cell Biol.*, **10**:405–455.

Silber, A., Newman, W., Reimann, K.A., Hendricks, E., Walsh, D. and Ringler, D.J. (1994). Kinetic expression of endothelial adhesion molecules and relationship to leukocyte recruitment in two cutaneous models of inflammation. *Lab. Invest.*, **70**:163–175.

Smeets, E.F., de Vries, T., Leeuwenberg, J.F.M., Van den Eijnden, D.H., Buurman, W.A. and Neefjes, J.J. (1993). Phosphorylation of surface E-selectin and the effect of soluble ligand (Sialyl Lewis^X) on the half-life of E-selectin. *Eur. J. Immunol.*, **23**:147–151.

Smith, G.M., Whelan, J., Pescini, R., Ghersa, P., Delamarter, J.F. and Hooft van Huijsduijnen, R. (1993). DNA-methylation of the E-selectin promoter represses NF-κB transactivation. *Biochem. Biophys. Res. Commun.*, **194**:215–221.

Stenberg, P.E., McEver, R.P., Shuman, M.A., Jacques, Y.V. and Bainton, D.F. (1985). A platelet alpha-granule membrane protein (GMP-140) is expressed on the plasma membrane after activation. *J. Cell Biol.*, **101**:880–886.

Strubel, N.A., Nguyen, M., Kansas, G.S., Tedder, T.F. and Bischoff, J. (1993). Isolation and characterization of a bovine cDNA encoding a functional homolog of human P-selectin. *Biochem. Biophys. Res. Commun.*, **192**:338–344.

Subramaniam, M., Koedam, J.A. and Wagner, D.D. (1993). Divergent fates of P- and E-selectins after their expression on the plasma membrane. *Mol. Biol. Cell*, **4**:791–801.

Symon, F.A., Walsh, G.M., Watson, S.R. and Wardlaw, A.J. (1994). Eosinophil adhesion to nasal polyp endothelium is P-selectin-dependent. *J. Exp. Med.*, **180**:371–376.

Takeda, I., Kaise, S., Nishimaki, T. and Kasukawa, R. (1994). Soluble P-selectin in the plasma of patients with connective tissue diseases. *Int. Arch. Allergy Immunol.*, **105**:128–134.

Thanos, D. and Maniatis, T. (1992). The high mobility group protein HMG I(Y) is required for NF-κB-dependent virus induction of the human IFN-β gene. *Cell*, **71**:777–789.

Thornhill, M.H. and Haskard, D.O. (1990). IL-4 regulates endothelial cell activation by IL-1, tumor necrosis factor, or IFN-γ. *J. Immunol.*, **145**:865–872.

Traencker, E.B-M., Wilk, S. and Baeuerle, P.A. (1994). A proteasome inhibitor prevents activation of NF-κB and stabilizes a newly phosphorylated form of IκB-α that is still bound to NF-κB. *EMBO J.*, **13**:5433–5441.

Trowbridge, I.S., Collawn, J.F. and Hopkins, C.R. (1993). Signal-dependent membrane protein trafficking in the endocytic pathway. *Annu. Rev. Cell Biol.*, **9**:129–161.

Ushiyama, S., Laue, T.M., Moore, K.L., Erickson, H.P. and McEver, R.P. (1993). Structural and functional characterization of monomeric soluble P-selectin and comparison with membrane P-selectin. *J. Biol. Chem.*, **268**:15229–15237.

Van Huijsduijnen, R.H., Whelan, J., Pescini, R., Becker-Andre, M., Schenk, A-M. and Delamarter, J.F. (1992). A T-cell enhancer cooperates with NF-κB to yield cytokine induction of E-selectin gene transcription in endothelial cells. *J. Biol. Chem.*, **267**:22385–22391.

Von Asmuth, E.J.U., Smeets, E.F., Ginsel, L.A., Onderwater, J.J.M., Leeuwenberg, J.F.M. and Buurman, W.A. (1992). Evidence for endocytosis of E-selectin in human endothelial cells. *Eur. J. Immunol.*, **22**:2519–2526.

Weller, A., Isenmann, S. and Vestweber, D. (1992). Cloning of the mouse endothelial selectins. Expression of both E- and P-selectin is inducible by tumor necrosis factor. *J. Biol. Chem.*, **267**:15176–15183.

Whelan, J., Ghersa, P., Van Huijsduijnen, R.H., Gray, J., Chandra, G., Talabot, F. and Delamarter, J.F. (1991). An NFκB-like factor is essential but not sufficient for cytokine induction of endothelial leukocyte adhesion molecule 1 (ELAM-1) gene transcription. *Nucl. Acids Res.*, **19**:2645–2653.

Whitley, M.Z., Thanos, D., Read, M.A., Maniatis, T. and Collins, T. (1994). A striking similarity in the organizations of the E-selectin and beta interferon gene promoters. *Mol. Cell. Biol.*, **14**:6464–6475.

3 Regulated Proteolysis of L-selectin

Grace I. Migaki and Takashi Kei Kishimoto

Department of Immunology, Boehringer Ingelheim Pharmaceuticals, Inc., Ridgefield, CT. 06877
Address correspondence to: Takashi Kei Kishimoto, Ph.D., Boehringer Ingelheim Pharmaceuticals, Department of Immunology, R6-5, 900 Ridgebury Rd. Box 368, Ridgefield, CT 06877 USA
Express mail address: Boehringer Ingelheim Pharmaceuticals, Department of Immunology, R6-5, 175 Briar Ridge Rd, Ridgefield, CT 06877 USA
Tel: (203) 798-4651, Fax: (203) 791–6196

Induction of adhesion in biological systems can be effected *via* a number of mechanisms, from qualitative changes in receptor activity (Dustin and Springer, 1989; Spertini *et al.*, 1991b) to quantitative upregulation of rapidly mobilized pools of receptors (Todd *et al.*, 1984; McEver *et al.*, 1989). Due to the transient nature of leukocyte adhesion interactions, mechanisms that cause de-adhesion or downregulation of adhesion protein expression are equally important for regulating cellular adhesion. Both E- and P-selectin have a relatively short half-life on stimulated endothelial cells *in vitro*. E-selectin is gradually lost over a period of 8–24 hours after Il-1 or TNF stimulation of cultured endothelial cells, while P-selectin is cleared within 30 min following thrombin or histamine stimulation of primary endothelial cell cultures. Both E- and P-selectin are downregulated primarily through receptor internalization (Green *et al.*, 1994; Subramaniam *et al.*, 1993). L-selectin is constitutively expressed on the cell surface of myeloid cells and a large subset of lymphocytes (Lewinsohn *et al.*, 1987). L-selectin expression can be downmodulated at the transciptional level during lymphocyte differentiation from a naive to memory cell phenotype. In addition L-selectin expression can be rapidly downregulated by a proteolytic activity which cleaves L-selectin at an extracellular site proximal to the cell membrane. The proteolytic activity is unusual in its rapid kinetics, resistance to most protease inhibitors, and in its relaxed sequence specificity. Thus L-selectin belongs to an emerging class of biologically diverse and important cell surface proteins whose expression can be rapidly downregulated by membrane-proximal proteolysis. This review will focus on the mechanism and consequences of regulated proteolysis of L-selectin.

REGULATED EXPRESSION OF L-SELECTIN

Activation-dependent Downregulation of Lymphocyte L-selectin

It was originally observed that activated T cell clones (Dailey *et al.*, 1985) and germinal center B cell blasts (Reichert *et al.*, 1983) were negative for L-selectin expression and that this phenotype correlated with the inability of these cells to bind HEV and to migrate specifically to peripheral lymph node tissue. *In vitro* analysis revealed that although submitogenic stimulation led to an increase in L-selectin expression, lymphocytes lost L-selectin expression (up to 75%) over a course of 3–6 days in mixed lymphocyte cultures and in response to stimulation with con A mitogen and LPS (Hamann *et al.*, 1988; Jung *et al.*, 1988). Furthermore, it was determined that the loss of L-selectin expression was due to downregulation of L-selectin expression rather than selective proliferation of a small subset of L-selectin-negative resting lymphocytes (Jung *et al.*, 1988). L-selectin expression can be restored on some of these cells by allowing day 3 con A blasts to recover for 3 additional days without further stimulation.

L-selectin expression on activated lymphocytes is likely to involve regulation at a transcriptional level as well as rapid proteolysis of L-selectin from the cell surface. Mitogen-stimulated proliferation of T lymphocytes induces a transient increase in L-selectin expression by day 3 post-stimulation, followed by a steady and marked decline of L-selectin expression by day 10. L-selectin mRNA levels parallel these cell surface changes in L-selectin expression (Kaldjian and Stoolman, 1995). Activation of protein kinase C by phorbol esters also increased L-selectin message levels over a course of several days. This increase in both L-selectin cell surface expression (Bührer *et al.*, 1990) and mRNA levels (Kaldjian and Stoolman, 1995) on PMA-stimulated lymphocytes was blocked by co-treatment with calcium ionophore. These results suggests that protein kinase C upregulates L-selectin message early in T cell proliferation, and that subsequent chronic elevation of intracellular calcium activates calcineurin which leads to downregulation of L-selectin mRNA transcription (Kaldjian and Stoolman, 1995).

Although protein kinase C and calcium influx have opposing effects on L-selectin message levels, both phorbol esters and calcium ionophores can induce a rapid downregulation of L-selectin from the cell surface (Jung and Dailey, 1990; Kishimoto *et al.*, 1990; Spertini *et al.*, 1991a; Bührer *et al.*, 1990). Pretreatment of lymphocytes with a protein kinase C inhibitor, H7, blocked phorbol ester-induced rapid downregulation of L-selectin (Jung and Dailey, 1990; Bührer *et al.*, 1990). However H7 does not inhibit the rapid downregulation of L-selectin in response to calcium ionophores, suggesting that involvement of protein kinase C is not essential (Bührer *et al.*, 1990). This downregulation of cell surface L-selectin occurs at a post-translational level over a time course of 30–60 minutes, which is rapid but still considerably slower than the kinetics of L-selectin down-regulation from the surfaces of stimulated neutrophils (see below). Interestingly L-selectin expression on gamma-delta T lymphocytes is more resistant to phorbol ester-induced downregulation than that of alpha-beta T cells (Walcheck and Jutila, 1994).

Activation-dependent Downregulation of Neutrophil L-selectin Expression

L-selectin cell-surface expression is also downmodulated on neutrophils during recruitment to sites of inflammation. Although peripheral blood neutrophils have uniformly high levels of L-selectin expression, cells harvested from the peritoneum 3 hours after thioglycollate challenge are largely L-selectin low or negative (Jutila *et al.*, 1989). Similarly neutrophils which have migrated to inflamed footpad and lung are largely L-selectin negative (Kishimoto *et al.*, 1989; Burns and Doerschuk, 1994). The recruitment of neutrophils to the inflamed peritoneum was observed to occur over a time course of hours,

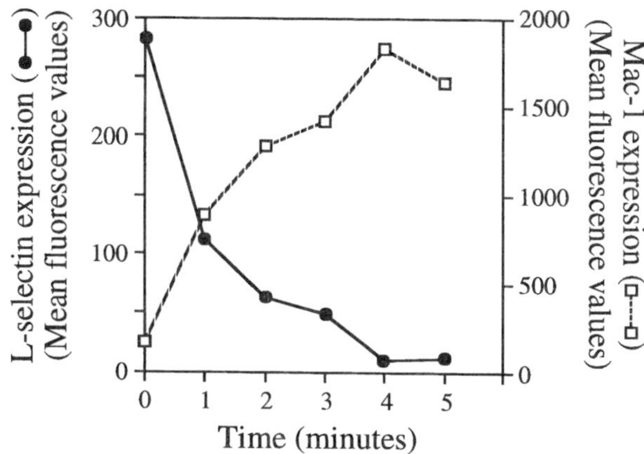

Figure 1 Inverse regulation of L-selectin and Mac-1 on C5a-stimulated neutrophils (adapted from Kishimoto *et al.* (1989), *Science*, **245**:1238–1241). Neutrophils were stimulated with C5a chemoattractant and analyzed at one minute intervals for cell surface expression of L-selectin (closed circles) and Mac-1 (open squares).

suggesting that L-selectin expression can be rapidly modulated *in vivo*. Like activated lymphocytes, activated neutrophils show defective migration *in vivo* (Jutila *et al.*, 1989).

Detailed kinetic analysis revealed the surprising finding that L-selectin can be downregulated from the cell surface of neutrophils over a time course of 1–5 minutes in response to the C5a chemoattractant (Kishimoto *et al.*, 1989; Jutila *et al.*, 1990) (Figure 1). The unusually rapid kinetics of L-selectin downregulation excludes any significant contribution from transcriptional regulation. L-selectin downregulation on neutrophils, monocytes and myeloid precursor cells was found to occur in response to a variety of neutrophil chemoattractants and activating factors, such as fMLP, C5a, LTB4, IL-8, TNF, GM-CSF, and calcium ionophore (A23187) (Kishimoto *et al.*, 1989; Jutila *et al.*, 1989; Griffin *et al.*, 1990; Berg and James, 1990). However, L-selectin downregulation on neutrophils was not affected by treatment with G-CSF, M-CSF, IL-1 or IFN-gamma (Griffin *et al.*, 1990). Neutrophils exposed to Il-1 stimulated HUVEC monolayers for thirty minutes also downregulate cell surface L-selectin (Smith *et al.*, 1991). The resultant loss of L-selectin may have been due to release of IL-8 or other neutrophil activating factors by the stimulated endothelial monolayers.

Activation-independent Downregulation of L-selectin

Paleconda *et al.* (1992) demonstrated that L-selectin can also be rapidly downregulated from leukocyte cell surfaces by crosslinking surface proteins with chemical crosslinking agents or by crosslinking L-selectin with immobilized MAb directed against L-selectin. Soluble L-selectin has been detected in the plasma of healthy adults whose peripheral blood lymphocytes exhibited no signs of activation (Palecanda *et al.*, 1992; Schleiffenbaum *et al.*, 1992). Interestingly it is calculated that the level of sL-selectin in normal serum is 10–25-fold greater than the total cell-associated L-selectin, suggesting that L-selectin shedding is an ongoing event (Schleiffenbaum *et al.*, 1992). Jutila and colleagues (Palecanda *et al.*, 1992) have proposed that engagement of L-selectin by its natural ligand on endothelial cells may directly cause L-selectin proteolysis in the absence of any additional activating signal.

Receptor engagement of L-selectin may induce conformational changes in membrane-bound L-selectin, or may induce the activation of the protease responsible for L-selectin processing. Recent evidence indicates that engagement and crosslinking of L-selectin can transduce signals to neutrophils, including activation of protein kinase-C, Ca^{++} flux, activation of MAP kinase and tyrosine phosphorylation (Crockett-Torabi *et al.*, 1995; Waddell *et al.*, 1994; Simon *et al.*, 1995; Waddell *et al.*, 1995; Laudanna *et al.*, 1994). Crosslinking of L-selectin on human neutrophils leads to increased tyrosine phosphorylation of several proteins. The effect is specific for L-selectin and is not inhibited by cytochalasin, thus reorganization of the actin cytoskeleton is probably not required for this response. Putative natural ligands for L-selectin, such as sulfatides and sulfated glycolipids, also induce a rapid, dose-dependent increase in tyrosine phosphorylation.

Recently, Diaz-Gonzalez *et al.* (1995) observed that certain non-steroidal anti-inflammatory drugs (NSAIDs) also induce L-selectin downregulation from the cell-surface of neutrophils and inhibited neutrophil-endothelial attachment without affecting other adhesion molecules such as CD11a, CD11b, CD31 and CD50. Neutrophils were treated with indomethacin, diclofenac, ketoprofen or aspirin. Quantitative evaluation of soluble L-selectin immunoprecipitated from the supernatants of treated cells indicated that L-selectin was efficiently downregulated in a time- and dose- dependent manner. In contrast, steroids had little or no effect on cell-surface expression of L-selectin on neutrophils. Crosslinking of other signalling surface structures, such as the CD16 Fc receptor, can also result in downregulation of L-selectin from the cell surface in the absence of additional activating signals (Bazil and Strominger, 1994).

MECHANISM OF L-SELECTIN DOWNREGULATION

Identification of L-selectin Cleavage Fragments and the Cleavage Site

The rapid downregulation of L-selectin from cell surfaces of activated neutrophils excludes any significant contribution of transcriptional regulation. To determine whether L-selectin was shed, internalized, or whether the detecting MAb epitope was merely masked on activated cells, the fate of $[^{125}I]$-surface labeled L-selectin was followed on activated neutrophils and lymphocytes (Kishimoto *et al.*, 1989; Berg and James, 1990; Jung and Dailey, 1990). Virtually all of the L-selectin was quantitatively recovered from the cell-free supernatants of activated cells. The shed soluble form of L-selectin has an apparent molecular weight approximately 4–12 kD less than the cell surface form of L-selectin suggesting that the large ectodomain of L-selectin is released relatively intact by limited proteolysis close to the site of membrane insertion (Kishimoto *et al.*, 1989; Jung and Dailey, 1990).

The cleavage fragments of L-selectin were further characterized with polyclonal antisera generated against the extracellular domain (JK923) and against the cytoplasmic domain (JK564) of L-selectin (Kahn *et al.*, 1994). Both antisera immunoprecipitated the full-length form of L-selectin from metabolically labeled PHA-stimulated lymphoblasts, peripheral blood neutrophils and L-selectin-transfected COS cells. However only the anti-cytoplasmic domain serum immunoprecipitated a 6 kD transmembrane cleavage

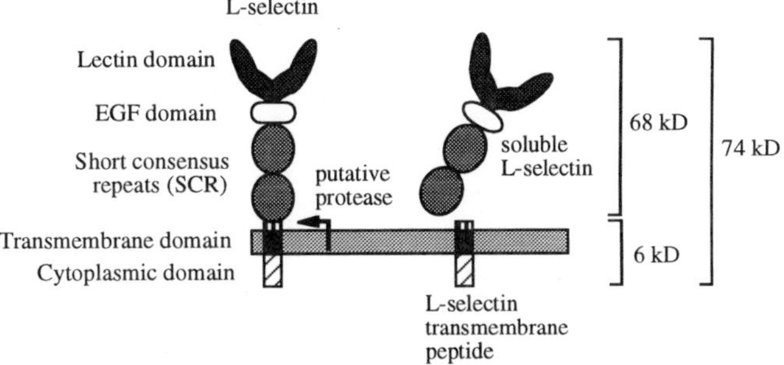

Figure 2 Cleavage fragments of L-selectin. Schematic representation of the cell surface form of L-selectin and its cleavage by a putative protease at a membrane-proximal site to yield a 68 kD soluble fragment and a 6 kD transmembrane fragment (size estimates are for the lymphocyte form of L-selectin).

fragment of L-selectin (Figure 2). The appearance of the 6 kD fragment correlated with the disappearance of the full-length form of L-selectin and the subsequent appearance of soluble L-selectin. A third polyclonal serum generated against the extracellular portion of the membrane proximal region (JK924) also immunoprecipitated the 6 kD fragment confirming that this fragment is a transmembrane peptide of L-selectin.

The putative cleavage site of L-selectin was determined by radiochemical sequence analysis of the 6 kD L-selectin transmembrane peptide (L-STMP) (Kahn *et al.*, 1994). Analysis of the 6 kD fragment isolated from cells metabolically labeled with [^{35}S]-methionine and [^{3}H]-phenylalanine revealed a putative cleavage site between Lys 321 and Ser 322 in the membrane-proximal region between the second short consensus repeat and the transmembrane domain (Figure 3). Cleavage at this site is predicted to yield a

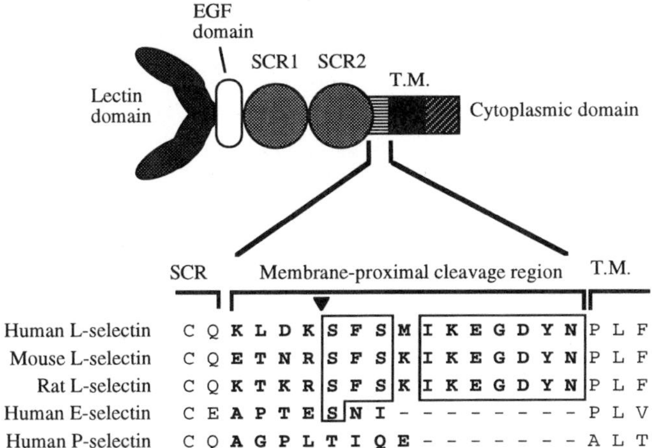

Figure 3 L-selectin cleavage site (adapted from Kahn *et al.* (1994) *J. Cell Biol.*, **125**:461–470). Schematic representation of L-selectin with the amino acid sequence of the membrane-proximal cleavage region shown below. Sequences of human, mouse, and rat L-selectin aligned with corresponding regions of human E-selectin and P-selectin. The putative cleavage site is indicated with an inverted triangle.

transmembrane peptide of 5.8 kD, which is in excellent agreement with the observed 6 kD species. This region is highly conserved among human, mouse and rat L-selectin; however, the region is highly divergent among E-selectin and P-selectin despite the high degree of homology that exists among their extracellular domains. This is consistent with the observation that only L-selectin is rapidly downregulated through membrane-proximal cleavage.

Site-directed Mutagenesis of the L-selectin Cleavage Region

The sequence specificity of L-selectin proteolysis was examined using site-directed mutagenesis (Migaki *et al.*, 1995; Chen *et al.*, 1995). The necessity of the cleavage domain was demonstrated by replacing the short cleavage domain of L-selectin with the corresponding membrane-proximal region of E-selectin to yield a chimeric protein resistant to membrane-proximal cleavage (Migaki *et al.*, 1995). This chimera, L-CDX (L-selectin cleavage domain exchange), was expressed at higher cell-surface levels than wildtype L-selectin, consistent with the apparent lack of downregulation from transfected cell surfaces. In addition, there was a marked absence of 68 kD soluble cleavage product and 6 kD transmembrane cleavage product immunoprecipitated from radiolabeled L-CDX transfectants. Soluble cleavage product was also absent in cell-free supernatants from L-CDX tranfected as demonstrated by trapping ELISA. Similarly analysis of L-selectin-P-selectin chimeras revealed that only those chimeric molecules retaining the membrane-proximal region of L-selectin were cleaved in a regulated manner (Chen *et al.*, 1995). Taken together, these results confirm that the membrane-proximal region of L-selectin is required for proteolysis.

Surprisingly, extensive mutation of the putative cleavage site revealed that L-selectin proteolysis shows extremely relaxed sequence-specificity surrounding the cleavage site (Migaki *et al.*, 1995; Chen *et al.*, 1995). Alanine scanning mutations of the cleavage site at both the P1 and P1′ positions, as well as mutations of those amino acid residues conserved across species were found to have little effect on production of both soluble and transmembrane cleavage product in radiolabeled COS cell transfectants and soluble L-selectin protein in lymphoid cell transfectants. The lysine at the P1 position is suggestive of a trypsinlike specificity, yet mutating the basic lysine residue to an acidic glutamic acid does not abolish proteolysis (Migaki *et al.*, 1995; Chen *et al.*, 1995). In all point mutations, the size of the 6 kD transmembrane cleavage product was indistinguishable from the corresponding fragment from wildtype L-selectin (Migaki *et al.*, 1995). Interestingly, mutations of multiple consecutive amino acids to alanine sequences, including a serine-phenylalanine-serine motif adjacent to the cleavage site did not appreciably affect proteolysis (Migaki *et al.*, 1995; Chen *et al.*, 1995). However, short deletions of consecutive amino acids within the cleavage region drastically decreased the amount of soluble cleavage product detected in cell-free supernatants (Migaki *et al.*, 1995; Chen *et al.*, 1995) and the amount of the 6 kD transmembrane cleavage product immunoprecipitated from cell lysates (Migaki *et al.*, 1995). Fluorescence microscopy and FACS analysis indicated that the deletion mutants were expressed at higher cell-surface levels than wildtype L-selectin, consistent with the lack of proteolysis from the cell surface. Two deletion mutants, I327-N332 and M325-N332 retain the native cleavage site, yet inhibit L-selectin proteolysis as well as deletion mutants that exclude the native P1 and P1′

residues (Migaki *et al.*, 1995; Chen *et al.*, 1995). Restoring the amino acids between I327 and N332 with five alanine residues restores L-selectin proteolysis (Migaki *et al.*, 1995). In addition, radiochemical sequence analysis of the 6 kD tramsmembrane cleavage product from K327EGDY/AAAAA restoration indicates that cleavage still occurs at the native cleavage site between Lys321 and Ser322. Thus, secondary and tertiary structural elements as well as the physical length of the cleavage domain may regulate proteolysis of L-selectin.

L-selectin Cleavage is Resistant to a Variety of Protease Inhibitors

Low doses of exogenous chymotrypsin were observed to mimic the downregulation of L-selectin, yet a variety of chymotrypsin inhibitors had no effect on L-selectin regulation on activated leukocytes (Jutila *et al.*, 1991). The downregulation of L-selectin was further found to be resistant to a broad panel of protease inhibitors, including general inhibitors of serine proteases, metalloproteases, aspartic proteases, and cysteine proteases (Campanero *et al.*, 1991; Shipp *et al.*, 1991; Bazil and Strominger, 1994; Kishimoto *et al.*, 1995). Interestingly, one chymotrypsin inhibitor, TPCK, inhibited Mac-1 upregulation and neutrophil oxidative burst but had no effect on L-selectin downregulation. Thus the precise class of protease involved in L-selectin downregulation could not be determined through the use of conventional protease inhibitors.

L-selectin Cleavage is Inhibitable by Hydroxamic Acid-based Metalloprotease Inhibitors

To aid in the screening of potential inhibitors, an L-selectin-alkaline phosphatase reporter construct (L-AP) consisting of the ectodomain of human placental alkaline phosphatase fused to the cleavage, transmembrane and cytoplasmic domains of L-selectin was developed (Feehan *et al.*, 1996). The L-AP reporter is cleaved from the cell surface and the released alkaline phosphatase activity can be detected with a chemiluminescent substrate. The release of the L-AP reporter is inhibited by a hydroxamic acid-based metalloprotease inhibitor, KD-IX-73-4 in a dose dependent manner. KD-IX-73-4 was originally synthesized as an inhibitor of zinc-dependent matrix metalloproteases, such as collagenase and gelatinase. The diastereomer, KD-IX-73-5, is less active against matrix metalloproteases and at least 25-fold less potent than KD-IX-73-4 in inhibiting cleavage of L-AP. In addition, the hydroxamic acid-based peptide was also found to inhibit wildtype L-selectin proteolysis from the surfaces of PMA-activated peripheral blood lymphocytes, PHA-stimulated lymphoblasts and fMLP- and PMA-stimulated neutrophils. The hydroxamic acid-based metalloprotease inhibitors do not affect general neutrophil activation, as judged by a lack of effect on Mac-1 mobilization and neutrophil oxidative burst. However, treatment with the hydroxamic acid-based compound inhibited L-selectin downregulation in a dose-dependent manner with an IC50 of approximately 4 uM. Analysis of the proteolytic cleavage products of L-selectin confirmed that KD-IX-73-4 inhibits formation of both the 6kD transmembrane- and 68kD soluble-cleavage products in PHA lymphoblasts and in neutrophils stimulated with PMA. Thus L-selectin proteolysis appears to involve a metal-dependent protease, possibly a novel member of the metalloprotease family.

FUNCTIONAL SIGNIFICANCE OF L-SELECTIN PROTEOLYSIS

Monoclonal antibodies directed against L-selectin and Mac-1 both attenuate neutrophil recruitment to the inflamed peritoneum, suggesting that both adhesion molecules are involved in neutrophil trafficking to sites of inflammation (Lewinsohn *et al.*, 1987; Rosen and Gordon, 1987; Jutila *et al.*, 1989). However Jutila *et al.* (1989) observed that neutrophils that had migrated into the peritoneum showed profound upregulation of Mac-1 and downregulation of L-selectin. Smith and colleagues (Lawrence *et al.*, 1990) also demostrated that anti-CD18 MAb profoundly affects neutrophil transendothelial migration but had no effect on neutrophil adhesion to stimulated endothelial cells under conditions of hydrodynamic flow which mimics the shear stress found in post-capillary venules. Thus the observation that L-selectin and Mac-1 are inversely regulated within minutes after chemotactic stimulation led to the suggestion that L-selectin may be involved in the intial adhesion event between circulating neutrophils and the inflamed endothelium, and that chemotactic factors released at the site of inflammation may induce downregulation of L-selectin and engagement of Mac-1, allowing extravasation to occur (Kishimoto *et al.*, 1989). Subsequent studies demonstrated that L-selectin can mediate, in part, neutrophil rolling on stimulated endothelium both *in vitro* and *in vivo* (Von Andrian *et al.*, 1991; Ley *et al.*, 1991; Smith *et al.*, 1991). L-selectin shedding, therefore, was postulated to be a "de-adhesion" mechanism that facilitates neutrophil migration following binding, or the release of the neutrophil from the endothelium after diapedesis.

Proteolysis of L-selectin may provide a means for leukocyte detachment from endothelial cells as they approach inflammatory sites, or prior to transendothelial migration. Smith *et al.* (1991) observed that the chemotactic stimulation of neutrophils induced by a soluble factor or factors released by stimulated HUVEC monolayers resulted in the downregulation of L-selectin from neutrophil cell-surfaces. The level of L-selectin expression on neutrophil surfaces closely correlated with their ability to adhere to HUVEC monolayers at a shear stress of 1.85 dynes/cm2.

In addition, shedding of L-selectin may be a protective mechanism that prevents activated neutrophils from entering secondary sites that are unrelated to an initial insult. The cleavage of L-selectin from the surface of murine neutrophils by low doses of chymotrypsin had no effect on neutrophil ability to adhere to plastic, migrate to C5a or regulate Mac-1 (Jutila *et al.*, 1991). However, chymotrypsin-treated neutrophils injected back into the animal lost their ability to home to inflamed peritoneum. Neutrophils from mice genetically-deficient in L-selectin expression exhibit a significant defect in their ability to migrate into an inflamed peritoneum 24 and 48 hours after administration of thioglycollate (Arbones *et al.*, 1994; Tedder *et al.*, 1995). Delayed-type hypersensitivity reactions, footpad swelling response to SRBC challenge and contact hypersensitivity were all reduced significantly in these L-selectin knock-out animals (Tedder *et al.*, 1995). These results suggest that downmodulation of L-selectin expression can profoundly influence neutrophil trafficking.

Inhibition of L-selectin Shedding Affects Neutrophil Rolling Behavior *in vitro*

The physiological consequence of inhibiting L-selectin proteolysis with hydroxamic acid-based inhibitors was tested in an *in vitro* model of neutrophil rolling under hydrodynamic flow (Walcheck *et al.*, 1996). The hydroxamic acid-based inhibitor, KD-IX-73-4, had a profound affect on neutrophil rolling behavior by prolonging transient attachment times to immobilized L-selectin ligand, PNAd, at a wall shear stress of 2 dynes/cm^2. The neutrophil accumulation rate increased from 6.3 cells/field/min for control neutrophils to 28.5 cells/field/min for neutrophils treated with 25 μg/mL of KD-IX-73-4. Neutrophils treated with the hydroxamic acid-based compound rolled at significantly lower velocities (9.2 μm/sec) than control cells (27.7 μm/sec) under conditions of flow. Control neutrophils primarily exhibit a fast rolling phenotype interrupted by short periods of slow rolling. However, neutrophils treated with hydroxamic acid-based compounds exhibited dramatically increased periods of slow rolling or transient arrest disrupted by short periods of fast rolling. These studies indicate that L-selectin is cleaved within seconds — the timecourse of these rolling interactions — and that proteolysis of L-selectin contributes significantly to the overall velocity of leukocyte rolling.

The significant effect of hydroxamic acid-based inhibitors on neutrophils rolling is experimental evidence for the occurrence of L-selectin proteolysis during the process of L-selectin-dependent rolling. L-selectin shedding during rolling may enable leukocytes to quickly scan a length of endothelium and rapidly detach if an appropriate chemokine signal is not encountered. In addition, ligation-dependent shedding of L-selectin may also limit L-selectin-dependent neutrophil-neutrophil interactions that lead to neutrophil accumulation under flow (Bargatze *et al.*, 1994) and neutrophil aggregation (Simon *et al.*, 1992).

A Biological Role for Circulating L-selectin

The continuous shedding of L-selectin during neutrophil rolling and leukocyte trafficking in response to subclinical stimuli may partially account for the high levels of circulating L-selectin found in normal human serum (Palecanda *et al.*, 1992; Schleiffenbaum *et al.*, 1992). ELISA analysis indicated a mean sL-selectin level of 1.6 to 2.0 μg/mL present in serum from normal human donors (Schleiffenbaum *et al.*, 1992). Additionally, semi-purified sL-selectin was able to inhibit L-selectin-specific attachment of lymphocytes to cytokine-activated endothelium in a dose-dependent manner with maximal inhibition achieved at concentrations of 12–15 μg/mL (Schleiffenbaum *et al.*, 1992). Although it is unlikely that the level of sL-selectin in normal circulation is high enough to impact L-selectin-mediated adhesion, it is possible that during the course of inflammation the concentration of sL-selectin in microenvironments may achieve levels high enough to affect adhesion. In certain disease states, such as chronic myeloid leukemia, soluble L-selectin in serum can reach close to mg/ml levels (Zetterberg and Richter, 1993).

Downregulation of Other Cell Surface Proteins by Regulated Proteolysis

L-selectin is a member of an emerging class of diverse proteins whose cell surface expression can be downmodulated by regulated proteolysis (Ehlers and Riordan, 1991; Bazil, 1995). Other leukocyte membrane-bound proteins such as CD14 (Bazil and Strominger, 1991), CD16 (Bazil and Strominger, 1994), CD43 (Bazil and Strominger, 1993; Campanero *et al.*, 1991) , CD44 (Bazil and Horejsí, 1992; Campanero *et al.*, 1991; Rieu *et al.*, 1992), ICAM-3 (Del Pozo *et al.*, 1994), IL-6 receptor (Mullberg *et al.*, 1994;

Mullberg *et al.*, 1993), and TNF-receptor (Porteu and Nathan, 1990) have also been reported to be shed upon leukocyte activation. However the rapid kinetics of L-selectin shedding are unusual and differentiate it from the regulation of many of these other surface proteins. For example, CD43 and CD44 are downregulated over a period of 15–60 min on phorbol ester-stimulated neutrophils (Bazil and Horejsí, 1992; Campanero *et al.*, 1991; Rieu *et al.*, 1992; Bazil and Strominger, 1993). In contrast L-selectin is downregulated in minutes on the same cells (Kishimoto *et al.*, 1989). Furthermore L-selectin shedding on neutrophils follows similarly rapid kinetics in response to chemotactic agents, such as fMLP; whereas, fMLP has little or no effect on CD43 and CD44 expression. The requirement for a potent stimulus such as phorbol esters suggests the possibility that proteases associated with secondary granules must be released for downregulation of CD43 and CD44. Moreover L-selectin shedding is resistant to common protease inhibitors. However, cleavage of CD14, CD16, CD43 and CD44 has been shown to be inhibitable when cells are treated with serine protease and metalloprotease inhibitors (Bazil and Horejsí, 1992; Campanero *et al.*, 1991; Rieu *et al.*, 1992; Bazil and Strominger, 1994; Bazil and Strominger, 1991). In contrast L-selectin proteolysis is resistant to serine protease inhibitors and common metalloprotease inhibitors, such as EDTA, phosphoramidon, and phenanthroline. Thus the mechanism that regulates L-selectin proteolysis is distinct from that of many other leukocyte cell-surface antigens.

The molecular basis of membrane-proximal proteolysis has been examined for only a few proteins. However it is becoming apparent that proteolysis of L-selectin shares certain features with that of such biologically important molecules as TGF-α (Pandiella and Massague, 1991; Arribas and Massague, 1995), TNF-α (Mohler *et al.*, 1994; Gearing *et al.*, 1994; McGeehan *et al.*, 1994), IL6-receptor (Mullberg *et al.*, 1994; Mullberg *et al.*, 1993), angiotensin converting enzyme (ACE; (Ramchandran *et al.*, 1994; Ehlers *et al.*, 1991)), and beta amyloid precursor protein (β-APP; (Ramchandran *et al.*, 1994; Sahasrabudhe *et al.*, 1992)). Interestingly, rabbit ACE (Ramchandran *et al.*, 1994) and human β-APP (Sisodia, 1992; Sahasrabudhe *et al.*, 1992) are both cleaved at sites homologous to the P1 and P1′ positions of the L-selectin cleavage site. However, mutation of the P1 and P1′ positions have surprisingly little effect on the efficiency of proteolysis. Similarly a Ser-Phe-Ser motif adjacent to the cleavage site of L-selectin is conserved among human, mouse and rat L-selectin, and a related motif is found proximal to the transmembrane regions of other downregulated proteins, including ACE, CD16-II, and TNF-RII. Yet mutagenesis of the Ser-Phe-Ser motif had no significant effect on L-selectin proteolysis (Migaki *et al.*, 1995; Chen *et al.*, 1995). The apparent lack of sequence specificity in the cleavage region of L-selectin is reminiscent of the relaxed specificity described for other molecules regulated by membrane-proximal proteolysis including β-APP (Sisodia, 1992; Sahasrabudhe *et al.*, 1992), TGF-α (Pandiella *et al.*, 1992), IL6-R (Mullberg *et al.*, 1993) and ACE (Ramchandran *et al.*, 1994).

Short deletions in the cleavage region of L-selectin profoundly inhibited proteolysis. Similar results were observed for the IL6-R (Mullberg *et al.*, 1993). However, large deletions in the cleavage regions of β-APP and ACE exhibit no effect on their proteolysis (Sisodia, 1992). The cysteine residue at position 316 of L-selectin contributes to an intrachain disulfide bridge which may provide steric or structural interference with the protease, thus preventing cleavage at upstream sequences in the deletion mutants. In general the membrane-proximal proteolysis of cell surface proteins may depend on secondary and tertiary structural characteristics, such as an alpha-helical conformation of the cleavage domain, as well as the physical distance of the hydrolyzed bond from the membrane (Sisodia, 1992; Sahasrabudhe *et al.*, 1992; Migaki *et al.*, 1995; Chen *et al.*, 1995; Mullberg *et al.*, 1993).

Although L-selectin proteolysis is resistant to numerous protease inhibitors, it has been recently demonstrated that L-selectin cleavage is inhibited by a hydroxamic-acid based metalloprotease inhibitor, KD-IX-73-4 (Feehan *et al.*, 1996; Walcheck *et al.*, 1996). Similar hydroxamic acid-based peptides have been demonstrated to affect processing of membrane-bound TNF (Mohler *et al.*, 1994; Gearing *et al.*, 1994; McGeehan *et al.*, 1994) and TNF-R (Crowe *et al.*, 1995). However, TNF-alpha processing can also be inhibited by EDTA (Mohler *et al.*, 1994), whereas L-selectin proteolysis is resistant to EDTA treatment (Bazil and Strominger, 1994; Kishimoto *et al.*, 1995). In a parallel approach, Arribas and Massague (1995) mutagenized CHO cells and selected for those that were unable to cleave membrane anchored TGF-α. The mutants do not appear to have any obvious defects in protein synthesis or transport to the cell surface, thus the defect appears to be at the level of proteolysis. Interestingly, β-APP transfected into these CHO cell mutants is also resistant to proteolysis. Recent data have revealed that these mutants also do not cleave L-selectin indicating that these systems may share a common step in the proteolytic process (see note). It is still too early to predict whether a single protease activity or multiple related activities are responsible for the proteolysis of such biologically diverse proteins. However the field is rapidly converging on isolating the protease activities involved. Identification of such proteases may provide significant insight into the regulation of inflammation (L-selectin, TNF-receptor, IL-6 receptor), hypertension (ACE), developmental regulation (TGF-α) or the pathogenesis of Alzheimer's disease (β-APP).

Note

Recently three other groups have independently demonstrated that hydroxamic acid-based metalloprotease inhibitors block L-selectin shedding [Pierce, G., Murphy, G., and Ager, A. (1996). Metalloproteinase-mediated regulation of L-selectin levels on leucocytes. *J. Biol. Chem.*, *271*, 11634–11640; Arribas, J., Coodly, L., Vollmer, P., Kishimoto, T.K., Rose-John, S., and Massague, J. (1996). Diverse cell surface protein ectodomains are shed by a system sensitive to metalloprotease inhibitors. *J. Biol. Chem.* *271*, 11376–11382; and Bennett, T.A., Lynam, E.B., Sklar, L.A., and Rogelj, S. (1996). Hydroxamate-based metalloprotease inhibitor blocks shedding of L-selectin adhesion molecule from leukocytes. *J. Immunol.*, *156*, 3093–3097]. Arribas *et al.* also showed that CHO cell mutants defective in proTGF-α processing are also defective in the ability to cleave L-selectin and IL-6 receptor. Bennett *et al.* demonstrated that inhibition of L-selectin shedding affects neutrophil aggregation.

References

Arbones, M.L., Ord, D.C., Ley, K., Ratech, H., Maynard-Curry, C., Otten, G., Capon, D.J. and Tedder, T.F. (1994). Lymphocyte homing and leukocyte rolling and migration are impaired in L-selectin-deficient mice. *Immunity*, **1**:247–260.

Arribas, J. and Massague, J. (1995). Transforming growth factor-alpha and beta-amyloid precursor protein share a secretory mechanism. *J. Cell Biol.*, **128**:433–441.

Bargatze, R.F., Kurk, S., Butcher, E.C. and Jutila, M.A. (1994). Neutrophils roll on adherent neutrophils bound to cytokine-induced endothelial cells *via* L-selectin on the rolling cells. *J. Exp. Med.*, **180**:1785–1792.

Bazil, V. (1995). Physiological enzymatic cleavage of leukocyte membrane molecules. *Immunol. Today*, **16**:135–140.

Bazil, V. and Horejsí, V. (1992). Shedding of the CD44 adhesion molecule from leukocytes induced by anti-CD44 monoclonal antibody simulating the effect of a natural receptor ligand. *J. Immunol.*, **149**:747–753.

Bazil, V. and Strominger, J.L. (1991). Shedding as a mechanism of down-modulation of CD14 on stimulated human monocytes. *J. Immunol.*, **147**:1567–1574.

Bazil, V. and Strominger, J.L. (1993). CD43, the major sialoglycoprotein of human leukocytes, is proteolytically cleaved from the surface of stimulated lymphocytes and granulocytes. *Proc. Natl. Acad. Sci. USA*, **90**:3792–3796.

Bazil, V. and Strominger, J.L. (1994). Metalloprotease and serine protease are involved in cleavage of CD43, CD44, and CD16 from stimulated granulocytes. Induction of cleavage of L-selectin *via* CD16. *J. Immunol.*, **152**:1314–1322.

Berg, M. and James, S.P. (1990). Human neutrophils release the Leu-8 lymph node homing receptor during cell activation. *Blood*, **76**:2381–2388.

Burns, A.R. and Doerschuk, C.M. (1994). Quantitation of L-selectin and CD18 expression on rabbit neutrophils during CD18-independent and CD18-dependent emigration in the lung. *J. Immunol.*, **153**:3177–3188.

Bührer, C., Berlin, C., Thiele, H-G. and Hamann, A. (1990). Lymphocyte activation and expression of the human leucocyte-endothelial cell adhesion molecule 1 (Leu-8/TQ1 antigen). *Immunol.*, **71**:442–448.

Campanero, M.R., Pulido, R., Alonso, J.L., Pivel, J.P., Pimentel-Muinos, F.X., Fresno, M. and Sanchez-Madrid, F. (1991). Down-regulation by tumor necrosis factor-alpha of neutrophil cell surface expression of the sialophorin CD43 and the hyaluronate receptor CD44 through a proteolytic mechanism. *Eur. J. Immunol.*, **21**:3045–3048.

Chen, A., Engel, P. and Tedder, T.F. (1995). Structural requirements regulate endoproteolytic release of the L-selectin (CD62L) adhesion receptor from the cell surface of leukocytes. *J. Exp. Med.*, **182**:519–530.

Crockett-Torabi, E., Sulenbarger, B., Smith, C.W. and Fantone, J.C. (1995). Activation of human neutrophils through L-selectin and Mac-1 molecules. *J. Immunol.*, **154**:2291–2302.

Crowe, P.D., Walter, B.N., Mohler, K.M., Otten-Evans, C., Black, R.A. and Ware, C.F. (1995). A metalloprease inhibitor blocks shedding of the 80-kD TNF receptor and TNF processing in T lymphocytes. *J. Exp. Med.*, **181**:1205–1210.

Dailey, M.O., Gallatin, W.M. and Weissman, I.L. (1985). The *in vivo* behavior of T cell clones: altered migration due to loss of the lymphocyte surface homing receptor. *J. Mol. Cell. Immunol.*, **2**:27–36.

Del Pozo, M.A., Pulido, R., Munoz, C., Alvarez, V., Humbria, A., Campanero, M.R. and Sanchez-Madrid, F. (1994). Regulation of ICAM-3 (CD50) membrane expression on human neutrophils through a proteolytic shedding mechanism. *Eur. J. Immunol.*, **24**:2586–2594.

Diaz-Gonzalez, F., Gonzalez-Alvaro, I., Campanero, M.R., Mollinedo, F., del, Pozo, M.A., Munoz, C., Pivel, J.P. and Sanchez-Madrid, F. (1995). Prevention of *in vitro* neutrophil-endothelial attachment through shedding of L-selectin by nonsteroidal antiinflammatory drugs. *J. Clin. Invest.*, **95**:1756–1765.

Dustin, M.L. and Springer, T.A. (1989). T cell receptor cross-linking transiently stimulates adhesiveness through LFA-1. *Nature*, **341**:619–624.

Ehlers, M.R.W., Chen, Y-N.P. and Riordan, J.F. (1991). Spontaneous solubilization of membrane-bound human testis angiotensin-converting enzyme expressed in Chinese hamster ovary cells. *Proc. Natl. Acad. Sci. USA*, **88**:1009–1013.

Ehlers, M.R.W. and Riordan, J.F. (1991). Membrane proteins with soluble counterparts: Role of proteolysis in the release of transmembrane proteins. *Biochem.*, **30**:10065–10074.

Feehan, C., Darlak, K., Kahn, J., Walcheck, B., Spatola, A.F. and Kishimoto, T.K. (1996). Shedding of the lymphocyte L-selectin adhesion molecule is inhibited by a hydroxamic acid-based protease inhibitor. *J. Biol. Chem.*, **271**:7019–7024.

Gearing, A.J.H., Beckett, P., Christodoulou, M., Churchill, M., Clements, J., Davidson, A.H., Drummond, A.H. *et al.* (1994). Processing of tumor necrosis factor-alpha precursor by metalloproteinases. *Nature*, **370**:555–557.

Green, S.A., Setiadi, H., McEver, R.P. and Kelly, R.B. (1994). The cytoplasmic domain of P-selectin contains a sorting determinant that mediates rapid degradation in lysosomes. *J. Cell Biol.*, **124**:435–448.

Griffin, J.D., Spertini, O., Ernst, T.J., Belvin, M.P., Levine, H.B., Kanakura, Y. and Tedder, T.F. (1990). Granulocyte-macrophage colony-stimulating factor and other cytokines regulate surface expression of the leukocyte adhesion molecule-1 on human neutrophils, monocytes, and their precursors. *J. Immunol.*, **145**:576–584.

Hamann, A., Jablonski-Westrich, D., Scholz, K-U., Duijvestijn, A., Butcher, E.C. and Thiele, H-G. (1988). Regulation of lymphocyte homing. I. Alterations in homing receptor expression and organ-specific high endothelial venule binding of lymphocytes upon activation. *J. Immunol.*, **140**:737–743.

Jung, T.M., Gallatin, W.M., Weissman, I.L. and Dailey, M.O. (1988). Down-regulation of homing receptors after T cell activation. *J. Immunol.*, **141**:4110–4117.

Jung, T.M. and Dailey, M.O. (1990). Rapid modulation of homing receptors (gp90Mel-14) induced by activators of protein kinase C. *J. Immunol.*, **144**:3130–3136.

Jutila, M.A., Rott, L., Berg, E.L. and Butcher, E.C. (1989). Function and regulation of the neutrophil MEL-14 antigen *in vivo*: Comparison with LFA-1 and Mac-1. *J. Immunol.*, **143**:3318–3324.

Jutila, M.A., Kishimoto, T.K. and Butcher, E.C. (1990). Regulation and lectin activity of the human neutrophil peripheral lymph node homing receptor. *Blood*, **76**:178–183.

Jutila, M.A., Kishimoto, T.K. and Finken, M. (1991). Low-dose chymotrypsin treatment inhibits neutrophil migration into sites of inflammation *in vivo*: Effects on Mac-1 and MEL-14 adhesion protein expression and function. *Cell Immunol.*, **132**:201–214.

Kahn, J., Ingraham, R.H., Shirley, F., Migaki, G.I. and Kishimoto, T.K. (1994). Membrane proximal cleavage of L-selectin: Identification of the cleavage site and a 6-kD transmembrane peptide fragment of L-selectin. *J. Cell Biol.*, **125**:461–470.

Kaldjian, E.P. and Stoolman, L.M. (1995). Regulation of L-selectin mRNA in Jurkat cells: Opposing influences of calcium- and protein kinase C-dependent signaling pathways. *J. Immunol.*, **154**:4351–4362.

Kishimoto, T.K., Jutila, M.A., Berg, E.L. and Butcher, E.C. (1989). Neutrophil Mac-1 and MEL-14 adhesion proteins inversely regulated by chemotactic factors. *Science*, **245**:1238–1241.

Kishimoto, T.K., Jutila, M.A. and Butcher, E.C. (1990). Identification of a human peripheral lymph node homing receptor: A rapidly down-regulated adhesion molecule. *Proc. Natl. Acad. Sci. USA*, **87**:2244–2248.

Kishimoto, T.K., Kahn, J., Migaki, G., Mainolfi, E., Shirley, F., Ingraham, R. and Rothlein, R. (1995). Regulation of L-selectin expression by membrane proximal proteolysis. *Agents and Actions — Supplement*, **47**:121–134.

Laudanna, C., Constantin, G., Baron, P., Scarpini, E., Scarlato, G., Cabrini, G., Dechecchi, C., Rossi, F., Cassatella, M.A. and Berton, G. (1994). Sulfatides trigger increase of cytosolic free calcium and enhanced expression of tumor necrosis factor-alpha and interleukin-8 mRNA in human neutrophils. Evidence for a role of L-selectin as a signaling molecule. *J. Biol. Chem.*, **269**:4021–4026.

Lawrence, M.B., Smith, C.W., Eskin, S.G. and McIntire, L.V. (1990). Effect of venous shear stress on CD18-mediated neutrophil adhesion to cultured endothelium. *Blood*, **75**:227–237.

Lewinsohn, D.M., Bargatze, R.F. and Butcher, E.C. (1987). Leukocyte-endothelial cell recognition: Evidence of a common molecular mechanism shared by neutrophils, lymphocytes, and other leukocytes. *J. Immunol.*, **138**:4313–4321.

Ley, K., Gaehtgens, P., Fennie, C., Singer, M.S., Lasky, L.A. and Rosen, S.D. (1991). Lectin-like cell adhesion molecule 1 mediates leukocyte rolling in mesenteric venules *in vivo*. *Blood*, **77**:2553–2555.

McEver, R.P., Beckstead, J.H., Moore, K.L., Marshall, C.L. and Bainton, D.F. (1989). GMP-140, a platelet alpha-granule membrane protein, is also synthesized by vascular endothelial cells and is localized in Weibel-Palade bodies. *J. Clin. Invest.*, **84**:92–99.

McGeehan, G.M., Becherer, J.D., Bast Jr., R.C., Boyer, C.M., Champion, B., Connolly, K.M., Conway, *et al.* (1994). Regulation of tumor necrosis factor-alpha processing by a metalloproteinase inhibitor. *Nature*, **370**:558–561.

Migaki, G.I., Kahn, J. and Kishimoto, T.K. (1995). Mutational analysis of the membrane-proximal cleavage site of L-selectin: Relaxed sequence specifity surrounding the cleavage site. *J. Exp. Med.*, **182**:549–557.

Mohler, K.M., Sleath, P.R., Fitzner, J.N., Cerretti, D.P., Alderson, M., Kerwar, S.S., Torrance, D.S., *et al.* (1994). Protection against a lethal dose of endotoxin by an inhibitor of tumor necrosis factor processing. *Nature*, **370**:218–220.

Mullberg, J., Schooltink, H., Stoyan, T., Gunther, M., Graeve, L., Buse, G., Mackiewicz, A., Heinrich, P.C. and Rose-John, S. (1993). The human soluble interleukin-6 receptor is generated by shedding. *Eur. J. Immunol.*, **23**:473–480.

Mullberg, J., Oberthur, W., Lottspeich, F., Mehl, E., Dittrich, E., Graeve, L., Heinrich, P.C. and Rose-John, S. (1994). The soluble human IL-6 receptor: Mutational characterization of the proteolytic cleavage site. *J. Immunol.*, **152**:4958–4968.

Palecanda, A., Walcheck, B., Bishop, D.K. and Jutila, M.A. (1992). Rapid activation-independent shedding of leukocyte L-selectin induced by cross-linking of the surface antigen. *Eur. J. Immunol.*, **22**:1279–1286.

Pandiella, A., Bosenberg, M.W., Huang, E.J., Besmer, P. and Massague, J. (1992). Cleavage of membrane-anchored growth factors involves distinct protease activities regulated through common mechanisms. *J. Biol. Chem.*, **267**:24028–24033.

Pandiella, A. and Massague, J. (1991). Cleavage of the membrane precursor for transforming growth factor alpha is a regulated process. *Proc. Natl. Acad. Sci. USA*, **88**:1726–1730.

Porteu, F. and Nathan, C. (1990). Shedding of tumor necrosis factor receptors by activated human neutrophils. *J. Exp. Med.*, **172**:599–607.

Ramchandran, R., Sen, G.C., Misono, K. and Sen, I. (1994). Regulated cleavage-secretion of the membrane-bound angiotensin-converting enzyme. *J. Biol. Chem.*, **269**:2125–2130.

Reichert, R.A., Gallatin, W.M., Weissman, I.L. and Butcher, E.C. (1983). Germinal center B cells lack homing receptors necessary for normal lymphocyte recirculation. *J. Exp. Med.*, **157**:813–827.

Rieu, P., Porteu, F., Bessou, G., Lesavre, P. and Halbwachs-Mecarelli, L. (1992). Human neutrophils release their major membrane sialoprotein, leukosialin (CD43), during cell activation. *Eur. J. Immunol.*, **22**:3021–3026.

Rosen, H. and Gordon, S. (1987). Monoclonal antibody to the murine type 3 complement receptor inhibits adhesion of myelomonocytic cells *in vitro* and inflammmatory cell recruitment *in vivo*. *J. Exp. Med.*, **166**:1685–1701.

Sahasrabudhe, S.R., Spruyt, M.A., Muenkel, H.A., Blume, A.J., Vitek, M.P. and Jacobsen, J.S. (1992). Release of amino-terminal fragments from amyloid precursor protein reporter and mutated derivatives in cultured cells. *J. Biol. Chem.*, **267**:25602–25608.

Schleiffenbaum, B., Spertini, O. and Tedder, T.F. (1992). Soluble L-selectin is present in human plasma at high levels and retains functional activity. *J. Cell Biol.*, **119**:229–238.

Shipp, M.A., Stefano, G.B., Switzer, S.N., Griffin, J.D. and Reinherz, E.L. (1991). CD10 (CALLA)/ neutral endopeptidase 24.11 modulates inflammatory peptide-induced changes in neutrophil morphology, migration, and adhesion proteins and is itself regulated by neutrophil activation. *Blood*, **78**:1834–1841.

Simon, S.I., Chambers, J.D., Butcher, E. and Sklar, L.A. (1992). Neutrophil aggregation is β_2-integrin- and L-selectin-dependent in blood and isolated cells. *J. Immunol.*, **149**:2765–2771.

Simon, S.I., Burns, A.R., Taylor, A.D., Gopalan, P.K., Lynam, E.B., Sklar, L.A. and Smith C.W. (1995). L-selectin (CD62L) cross-linking signals neutrophil adhesive functions via the Mac-1 (CD11b/CD18) beta$_2$-integrin. *J. Immunol.*, **155**:1502–1514.

Sisodia, S.S. (1992). Beta-amyloid precursor protein cleavage by a membrane-bound protease. *Proc. Natl. Acad. Sci. USA*, **89**:6075–6079.

Smith, C.W., Kishimoto, T.K., Abbassi, O., Hughes, B.J., Rothlein, R., McIntire, L.V., Butcher, E.C. and Anderson, D.C. (1991). Chemotactic factors regulate lectin adhesion molecule-1 (LECAM-1)-dependent neutrophil adhesion to cytokine-stimulated endothelial cells *in vitro*. *J. Clin. Invest.*, **87**:609–618.

Spertini, O., Freedman, A.S., Belvin, M.P., Penta, A.C., Griffin, J.D. and Tedder, T.F. (1991a). Regulation of leukocyte adhesion molecule-1 (TQ1, Leu-8) expression and shedding by normal and malignant cells. *Leukemia*, **5**:300–308.

Spertini, O., Kansas, G.S., Munro, J.M., Griffin, J.D. and Tedder, T.F. (1991b). Regulation of leukocyte migration by activation of the leukocyte adhesion molecule-1 (LAM-1) selectin. *Nature*, **349**:691–694.

Subramaniam, M., Koedam, J.A. and Wagner, D.D. (1993). Divergent fates of P- and E-selectins after their expression on the plasma membrane. *Mol. Biol. Cell*, **4**:791–801.

Tedder, T.F., Steeber, D.A. and Pizcueta, P. (1995). L-selectin-deficient mice have impaired leukocyte recruitment into inflammatory sites. *J. Exp. Med.*, **181**:2259–2264.

Todd, R.F.I., Arnaout, M.A., Rosin, R.E., Crowley, C.A., Peters, W.A. and Babior, B.M. (1984). Subcellular localization of the large subunit of Mo1 (Mo1 alpha; formerly gp 110), a surface glycoprotein associated with neutrophil adhesion. *J. Clin. Invest.*, **74**:1280–1290.

Von Andrian, U.H., Chambers, J.D., McEvoy, L.M., Bargatze, R.F., Arfors, K-E. and Butcher, E.C. (1991). Two-step model of leukocyte-endothelial cell interaction in inflammation: Distinct roles for LECAM-1 and the leukocyte β_2 integrins *in vivo*. *Proc. Natl. Acad. Sci. USA*, **88**:7538–7542.

Waddell, T.K., Fialkow, L., Chan, C.K., Kishimoto, T.K. and Downey, G.P. (1994). Potentiation of the oxidative burst of human neutrophils. A signaling role for L-selectin. *J. Biol. Chem.*, **269**:18485–18491.

Waddell, T.K., Fialkow, L., Chan, C.K., Kishimoto, T.K. and Downey, G.P. (1995). Signalling functions of L-selectin — enhancement of tyrosine phosphorylation and activation of MAP kinase. *J. Biol. Chem.*, **270**:15403–15411.

Walcheck, B., Kahn, J., Fisher, J.M., Wang, B.B., Fisk, R.S., Payan, D.G., Feehan, C., Betageri, R., Darlak, K., Spatola, A.F., and Kishimoto, T.K. (1996). Neutrophil rolling altered by inhibition of L-selectin shedding *in vitro*. *Nature*, **380**:720–723.

Walcheck, B. and Jutila, M.A. (1994). Bovine gamma delta T cells express high levels of functional peripheral lymph node homing receptor (L-selectin). *International Immunol.*, **6**:81–91.

Zetterberg, E. and Richter, J. (1993). Correlation between serum level of soluble L-selectin and leukocyte count in chronic myeloid and lymphocytic leukemia and during bone marrow transplantation. *Eur. J. Haematol.*, **51**:113–119.

4 The Selectins as Rolling Receptors

Klaus Ley*

**Department of Biomedical Engineering, University of Virginia School of Medicine, Charlottesville, VA 22908*
Address for correspondence: Klaus Ley, M.D., University of Virginia School of Medicine, Department of Biomedical Engineering, Health Sciences Center Box 377, Charlottesville, VA 22908, Tel: (804) 924-1722, Dept. office phone: (804) 924-5101, Fax: (804) 982-3870

Leukocyte recruitment in inflammation is initiated by the capture of leukocytes from the blood stream followed by their rolling along the endothelium of postcapillary venules. Rolling is a slow downstream motion of leukocytes caused by a balance between shear forces exerted by the streaming blood and transient bonds between surface receptors on leukocytes and endothelial cells. Selectin-mediated leukocyte rolling is an essential component of the recruitment paradigm, because blockade of selectin function can inhibit the inflammatory process (55, 149, 162, 232). The selectins (174, 232, 248), highly conserved transmembrane molecules with an N-terminal lectin domain, are expressed on leukocytes and activated endothelial cells and play a crucial part in mediating leukocyte rolling. This chapter summarizes the current insight into selectin-mediated leukocyte rolling gained from antibody blocking studies, *in vitro* and *in vivo* reconstitution assays, gene-targeted mice, and modeling studies.

EARLY INVESTIGATIONS OF LEUKOCYTE ROLLING

The phenomenon of leukocyte rolling was described more than 150 years ago in amphibian tissues investigated by intravital microscopy. The first explicit description of this phenomenon was reported by R. Wagner in 1839 (267). The interpretation of the physiological significance of leukocyte rolling, adhesion and transmigration remained elusive well into the second half of the nineteenth century. In his famous textbook, Rudolf Virchow maintained that leukocyte emigration served a function in the nourishment of tissues (259). The link between leukocyte trafficking and inflammation was identified through the work of Julius Cohnheim and Elie Metchnikoff (61, 62, 178). After great initial strides including the development of the first effective vaccinations, progress in inflammation research slowed down in the first decade of the twentieth century, and interest in leukocyte rolling and adhesion diminished. Modern quantitative investigation of leukocyte rolling and adhesion by intravital microscopy resumed with the work of Atherton and Born in 1972 (21, 23).

RHEOLOGY OF LEUKOCYTE ROLLING

Blood Flow in Microvessels

The shape of most microvessels can be approximated as cylindrical tubes. Laminar flow of incompressible Newtonian fluids (with a uniform and constant viscosity) through cylindrical tubes is governed by Poiseuille's equation, which essentially states that the volume flow rate is proportional to the pressure difference between the beginning and end of the tube, inversely proportional to the tube length and fluid viscosity, and proportional to the fourth power of the tube radius or diameter. This equation accurately predicts that the resistance to flow through capillary vessels is very large. However, the presence of cells, mainly erythrocytes, suspended in plasma greatly complicates the flow behavior of blood in microvessels, because its viscosity is neither uniform nor constant (58). In fact, red blood cells accumulate in the center of microvessels, leaving a cell-free marginal layer. These factors lead, in essence, to a blunted velocity profile in microvessels, which has a steeper velocity gradient close to the vessel wall than would be seen for laminar flow of a Newtonian fluid. Velocity profiles measured *in vivo* have a large region close to the center of the vessel in which the velocity is almost uniform (242). Based on such measurements, the wall shear rate (i.e., the change of axial flow velocity with radial position at the vessel wall) for blood-perfused microvessels *in vivo* has been estimated to be about 2.1 times larger than that expected for a Newtonian fluid (215) at the same volume flow rate.

Wall Shear Stress

The wall shear stress τ_w is defined as the shear force acting per unit area of vessel wall and has the dimension of a pressure, e.g. dyn/cm^2, or Pascal (Pa, 1 Pa = 10 dyn/cm^2). Wall shear stress is directly proportional to wall shear rate γ_w, which is the gradient of axial flow velocity in radial direction close to the wall and has dimensions of a velocity per unit length (radius), or s^{-1}. The ratio of wall shear stress and wall shear rate defines the viscosity μ: $\mu = \tau/\gamma$. If the viscosity is known, wall shear stress can be estimated from velocity and diameter measurements. Using 0.01 dyn·s/cm^2 (1 centipoise) for plasma viscosity, such measurements yield wall shear stresses between 2 and about 30 dyn/cm^2 in venules, and up to 300 dyn/cm^2 in arterioles (215). Wall shear stress is the most important physical parameter modulating leukocyte-endothelial interactions in the presence of flow. Hence, *in vitro* models exposing the leukocytes to shear stress have been developed using various designs of flow chambers (see below).

Estimates of Mean Blood Flow Velocity

Mean blood flow velocity v_b is defined as the volume flow passing through a microvessel, divided by its cross-sectional area. In most cases, mean blood flow velocity cannot be determined directly, but must be inferred from other velocities which can be measured *in vivo*. Two parameters commonly measured in the microcirculation are *centerline velocity* v_{CL} and *individual cell velocities*. Centerline velocity is usually determined using a dual or multiple photodiode system and auto-tracking cross correlation (273). This method averages the velocities of particles projected in the centerline of the blood vessel. Depending on the width of the sensor diodes and other parameters, the ratio v_{CL}/v_b varies

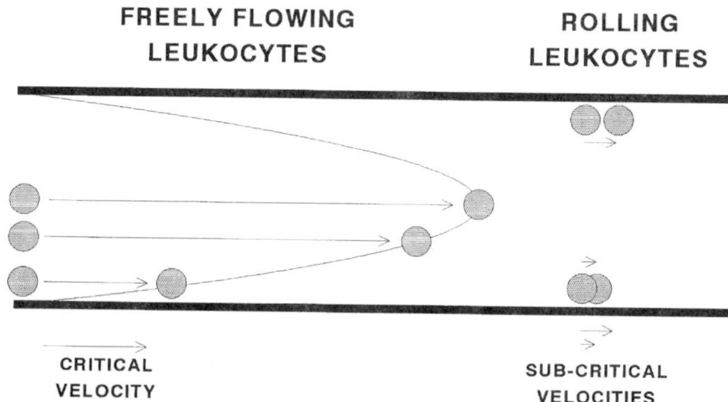

Figure 1 The concept of critical velocity to unequivocally identify rolling leukocytes. In a microvessel, leukocytes travel at various velocities (indicated by arrows) as a consequence of the existing profile of flow velocities (here shown as a parabola). Because of their finite size, the center of leukocytes cannot approach the vessel wall (indicated by solid lines) beyond a distance equal to their own radius. Therefore, at any given vessel size and blood flow velocity, a lower limit for leukocyte velocity exists (critical velocity). In this schematic, critical velocity is shown as the velocity of a fluid element traveling at a distance of one leukocyte radius from the vessel wall. Actual critical velocity is smaller than this value because of additional drag caused by the cell moving close to the wall (see Figure 2). Rolling leukocytes are characterized by sub-critical velocities which are caused by molecular interactions with the vessel wall.

(211), but under most experimental conditions the ratio is close to 1.6 (27, 161, 165). Individual cell velocities can be measured using fluorescence-labeling techniques of platelets, red or white blood cells. To obtain sufficient time resolution, stroboscopic measurements are usually required, using one or two xenon arc flash pulses per video field (60–120 flashes per second) (161, 242). Since blood cells may be traveling at different radial positions within the microvessel, their individual velocities can vary between a maximum, which is usually reached in the center of the vessel, and a minimum close to the wall (Figure 1). Mean blood flow velocity can be estimated from experimentally determined maximum cell velocity v_{max} as $v_b = v_{max}/(2-\epsilon)$, where ϵ is the ratio of cell to microvessel diameter (85, 161). However, this estimate does not take into account the experimentally determined bluntness of the velocity profile. Alternative ways to estimate v_b from v_{max} exist (97, 275), and the different models yield values which fall within 10% of each other within the range of realistic microvessel geometries.

Critical Velocity

In a very large tube, a very small particle can travel at a very low velocity if it is close to the wall. Since blood cell dimensions are of the same order of magnitude as the diameter of the microvessels through which they pass, the minimum or critical velocity v_{crit} they can assume in the absence of adhesive interactions is different from zero (85). The minimum distance between the center of mass of a blood cell and the vessel wall is dictated by the cell radius (Figure 1). In order to unequivocally identify leukocytes that interact with the vessel wall, it is helpful to estimate v_{crit}, which is significantly lower than the velocity a fluid element would have in the same radial position in absence of a blood cell because of additional drag introduced by the portion of the cell close to the wall. The

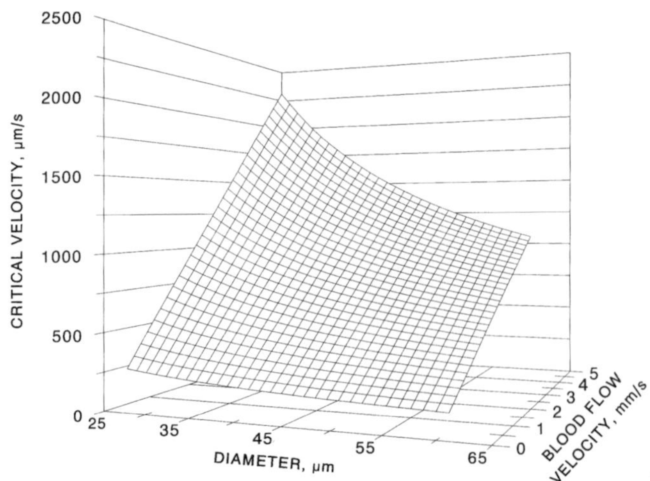

Figure 2 Calculated critical velocity as a function of microvessel diameter and blood flow velocity. Critical velocity is a linear function of blood flow velocity and increases with decreasing microvessel diameter. The front edge of the area plot depicts critical velocity for vessels between 25 and 60 μm diameter at a blood flow velocity of 0.5 mm/s, the rear edge shows the same at a blood flow velocity of 5 μm/s. This graph is a based on the equation critical velocity = (blood flow velocity) · $\varepsilon \cdot (2-\varepsilon)$, where ε is the cell-to-vessel diameter ratio. This estimate of critical velocity is very close to direct measurements of critical velocity *in vivo* (197) and *in vitro* (Michael Lawrence, personal communication).

problem of a spherical particle flowing in a large tube or flow chamber has been solved theoretically by Goldman *et al.* (87). The value of the calculated critical velocity depends on mean blood flow velocity, the cell-to-microvessel diameter ratio ε, and the clearance between the sphere and the wall. For a distance of 18 nm, which corresponds to the approximate length of typical adhesion molecules, v_{crit} can be estimated as $v_{crit} = v_b \cdot \varepsilon \cdot (2-\varepsilon)$ (161). The calculated critical velocity for a range of microvessel diameters and blood flow velocities is shown in Figure 2 for a 7 μm diameter sphere traveling at a clearance of 18 nm. It is important to note that this estimate of v_{crit} is only approximate because leukocytes are not perfect spheres, the microvessel is not a smooth cylinder, the velocity profile is not parabolic, and experimental uncertainties exist in determining diameters and velocities. Nevertheless, the critical velocity estimated as shown separates rolling, i.e. interacting, leukocytes remarkably well from freely flowing leukocytes. The concept of critical velocity is also applicable to flow chamber studies *in vitro* (137).

Rheological Margination in Microvessels

The interaction between leukocytes and endothelial cells is initiated by close mechanical contact between the two cell types, a process called *margination*. The mechanisms of margination are operative even in the absence of inflammation, and (with one exception) rheological margination is not increased by the inflammatory process. Several rheological mechanisms of margination have been demonstrated by direct *in vivo* observations and *in vitro* model experiments (48, 88, 190, 222). The principal mechanisms of margination are based on the interaction of leukocytes with erythrocytes flowing in the same microvessel. In postcapillary venules, whose internal diameter typically increases from <10 μm to

~30 μm over a short distance, deformed erythrocytes can "push" leukocytes to a marginal position by virtue of their smaller hydrodynamic cross section and thus higher flow velocity within the paraboloid velocity profile (222). In larger microvessels (diameters 30–100 μm), low rates of blood flow can promote erythrocyte aggregation which causes leukocyte margination, because the larger red cell aggregates tend to occupy the central portions of the venular cross section (88, 190). This mechanism is more pronounced at higher concentrations of acute-phase proteins including fibrinogen. Since acute phase protein production is increased through the action of interleukin-6 and other inflammatory cytokines, aggregation-dependent margination may be enhanced in inflammation. However, extreme variability of red blood cell aggregation tendency among mammalian species (e.g., very high in cat, almost absent in rabbit (200)) suggests that aggregation-dependent margination is not critical for the inflammatory process. A third mechanism of margination may be operative in venules downstream from a confluent branch point: The blood delivered by the branch with lower flow rate is distributed to a marginal region (59) which tends to enhance the probability of leukocytes to interact with the endothelium (48).

It is important to note that none of the mechanisms of rheological margination is a rate-limiting factor for the induction of leukocyte rolling or inflammation, because leukocytes exit from capillaries whose diameter is smaller than their own, ensuring that every leukocyte exiting a capillary will necessarily be in contact with the blood vessel wall. There is no known condition associated with a deficiency of inflammatory function due to insufficient rheological margination.

Leukocyte Rolling Velocity

In a systematic fluorescence microscopic study using the critical velocity to identify all rolling leukocytes, the most frequent velocity range of rolling leukocytes was found to be between 20 and 40 μm/s in different shear rate classes covering the physiological range (161). Other intravital microscopic studies in a variety of tissues have yielded essentially similar results, showing a wide range of rolling velocities between 5 and more than 100 μm/s (23, 40, 83, 109, 173, 206, 223). The fraction of rolling leukocytes, expressed as percentage of all leukocytes passing the investigated vessel, drops with increasing wall shear rate (78, 161), but some leukocytes exhibit quite robust rolling. Investigation of the shapes of rolling leukocytes *in vivo* suggests that only the most deformable cells are able to roll at higher shear rates (78). Deformation of rolling leukocytes continues to increase throughout the physiological range up to the highest wall shear stress values observed in venules (69).

Rolling velocity is remarkably invariant with increasing shear stress. Although fewer cell roll, those that do can maintain velocities below 50 μm/s at wall shear stresses up to 30 dyn/cm^2 (78, 161). The invariance of rolling velocity suggests that the rate-limiting step of rolling may be found in the detachment of the molecular bonds causing rolling. This finding can be accommodated by models assuming that bonds with a short lifetime continuously form and break between leukocytes and endothelial cells (96, 254).

Rolling leukocytes accumulate along microvessel walls, even though they are not firmly adherent, because their low velocity generates an increase in transit or residence time in the microvessel. At a given rate of leukocyte flux (cells passing per unit of time), slowly moving leukocytes are more likely to be present in any given segment of microvessel at any point in time. This is reflected by an increased microvascular concentration of leukocytes, which, in the absence of firm adhesion, can be shown to be proportional to the fraction of rolling leukocytes (77).

In vitro experiments show that leukocyte rolling velocity is dependent on the receptor site density. Also, variation of leukocyte rolling velocity on homogeneous substrates with uniform site density suggests that the number of bonds formed at any moment in time is small and that cell movement is determined by the stochastic probability distribution of the number of bonds formed (96). In addition, different adhesion molecules yield different typical rolling velocities when used in reconstituted systems at the same site density (136–138). Recently, similar results have been obtained *in vivo* using selectin-deficient mice (290, 291).

Molecular Mechanics of Leukocyte Rolling

Some considerations of the biomechanics of leukocyte rolling yield insights into the nature of the bonds that can mediate rolling interactions. The most important force for receptor-ligand interactions is the van der Waals component, which has a short range over distances of the order of 1 nm. At a typical critical velocity of 1, 000 μm/s, a leukocyte travels a distance of 1 nm within one microsecond, indicating that selectins (or other rolling receptors, see below) must have extremely high forward reaction rates (on-rates). The process of leukocyte capturing may be initiated by the formation of a single molecular bond. Once the first bond is formed, the leukocyte will pivot about the attachment point, leading to closer contact with the endothelium and the formation of more bonds. When stable rolling is established, bonds will continuously form at the leading edge and detach at the trailing edge of the rolling cell. Leukocyte deformation causes an increase of the effective contact area between the leukocyte and the endothelial cell (78), and this contact area appears to be 4 to 5 μm long measured in axial direction by intravital microscopy (69). At a typical rolling velocity of 20 μm/s, this means that any given part of the leukocyte is in continuous contact with the endothelium for 1/4 second. The individual bond lifetime can be much shorter than this, because multiple bonds are likely to be engaged at any moment in time. Based on published values for the mechanical strength of antigen-antibody bonds and calculated leukocyte adhesion forces (around 1 nN per cell (101, 207, 221)), it has been estimated that approximately 30 bonds are necessary to bear the load imposed on the leukocyte by the wall shear stress (101). It has been hypothesized that the bond lifetime may decrease with increasing stress and resulting strain (254). This predicition is supported by a recent experimental study (7) in which the lifetime of a bond between P-selectin and its cellular ligand has been measured directly in a flow chamber system. These data suggest that the rate of dissociation (k_{off}) for P-selectin bonds is about 1 s^{-1} in the unstressed state, increasing to 3.5 s^{-1} at a force of 110 pN applied on the bond.

EXPERIMENTAL APPROACHES TO INVESTIGATE LEUKOCYTE ROLLING

Intravital Microscopy

Various transparent tissues such as the rat or rabbit mesentery (288) or omentum (221) can easily be exposed and mounted for microscopic observation of the living tissue. Other preparations such as the hamster cheek pouch (73), the tenuissimus (47), cremaster

(24) or spinotrapezius (123) muscle, or the spleen (223) require some dissection. Chronically implanted chambers like the rabbit ear chamber (4) or the mouse, rat or hamster dorsal skin window (74, 193) allow observations of leukocyte rolling in the unanesthetized animal. A few preparations require neither surgery nor anesthesia, i.e. the bat wing (276), the auricle of the hairless mouse (173), or the hindlimb skin of trained rats (108). However, the optical transparency and resolution of these preparations is inferior to e.g. the mesentery or cremaster muscle, and precise measurements of microvascular diameters are difficult or impossible.

Rolling leukocyte flux (number of cells passing per minute) and velocity can be measured in transilluminated specimens, but more detailed analysis of the distribution and velocities of rolling as well as freely flowing leukocytes requires stroboscopic video microscopy after intravital labeling of leukocytes with fluorescent dyes (148, 241). Parenchymatous or opaque organs such as liver (170, 260) or lung (163) are also accessible to epifluorescence intravital microscopy. Differential staining of leukocyte species *in vivo* allows the identification of polymorphonuclear granulocytes under optimal optical conditions in very transparent tissues such as the mesentery (244).

Various parameters are used to characterize leukocyte rolling *in vivo* (Table 1). Although leukocyte rolling flux and mean leukocyte rolling velocity are the most commonly reported quantities, these are not necessarily the best representations of leukocyte rolling, because rolling is sensitive to hemodynamic variations (78, 161) and to the number of circulating leukocytes (154). These factors are accounted for in a parameter called normalized leukocyte rolling flux fraction (see Table 1).

Recently, the introduction of *ex vivo* fluorescence-labeled leukocytes and other cells into the mesenteric microcirculation has been described (156, 264). This allows to conduct reconstitution experiments *in vivo*. Using this approach, it has been shown that transfection of L-selectin into a non-adherent cell line confers the ability to roll in venules (156, 264) (see below). Other specialized uses of intravital microscopy include the use of micro-infusion to administer minute amounts of putative inhibitors of leukocyte rolling without affecting the systemic circulation of the animal (152, 155, 198).

With the recent availability of gene-targeted mice lacking selectins (12, 131, 171), mouse models of leukocyte rolling are becoming increasingly important to gain comprehensive insights into *in vivo* mechanisms of rolling and possible interactions between different sets adhesion molecules. Although the mouse mesentery can be used to study leukocyte rolling (12, 171), the mouse cremaster muscle (24) offers far superior optical resolution and a much more extensive network of microvessels (128, 159).

Flow Chambers

The highly dynamic nature of selectin-mediated adhesion results in their unique ability to mediate cell attachment in the presence of shear stress, much more so than adhesion mediated by other adhesion receptors. Therefore, assays involving exposure to shear stress increase the specificity and sensitivity for detection of selectin-mediated adhesion events.

Defined shear stress can be generated *in vitro* using flow chambers. In particular, such systems have successfully been used to reconstitute adhesion systems using purified or recombinant selectin molecules (7, 135, 137, 138) or selectin ligands (136). Molecules can either be incorporated into planar lipid bilayers or directly be coated on plastic.

Table 1 Parameters used to characterize leukocyte rolling *in vivo*

Parameter	*unit*	*definition*	*characteristics*	*Examples for use*
Rolling leukocyte flux	s^{-1}	number of rolling leukocytes passing per unit of time	varies with systemic leukocyte concentration dependent on microvessel diameter, flow velocity (shear stress), independent of rolling velocity	(21, 23, 78, 152, 244) (81, 83, 126, 151, 171) (108, 155, 157, 246, 287)
Rolling leukocyte flux fraction	%	number of rolling leukocytes as fraction of all passing leukocytes (total leukocyte flux)	independent of rolling velocity and systemic leukocyte concentration. Total leukocyte flux determined by stroboscopic fluorescence video microscopy or estimated from systemic leukocyte concentration and microvessel flow rate. Dependent on microvessel diameter and velocity (shear stress)	(12, 78, 156, 159, 160) (19, 31, 151, 154, 161)
Normalized rolling leukocyte flux fraction	%	rolling leukocyte flux fraction, empirically normalized for microvessel diameter and velocity	Independent of systemic leukocyte concentration, largely independent of microvessel diameter and blood flow velocity. Used to identify changes of adhesive properties between leukocytes and endothelium (as opposed to alterations of rolling caused by hemodynamics)	(128, 159)
Leukocyte rolling velocity distribution	μm/s	histogram of velocities of individual rolling leukocytes	represents population of rolling leukocytes. Sensitive to blood flow velocity (shear stress) and temporal variations of leukocyte rolling velocity. Labor-intensive interactive measurement.	(23, 108, 156, 161)
Mean leukocyte rolling velocity	μm/s	average velocity of all rolling leukocytes passing a point of observation	Sensitive to actual distribution of rolling velocities. Average rolling velocity of all leukocytes contained in a microvessel segment is lower (harmonic mean) than average velocity of all cells rolling past a point of observation (arithmetic mean).	(19, 20, 83, 126, 287) (81, 109, 125, 128)
Rolling leukocytes per 100 μm venule	$(100\ \mu m)^{-1}$	Number of rolling leukocytes present in a defined segment of venule	Varies with microvessel geometry (diameter), flow velocity, systemic leukocyte concentration, and leukocyte rolling velocity. "Composite" parameter reflecting "attendance" of microvessel segment by rolling leukocytes. Shares problems with rolling leukocyte flux.	(19, 20, 126, 287)
Rolling leukocyte volume fraction	%	rolling leukocytes present in microvessel segment as % of all leukocytes present	Independent of systemic leukocyte concentration, but sensitive to rolling velocity. Similar to rolling leukocytes per 100 μm venule, it reflects "attendance" of microvessel segment by rolling leukocytes	(77, 161)

In vitro flow chambers have also been used to investigate selectin-mediated leukocyte rolling and adhesion on activated, cultured endothelial cells (1, 2, 110, 167, 168, 227), immobilized platelets (56, 284), adherent leukocytes (28), and transfected cells (2, 181).

The flow chambers used in most assays are of the parallel-plate type (134) in which wall shear stress can easily be calculated from the flow rate and chamber geometry. Alternative designs include capillary tube flow chambers (28), which have the disadvantage of optical distortions. This can be overcome by using capillaries with rectangular cross sections (56, 63), which approximate parallel plate geometry. Another shear-stress generating apparatus consists of a rotating cone placed above a plate, which generates a stationary, relatively uniform shear field (150, 279). Parallel-plate chambers with specialized lateral geometry and hence variable cross-sectional area can be used to produce a range of shear stresses in a single experiment (252) and have successfully been used to study selectin-mediated cell adhesion (253, 286). A wide range of shear stresses can also be achieved using a radial flow chamber with a central inflow which produces a radial flow field (65).

PHYSIOLOGY OF LEUKOCYTE ROLLING

Induction of Rolling *in vivo*

Leukocyte rolling is not present under normal conditions in internal tissues such as the mesentery, but is rapidly induced upon exteriorization by surgical procedures (77). The procedures associated with exteriorizing internal tissues are sufficient to initiate a mild inflammatory response which appears to be largely triggered by mast cell degranulation (118, 157, 250). Experimental evidence for the time course of establishment of leukocyte rolling has been obtained by use of histological staining of whole mount preparations of the rabbit mesentery obtained immediately (<1 min) after exteriorization, which showed no elevation of intravenular leukocyte concentration above systemic concentration (77). Within 2 min, granulocytes and, to a lesser extent, mononuclear cells start to roll and hence accumulate in the venules of these whole mounts. Direct quantitation of leukocyte rolling in venules of the rat mesentery (157) and the mouse cremaster muscle (159) show a similar time course of induction with a broad peak at 20–60 min, after which time leukocyte rolling is maintained at a lower rate for several hours (21). Interestingly, P-selectin dependent granulocyte rolling *in vitro* after histamine stimulation of cultured endothelial cells follows a similar, but more rapid time course (110), suggesting that the expression pattern of P-selectin after mast cell degranulation *in vivo* dominates the kinetics of induction of leukocyte rolling (Figure 3). In dermal venules of the auricle of the hairless mouse (173) or the hindpaw of Lewis rats (108), leukocyte rolling appears to be prominent even without prior stimulation or surgical trauma. This may be a specialized feature of skin microvessels related to the barrier function of the skin as an interface constantly challenged with physical, chemical and microbial stimuli.

Granulocytes, Monocytes and Lymphocytes can Roll *in vivo*

Leukocytes rolling in mesenteric venules have been described as (mainly) granulocytes by intravital microscopy using both transillumination (4, 21, 23, 78) and fluorescence microscopy (244). However, rolling fractions of up to 80% of all leukocytes passing a

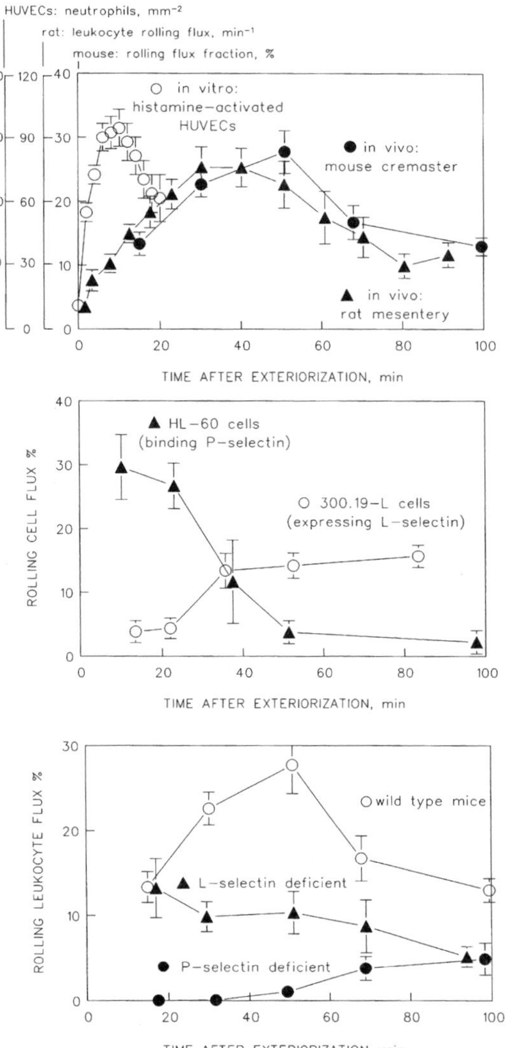

Figure 3 Cooperation between P- and L-selectin in mediating trauma-induced leukocyte rolling *in vivo*. **Top panel**: Time course of induction of leukocyte rolling (exteriorization at time 0) in rat mesentery venules (triangles) (157) and mouse cremaster muscle venules (filled circles) (159), compared with P-selectin dependent leukocyte rolling *in vitro* induced by histamine treatment of cultured endothelial cells (open circles) (110). The delayed response seen *in vivo* may be caused by diffusional barriers and the time course of mast cell degranulation, which is known to be responsible for trauma-induced leukocyte rolling (250). **Middle panel**: Rolling of HL-60 promyelocytes, which express ligands for P- and E-selectin, but no L-selectin (triangles), and L-selectin transfected 300.19 pre-B cells, which express L-selectin, but no ligands for P- or E-selectin (open circles) (160). The patterns of rolling of these cell types after exteriorization of the rat mesentery (at time 0) are roughly mirror images, indicating that P-selectin dominates early leukocyte rolling and L-selectin is important later. **Bottom panel**: Leukocyte rolling in venules of cremaster muscle of wild type (open circles), L-selectin deficient (triangles) and P-selectin deficient (filled circles) mice after exteriorization at time 0. P-selectin deficient mice show no leukocyte rolling in the first hour following exteriorization, while L-selectin deficient mice show markedly reduced rolling at later time points. Adapted from (159). These findings confirm the sequential role of P- and L-selectin in trauma-induced leukocyte rolling. In wild type mice, the number of rolling leukocytes at most time points is larger than the sum of rolling leukocytes seen in either mutant, suggesting cooperative interaction between P- and L-selectin.

given venule measured in the rat (161) or rabbit (151), species with granulocyte counts in the systemic circulation of well below 50%, suggest that other cells participate in rolling. Histological investigation of whole mounts of the rabbit mesentery or mouse cremaster muscle revealed a small but significant participation of mononuclear cells in rolling (77, 290). In both rat and mouse spleen microvessels, rolling of lymphocytes has been observed directly (223). Taken together, current evidence suggests that granulocytes constitute the predominant species of rolling leukocytes during the acute inflammatory response *in vivo*.

Preferential Leukocyte Rolling in Venules

In the majority of preparations, leukocyte rolling is confined to venules (4, 12, 21, 23, 78, 100, 151, 159, 161, 173, 221, 261), suggesting a preferential adhesiveness of venular endothelial cells for leukocytes. In addition, the fraction of rolling leukocytes is modulated by hemodynamic factors. Wall shear stress is proportional to blood flow velocity, and blood flow velocities are 3–5 times higher in arterioles than in venules of the same diameter (289). However, in the absence of exogenous stimulation, reducing flow in arterioles to venular levels does not induce leukocyte rolling, even when low flow is maintained for up to 90 min (161). Also, flow reversal through small microvessel networks leading to 'arteriolar' flow patterns in venules and 'venular' flow pattern in arterioles does not substantially alter preferential leukocyte rolling in venules (187). Taken together, these findings suggest that endothelial properties determine the extent of leukocyte rolling, while hemodynamic and rheological phenomena have modulating effects. This is supported by *in vitro* studies showing preferential granulocyte adhesion to microvascular endothelial cells thought to originate from venules (153). Under certain conditions like thermal injury (172), administration of oxidatively modified low density lipoproteins (140, 143, 145), or cigarette smoke (142), leukocytes have also been reported to roll in arterioles. We have recently observed P-selectin dependent leukocyte rolling in arterioles of TNF-α treated mouse cremaster muscle (manuscript in preparation).

Consequences and Possible Functions of Leukocyte Rolling

Leukocyte rolling is clearly essential for leukocyte recruitment from the flowing blood, because elimination of rolling (see below) also prevents leukocyte adhesion and emigration (126, 164, 243, 261, 262). Leukocyte rolling promotes firm attachment through activation-dependent adhesion receptors such as β_2 integrins, which cannot arrest leukocytes in the presence of shear stress (137).

Rolling leukocytes have vastly prolonged transit times, which can be 100-fold greater than those of red blood cells. As a consequence, non-activated leukocytes have a much better chance to sample the surface of an inflamed organ. Rolling cells can efficiently become activated by pro-inflammatory chemoattractants. Cytokines like interleukin-8 (103) and lipid mediators like platelet activating factor (166) are produced by activated endothelium, and these and other mediators may also diffuse into the vessel from extravascular tissue sources.

In the absence of activating chemoattractants, rolling leukocytes probably return to the systemic circulation in a fully functional state and are available for inflammatory responses at other sites. This situation is thought to be advantageous compared to possible release of activated leukocytes, which are known to become rigid due to actin polymerization (26, 202). Activated leukocytes have been shown to occlude capillaries by being unable to deform sufficiently to negotiate narrow passages (25, 75), which may result in ischemic tissue damage.

Leukocyte Sequestration in Lung and Liver

In the body, approximately 50% of the granulocytes circulate with transit times considerably longer than those of red cells or plasma, forming the so-called sequestered or marginated pool. This pool largely resides in the lungs (41, 163, 182) and, to a lesser extent, in the liver and spleen. It can be mobilized by administration of epinephrine or physical exercise (41, 79). Early investigators (258) assumed that leukocyte rolling may be responsible for the marginated pool of granulocytes. However, several lines of evidence indicate that the marginated pool of leukocytes is maintained through mechanisms different from those underlying leukocyte rolling. Polysaccharides (246) or antibodies (72) that block leukocyte rolling fail to mobilize the marginated pool, and P-selectin deficient mice, which have a severe deficiency of leukocyte rolling (159, 171), can normally mobilize neutrophils from the marginated pool upon epinephrine treatment (109). Furthermore, direct investigation of the subpleural microcirculation shows that delayed granulocyte transit through the lung microcirculation is mainly caused by plugging in capillaries rather than by rolling in venules (163, 280). In order to clearly distinguish the marginated pool from rolling leukocytes, the term 'sequestered pool' is advocated for physiological granulocyte retention in the lung (163).

SELECTINS AND LEUKOCYTE ROLLING

Calcium Dependence and Carbohydrate Inhibitors of Leukocyte Rolling

Leukocyte rolling *in vivo* is abolished by superfusion of tissues with calcium-chelating agents (21, 249), a finding that suggested that calcium is required for the molecular interaction underlying rolling. At the time of this discovery, however, rolling was not recognized as a receptor-mediated event and held to be due to unspecific "electrostatic attraction" between the leukocyte and the inflamed endothelium (3, 99, 256).

Before the selectins were implicated in leukocyte rolling, a number of sulfated polysaccharides were found to reduce or abolish leukocyte rolling (245). Specifically, dextran sulfate, xylan sulfate and heparin are effective blockers (151, 246), while chondroitin sulfate type C (chondroitin-6-sulfate) and heparan sulfate are partially effective, and other sulfated glycosaminoglycans (chondroitin-4-sulfates A and B, keratan sulfate) have no effect (151). Fucoidin, an algal fucose-4-sulfate polymer, inhibits leukocyte rolling (155), although another polyanionic carbohydrate, the polyphosphomannan PPME isolated from the yeast *Hansenula holstii*, does not (155). In addition, polycationic proteins (protamine) were found to reduce leukocyte rolling (246).

Taken together, these findings suggested that rolling may be caused by a lectin-like interaction, that charged carbohydrates may be involved in leukocyte binding, and that calcium is required for bond formation.

The Selectins: Discovery, Expression and Regulation

Although L-, E- and P-selectin had been studied in various assay systems for several years, this was not appreciated until molecular cloning and sequencing data revealed that these molecules form a closely related family. L-selectin, also known as CD62L, LECAM-1, leukocyte adhesion molecule-1, LAM-1, Leu-8, TQ-1 or DREG antigen in man and as $gp90^{Mel}$ or Mel-14 antigen in mouse, was already recognized as the peripheral lymphocyte homing receptor (86) and suspected to be a lectin (282). This molecule was known to mediate lymphocyte attachment to high endothelial venules of peripheral lymph nodes, a process that was inhibited by charged carbohydrates (237, 282, 283). E-selectin, or CD62E, Endothelial-Leukocyte Adhesion Molecule-1 (ELAM-1) was known to be an endothelial activation antigen (64), and antibodies to E-selectin blocked neutrophil adhesion (36). P-selectin, or CD62P, also known as Granule Membrane Protein-140 (GMP-140) or PADGEM, was known as an antigen on activated platelets (102, 175).

The close structural homology prompted the re-naming of these molecules to L-selectin (for leukocyte), E-selectin (for endothelial cell), and P-selectin (for platelet) (38). L-selectin is a 74–100 kDa molecule constitutively expressed on all granulocytes, monocytes and most circulating lymphocytes (93). E-selectin is a transcriptionally regulated endothelial cell glycoprotein expressed on cytokine-activated endothelial cells with a maximum at 4 hrs *in vitro* (36, 37), and a more sustained expression *in vivo* (119, 224). P-selectin is stored in α-granules of platelets and Weibel-Palade bodies of endothelial cells, from where it is released to the cell surface within minutes of exposure to thrombin or histamine (98). In addition, P-selectin is up-regulated in cytokine-activated endothelial cells *in vitro* (274) and *in vivo* (89, 219). Each of the three selectins contains an N-terminal lectin domain followed by a short domain with homology to epidermal growth factor (EGF) and two to nine consensus repeats with homology to complement regulatory proteins (reviewed in (174, 232, 248)). The selectins are monomeric, single membrane-spanning molecules with short cytoplasmic tails. Structure and function of the selectin molecules are described in detail elsewhere in this book.

Blocking Leukocyte Rolling *in vitro*

All three selectins have been shown to be involved in mediating leukocyte rolling and adhesion on cultured endothelial cells *in vitro*. Monoclonal antibodies recognizing E-selectin block granulocyte rolling on E-selectin transfectants in a parallel plate flow chamber system (2). In a similar system, P-selectin was shown to be responsible for neutrophil rolling on human umbilical vein endothelial cells stimulated with histamine (110). P-selectin is also responsible for rolling of neutrophils on platelets (56, 284). L-selectin has been shown to mediate adhesion of granulocytes to TNF-α activated human umbilical vein endothelial cells (229) and bovine microvascular endothelial cells cultured *in vitro* (45). In a parallel plate flow chamber assay, rolling of canine neutrophils on LPS-activated canine endothelial cells was shown to be inhibited by L-selectin antibodies (1). Monocyte adhesion to TNF-α stimulated, cultured endothelial cells under flow conditions is also blocked by L-selectin antibodies (230). This was confirmed in a defined flow field generated by a parallel plate flow chamber, showing that monocyte rolling on cultured endothelial cells activated with interleukin-4 is L-selectin dependent (167). T lymphocytes can also use selectin-dependent pathways for adhesion in the presence of flow (111). In particular,

P-selectin has been shown to support rolling of a subset of T-cells on TNF-activated human umbilical vein endothelial cells (168). L-selectin can mediate rolling of lymphocytes on purified peripheral lymph node addressin (136), a group of molecules involved in lymphocyte homing to peripheral lymph nodes. Taken together, there is solid evidence that all three selectins participate in mediating attachment of granulocytes, monocytes, and lymphocytes to inflamed endothelium under conditions of flow.

In some experimental systems, the pathways of E- and L-selectin mediated adhesion have not been clearly separable. In one study, antibodies to L-selectin and E-selectin each inhibited leukocyte attachment, yet the combination of both antibodies had no further effect beyond that seen with either antibody alone (122). This led the authors to conclude that E- and L-selectin pathways on adhesion are overlapping, and L-selectin may be a ligand binding endothelial E-selectin. This idea was further elaborated in binding studies using isolated L-selectin, which showed glycosylation-dependent binding of E-selectin to L-selectin isolated from granulocytes, but not to L-selectin from lymphocytes (210). However, other investigators found that a combination of E- and L-selectin antibodies did result in further reductions of attachment than either antibody alone (229). This and multiple other lines of evidence make it unlikely that granulocyte L-selectin is a physiologically important ligand for E-selectin (see below).

Reconstitution Experiments *in vitro*

As mentioned above, flow chamber systems are ideally suited to study leukocyte attachment in reconstituted systems. Attachment of leukocytes in the presence of shear stress has been studied on substrates containing purified selectins or selectin ligands. Completely cell-free reconstitution assays for selectin-mediated adhesion have not been described so far, although such systems exist for antibody-antigen binding (65). P-selectin incorporated into planar lipid bilayers or coated on plastic supports neutrophil rolling (137, 138). Similar data were obtained on E-selectin-containing substrates, which support granulocyte rolling at significantly lower velocities (138). A remarkable difference was observed in the shear stress resistance of rolling: the detachment rate of rolling cells was much lower on E-selectin than on P-selectin at the same site density, although the attachment rates were the same (137, 138).

Some of the ligands binding to selectins to mediate leukocyte rolling have been characterized in flow chamber systems. Neutrophil attachment to immobilized P-selectin is completely dependent on the function of a single glycoprotein, P-selectin glycoprotein ligand-1 (181). The ligand(s) for E-selectin in rolling interactions have not been fully identified (5), although glycolipids have been implicated in this interaction (6). L-selectin dependent rolling of lymphocytes on an affinity-purified ligand called peripheral lymph node addressin is significantly less efficient than P- or E-selectin-mediated rolling. This is reflected by larger rolling velocities even at high site densities (136). L-selectin dependent rolling on another ligand, mucosal addressin adhesion molecule-1 (MadCAM-1) was also found to occur at high velocities, suggesting weaker or more transient attachment (34). An interesting observation indicates that attachment of neutrophils to immobilized E-selectin requires L-selectin function, but once rolling is established, L-selectin antibodies have no effect (135). This finding may at least partially be explained by the ability of surface-bound granulocytes to support recruitment of more granulocytes via an L-selectin-dependent pathway (28).

Instead of immobilized adhesion receptors, transfected cell lines can be used to present adhesion molecules to suspended cells in flow chamber systems. A possible advantage of such systems is that the adhesion receptor may be presented in a more physiological cellular context. A disadvantage of transfected cells is that the site density cannot be controlled very well. Neutrophils have been shown to roll on P-selectin transfected CHO cells (181) and E-selectin transfected L-cells (2).

Blocking Leukocyte Rolling *in vivo*

Early microcirculatory studies demonstrated that antibodies blocking the common β chain (CD18) of the β_2 integrins inhibit the adhesion of neutrophils to microvascular endothelial cells, but not leukocyte rolling (13). This finding suggested that other adhesion molecules must be responsible for leukocyte rolling. The putative β_2 integrin-independent adhesion mechanism was prominent in the presence of shear stress and was initially defined as "dextran-sulfate-inhibitable adhesion mechanism" (150). Subsequently, antibody blocking studies have demonstrated the involvement of all three selectins in leukocyte rolling during the acute inflammatory response (see below). In addition, selectin antibodies have been effective at limiting leukocyte recruitment in a variety of inflammatory models in which the step at which leukocyte recruitment is blocked (i.e., rolling, firm adhesion or transmigration) cannot be directly investigated (51, 70, 139, 183, 184, 186). Since this chapter is focused on the selectins in their function as rolling receptors, these findings will not be discussed here in detail.

When interpreting results of antibody blocking experiments *in vivo*, it must be born in mind that antibodies can have effects apart from immunoneutralization of the targeted molecule. Some integrin- and selectin-antibodies can activate rather than block the ability of adhesion receptors to bind ligand (14, 42, 115, 120, 216, 228). Moreover, surface-bound antibodies may induce endocytosis and cause other secondary cellular effects. Cross-linking of adhesion receptors has been shown to induce activation signals in a variety of systems (10, 67, 133, 266, 269). With this caveat in mind, it is important to recognize that monoclonal antibodies are nevertheless one of the most versatile tools to investigate the role of leukocyte adhesion molecules *in vivo*.

P-selectin

Both constitutive and trauma-induced leukocyte rolling are initially dependent on the expression of P-selectin on the endothelial surface. Evidence for the importance of P-selectin for initiating leukocyte rolling is based on observations showing that P-selectin antibodies block both constitutive and trauma-induced rolling *in vivo* (72, 159, 193). Endothelial cell stimulation with secretagogues can induce an increase of leukocyte rolling *in vivo*, which is P-selectin dependent (127, 171). At reduced shear rates, a monoclonal antibody to P-selectin has been shown to increase the velocity of rolling leukocytes in cat mesenteric venules without affecting rolling leukocyte flux (40). A monoclonal P-selectin antibody inhibited most leukocyte rolling in mesenteric venules of the dog (72). Moreover, P-selectin antibody pretreatment prevents the rise of rolling leukocyte flux induced in rat mesenteric venules by hypoxanthine/xanthine oxidase, a superoxide anion radical-generating system, leaving basal leukocyte rolling unaffected (83). The insensitivity of 'basal' leukocyte rolling to P-selectin antibodies indicates that P-selectin-independent mechanisms of leukocyte rolling are operative *in vivo* at least under some experimental conditions. In fact, the P-selectin dependent component of leukocyte rolling appears to follow a clear time course in some models. Immediately following tissue trauma, rolling is largely P-selectin dependent (156, 159, 160), while L-selectin dependent mechanisms are prominent at later times (12, 156, 159, 160).

L-selectin

Micro-infusion of a soluble recombinant L-selectin-IgG fusion protein (271) or a polyclonal antiserum to L-selectin significantly reduces leukocyte rolling in venules of the rat mesentery by about 80% (152). Similarly, systemic infusion of a monoclonal antibody (both intact IgG or Fab fragments) recognizing and blocking an epitope in the lectin domain of L-selectin greatly reduces leukocyte rolling in rabbit mesenteric venules (261). L-selectin is known to be rapidly down-regulated from the cell surface upon granulocyte activation by receptor-dependent agonists or phorbol esters through proteolytic cleavage (93, 121, 147). This receptor down-regulation, or shedding, can be mimicked by mild protease treatment (113). Protease treatment of neutrophils effectively and dose-dependently reduces rolling of human granulocytes in mesenteric venules of rabbits pre-treated with intraperitoneal interleukin-1 (IL-1) to induce expression of L-selectin ligand(s) (262). Protease treatment is unspecific in that it also removes P-selectin glycoprotein ligand-1 (PSGL-1), and thus eliminates P-selectin mediated binding. In fact, isolated removal of PSGL-1 without removing L-selectin has been achieved by using O-sialoglycoprotease (OSGP) and has been shown to almost completely block P-selectin dependent rolling of neutrophils and HL-60 promyelocytes (160). In experiments conducted prior to the discovery of PSGL-1, both L- and P-selectin dependent adhesion mechanisms appear to have been eliminated through the use of unspecific proteases. The important role of L-selectin for leukocyte rolling *in vivo* has been confirmed in L-selectin deficient mice (12, 159), and it can be concluded from the available data that L-selectin is required for leukocyte rolling at inflammatory sites *in vivo*. In addition, rolling of resting lymphocytes on high endothelial venules of Peyer's patches has been shown to require L-selectin (29).

E-selectin

Leukocyte rolling via E-selectin has been difficult to demonstrate *in vivo*. Until recently, the only effect seen with E-selectin antibodies was a modest and very transient 50% reduction of leukocyte rolling in a rabbit mesenteric model following prolonged incubation with an IgM antibody (201). In a model of pulmonary inflammation (183), an E-selectin antibody limited tissue damage following IgG immune complex formation, but it is unclear whether this effect was related to inhibition of leukocyte rolling. In general, E-selectin antibodies alone appear to have limited effects on the inflammatory process *in vivo* ((44), and Barry Wolitzky, Hoffmann-La Roche, personal communication). This can probably be explained by the redundancy of function between E- and

P-selectin with respect to rolling (see below). Consistent with this notion, we have recently shown that leukocyte rolling in venules of TNF-α treated P-selectin deficient mice is completely blocked by an E-selectin antibody (128).

Reconstitution Experiments *in vivo*

In vivo reconstitution experiments using selectin-transfected cells are currently limited to one single model, the mesenteric microcirculation. This is because it is necessary to introduce a local arterial catheter to inject the transfected cells. Cell lines do not recirculate effectively through the heart and lungs, because the cells are larger than blood cells. Rolling of L-selectin transfected pre-B cells in rat mesenteric venules in the absence of intentional stimulation was seen to start at approximately 20 min after exteriorization of the rat mesentery and persisted for at least 2 hours (156, 160). Leukocyte rolling was also observed using a different L-selectin transfected cell line in the IL-1 treated rabbit mesentery (264). Both cell lines lack expression of ligands for E- or P-selectin, suggesting that L-selectin can mediate rolling *in vivo* independent of the other selectins. However, genuine L-selectin mediated leukocyte rolling is clearly less efficient than neutrophil rolling. This is reflected by the lower percentage (156, 264) and higher rolling velocity (156) of L-selectin transfectants *in vivo*. The time course of L-selectin dependent rolling after trauma shows that functional expression of L-selectin ligand(s) on venular endothelium occurs later than expression of P-selectin (see figure 3 above).

Transfected murine pre-B cells expressing a mutant of L-selectin which lacks the carboxy-terminal 11 amino acids of the 17 amino acids constituting the cytoplasmic domain of L-selectin (LΔcyto) completely lose the ability to roll on inflamed endothelium *in vivo* (116). This is paralleled by the inability of these cells to bind to high endothelial venules of peripheral lymph nodes in the presence of flow, although the lectin function is preserved in LΔcyto transfected cells as shown by their ability to bind PPME, a soluble ligand for L-selectin. The LΔcyto mutant also express the full range of extracellular L-selectin epitopes as defined by a series of monoclonal antibodies (115, 228), which suggests intact conformation of the extracellular portion of L-selectin. Interestingly, the phenotype of LΔcyto is reproduced by cytochalasin treatment of cells transfected with intact L-selectin, a procedure that disrupts actin filament formation. This indicates that cytoskeletal anchoring is important for L-selectin function (116). Recent findings indicate that L-selectin indeed interacts with cytoskeletal proteins via α-actinin (205). Cytoskeletal anchoring is also known to be important for the function of a variety of other adhesion molecules including integrins and E-selectin (60, 204, 285).

In order to directly address the contribution of the lectin domain versus other domains for L-selectin function *in vivo*, chimeric cDNAs of L- and P-selectin were constructed, expressed and characterized (115). When transfected into a pre-B cell line, P-selectin did not enable rolling in rat mesenteric venules. However, when the lectin domain of P-selectin was replaced by that of L-selectin, the transfectants rolled with a pattern and velocity indistinguishable from L-selectin transfectants. Similarly, a chimeric L-selectin in which only the EGF-domain was replaced by that of P-selectin conferred the ability to roll to transfected cells (117). These data indicate that the lectin portion of L-selectin is critical for its adhesive function underlying rolling *in vivo*, while other domains of the molecule can be replaced by the corresponding structures of P-selectin.

SELECTIN LIGANDS AND LEUKOCYTE ROLLING

Sialylated, Fucosylated and Sulfated Carbohydrates

Their N-terminal lectin domains enable the selectins to bind to specific carbohydrate structures. Thus, E-selectin has been shown to bind to sialyl-Lewis^x (sLe^x) (208, 268), sialyl-Lewis^a (sLe^a) (32), and related structures. These ligands lose binding activity when sialic acid is removed by neuraminidase treatment. P-selectin and L-selectin have also been shown to be able to bind to sLe^x (80, 213). Generally, small carbohydrate structures such as oligosaccharides bind to the selectins with low affinity (174, 217, 257), while high affinity binding requires the presence of specific glycoproteins (see below).

Sialylation

In vivo, conflicting results on leukocyte rolling have been obtained after systemic application of neuraminidase (22, 90, 91). Intravascular application of sialic acid has been shown to reduce trauma-induced leukocyte rolling *in vivo* (90). Also, systemic treatment with bacterial neuraminidases, enzymes which remove terminal sialic acid from oligosaccharides, reduces leukocyte rolling (22), while isolated perfusion with neuraminidase appears to increase leukocyte rolling upon subsequent reperfusion (91). Closer investigation of this problem necessitated treatment of isolated granulocytes, whose ability to roll in venules of IL-1 pretreated rabbit mesentery was found to be reduced by approx. 50–60% after neuraminidase treatment (264). In the untreated rat mesentery, neuraminidase treatment of exogenously administered human neutrophils almost completely blocked their rolling up to 30 minutes after exteriorization of the tissue, but had no effect at later times (160). This finding indicates that sialic acid moieties on granulocytes are indeed important for effective leukocyte rolling at least during the P-selectin dependent phase of rolling (160) and for rolling after endothelial activation with IL-1 (264). By contrast, selective treatment of the endothelial cells by micro-infusion of three different bacterial neuraminidases at activities sufficient to almost eliminate binding of the sialic acid-specific *limulus polyphemus* lectin did not reduce leukocyte rolling in venules of the untreated rat mesentery (Ley, Wintzer, Gaehtgens, unpublished results). This treatment apparently removed sialic acid residues from putative endothelial L-selectin ligand(s). Therefore, endothelial L-selectin ligands relevant for leukocyte rolling either do not require sialylation, or they may contain sialic acid in an unusual linkage that is not effectively removed by bacterial neuraminidases under the experimental conditions used.

Fucosylation

The importance of fucosylation of selectin ligands is most dramatically demonstrated by the properties of neutrophils isolated from a pair of patients with a defect in fucose metabolism (76). These granulocytes do not express sialyl Lewis^x on their surface (76) and have a vastly reduced ability to roll in venules of the exteriorized rabbit mesentery (263). These patients suffer from recurrent bacterial infections, which appear to be due to the deficiency of leukocyte adhesion (76). The critical role of fucosylation for selectin ligand function is further stressed by the requirement to co-transfect appropriate fucosyl transferases in order to render PSGL-1 functional in transfected cells (218, 234).

Sulfation

At least one of the molecules binding to L-selectin on specialized high endothelial venules in peripheral lymph nodes requires sulfation in addition to sialylation (106). Similarly, cultured human microvascular endothelial cells activated with TNF-α lose the ability to bind L-selectin transfected cells when they are incubated with chlorate, a treatment that greatly reduces sulfation, but not when they are treated with neuraminidase (286). Sulfatide, a sulfated glycolipid, has been proposed to be a ligand for P-selectin (15) and L-selectin (133), but the physiological importance of sulfatide binding remains to be established. Another class of sulfated glycolipids with an affinity for L- and P-selectin are sulfoglucuronyl glycosphingolipids (188). Fucoidin, a fucose-4-sulfate polymer, has long been known to bind to lymphocytes (46, 238) via L-selectin (282). It appears likely that these sulfated carbohydrates bind to the selectins via molecular mimicry, and they are probably not components of the physiological selectin ligands. Nevertheless, some high affinity selectin ligands appear to bear sulfate residues that are critical for function (see below).

Selectin binding and rolling of lymphocytes

Granulocytes and monocytes express sLe^x on surface glycoproteins and glycolipids, while most lymphocytes do not (82, 95, 240). However, a subset of memory T lymphocytes express sLe^x or related structures and have been shown to bind to E-selectin *in vitro* (5, 209). This ability strongly correlates with the expression of cutaneous lymphocyte antigen, CLA, a carbohydrate determinant that is fucosylated and sialylated. CLA expression predisposes lymphocytes to home to dermal sites via binding to E-selectin (33,209). An apparently distinct subset of T-cells binds to P-selectin (5, 129, 168). Approximately 15% of peripheral blood T cells can also roll on P-selectin when exposed to a defined shear field (5, 168). Certain B lymphocytes can also bind to E- and P-selectin, apparently via induction of synthesis of carbohydrate ligands (214).

Selectin binding and rolling of tumor cells

A wide variety of tumor cell lines have been shown to bind to E- and/or P-selectin (15, 16, 94, 107, 130, 169, 177, 220, 236, 253, 268). This includes breast carcinoma, squamous skin carcinoma, colon and pancreas carcinoma, neuroblastoma and small cell lung cancer. In some, but not all tumor cell lines (130) the ability to bind selectins correlates with expression of sialyl Lewis^x. In one study, tumor cell adhesion to TNF-stimulated endothelial cells was studied in a parallel plate flow chamber. In this system, E-selectin dependent rolling of mammary and colon carcinoma cells was demonstrated (253). Based on the above data, it appears possible that E- and P-selectin may play a role in carcinoma metastasis, carcinoma cell clearance by the immune system, or both.

High Affinity Glycoprotein Ligands for Selectins

Selectin ligands comprise a heterogeneous group of transmembrane glycoproteins that present oligosaccharide structures to the selectins. These include several O-glycans like P-selectin Glycoprotein Ligand-1 (PSGL-1) (181, 194, 218), Glycosylation-dependent Cell Adhesion Molecule-1 (GlyCAM-1) (132) and CD34 (30), and at least one N-glycan

(234). Beyond glycoproteins, selectins have been shown to bind to glycolipids (6, 251) and proteoglycans (195, 196). The physiological significance of glycolipids and proteoglycans for selectin-mediated cell adhesion remains to be determined.

P-selectin Glycoprotein Ligand-1

The best-characterized selectin ligand is P-selectin Glycoprotein Ligand-1 (PSGL-1), a widely expressed homodimer present on all lymphocytes, monocytes, eosinophils and neutrophils. Myeloicd-cell specific glycosylation is required for P-selectin binding by PSGL-1, and this can be mimicked by co-transfection of a fucosyl transferase in COS cells (218). PSGL-1 appears to account for all or most of the P-selectin dependent rolling of myeloid cells on P-selectin substrates *in vitro* (5, 181). N-terminal tyrosine sulfation sites are also critical for PSGL-1 function (174, 218), while N-linked carbohydrate chains are not (194). PSGL-1 is preferentially expressed on microvillous processes of neutrophils (181) and is specifically cleaved by O-sialoglycoprotease, an enzyme specific for glycoproteins with clustered O-glycans (194, 235).

E-selectin ligands

Initial reports suggested that oligosaccharides and glycolipids can effectively function as ligands for E-selectin (208, 251, 268), but more recent studies have shown that high-affinity E-selectin binding to myeloid cells is abolished by proteolytic treatment, indicating that glycoproteins serve as high-affinity E-selectin ligands. In fact, E-selectin is known to bind to PSGL-1 (17, 146, 218), although its binding affinity for PSGL-1 may be somewhat lower than that of P-selectin (218). Recently, a high affinity ligand for E-selectin has been identified in the murine system using affinity chromatography of myeloid cell extracts on an E-selectin-IgG column (146, 234). This ligand is called E-selectin ligand-1 (ESL-1) and shares a high degree of homology with a receptor for fibroblast growth factor (234). ESL-1 is expressed on a variety of cell types, but binding specificity for E-selectin is conferred by myeloid cell-specific glycosylation or, in experimental systems, by co-transfection of a fucosyl transferase. It is unclear at this time whether ESL-1 is responsible for most or all of the E-selectin dependent neutrophil binding observed in biological systems. Antibodies raised against ESL-1 significantly inhibit neutrophil adhesion to isolated E-selectin in a static *in vitro* assay (234).

An early study suggested that E-selectin may bind to L-selectin glycosylated in a neutrophil-specific manner, because E-selectin transfectants were able to bind to L-selectin purified from neutrophils but not to L-selectin purified from lymphocytes (210). However, neutrophil-derived L-selectin does not bind to E-selectin affinity columns (234), cells with little or no L-selectin can bind to E-selectin (39, 156), chemoattractant-induced shedding of L-selectin off isolated granulocytes does not impair their ability to adhere to E-selectin (43), and L-selectin cDNA co-transfected with fucosyl transferase does not support P- or E-selectin binding, while PSGL-1 cDNA does (218). These findings indicate that L-selectin is not a functionally important ligand for E-selectin.

L-selectin ligands

Two L-selectin ligands identified in high endothelial venules of peripheral lymph nodes are O-glycosylated mucins, CD34 (30) and glycosylation-dependent cell adhesion molecule-1

(GlyCAM-1), a secreted molecule (50). Neither CD34 nor GlyCAM-1 have been demonstrated to directly bind lymphocytes or neutrophils. Another L-selectin ligand has been identified in intestinal lymphatic organs. Mucosal Addressin Cell Adhesion Molecule MAdCAM-1 contains both mucin-like and immunoglobulin domains (49). As previously discussed for PSGL-1 and ESL-1, the function of MAdCAM-1 as an L-selectin ligand is glycosylation dependent, and only a subset of MAdCAM-1 molecules expressed in Peyer's patches of mice supports L-selectin dependent cell rolling and binding (29, 34). L-selectin dependent binding is related to the expression of a carbohydrate epitope recognized by mAb MECA-79 (32a). Although CD34 is expressed in microvascular endothelial cells of non-lymphatic organs, MECA-79 antigen is not expressed, and L-selectin-IgG fusion protein cannot be demonstrated to bind to inflamed post-capillary venules outside lymphatic organs. Heparan sulfate proteoglycans have been suggested as possible ligands for L-selectin expressed in activated endothelial cells (195). However, the presence of L-selectin binding glycosaminoglycans on the cell surface has not been demonstrated. Taken together, none of these candidate L-selectin ligands is directly implicated in leukocyte adhesion to extralymphatic endothelial cells at sites of inflammation. Interestingly, isolated neutrophils can roll on surface-bound neutrophils via an L-selectin dependent pathway (28), suggesting that neutrophils also express a ligand for L-selectin.

Selectin Ligands Important for Leukocyte Rolling *in vivo*

P-selectin Glycoprotein Ligand-1

PSGL-1 appears to account for all or most of the P-selectin dependent rolling of HL-60 myeloid cells and neutrophils *in vivo*. This is demonstrated by effective inhibition of rolling of HL-60 cells and neutrophils in rat mesenteric venules after pre-incubation with the mucin-specific enzyme O-sialoglycoprotease (OSGP) to remove PSGL-1 and other mucins from the cell surface (160). OSGP-treatment does not detectably diminish cell surface expression of sialyl Lewis^x (160), indicating that sialyl Lewis^x outside the context of the PSGL-1 glycoprotein is unable to mediate P-selectin dependent rolling in living microvessels. Recently, two monoclonal antibody recognizing PSGL-1 were developed (181), one of which (PL1) blocks PSGL-1 function. Treatment of human neutrophils or HL-60 promyelocytes with PL1 (as intact antibody or Fab fragments) also inhibits rolling of these cells in the rat mesentery during the P-selectin dependent, initial phase by more than 80% (197). The few cells that continue to roll do so in an intermittent fashion at high velocity. These findings clearly establish that at least 80% of the P-selectin dependent neutrophil rolling *in vivo* is mediated by its interaction with a single high affinity ligand, PSGL-1. Other high-affinity ligands for P-selectin have been hypothesized to exist that await full characterization (146).

L-selectin ligands

In order to investigate the functional expression of L-selectin ligand(s) on inflamed venules *in vivo*, the behavior of two cell lines transfected with cDNA for L-selectin was explored in two different systems *in vivo*. Two murine pre-B cell lines were unable to roll in venules of rat or rabbit mesentery, but when transfected with cDNA encoding for human L-selectin showed rolling in these models (156, 264). These findings indicate that functional L-selectin ligand(s) are expressed by inflamed endothelial cells. The time course over which L-selectin transfectants acquire the ability to roll after tissue trauma (within 30–60 min) (160) suggests that expression of L-selectin ligands is induced after tissue exteriorization. CD34 can bind to L-selectin and is expressed in microvascular endothelium (30). However, a polyclonal antiserum blocking CD34 function in other assays did not inhibit leukocyte rolling in mouse mesenteric venules (Klaus Ley and Larry Lasky, unpublished observations). At this time, the ligand responsible for L-selectin dependent binding of neutrophils to venular endothelium of non-lymphatic organs is unknown.

SELECTIN-DEFICIENT MICE GENERATED BY GENE-TARGETING

Gene-Targeted Mice

Gene-targeting in mouse embryonic stem (ES) cells has been used extensively over the past few years to generate mutations in specific genes (57). A replacement or insertion vector, designed to mutate the gene of interest, is constructed and incorporated into ES cells. These vectors also contain a selectable marker gene (usually neomycin resistance) which allows for the isolation of cells that were transfected successfully. ES cell clones which have undergone homologous recombination at the target locus are detected by Southern blotting or PCR. These cells are then injected into mouse blastocysts. These blastocysts are surgically implanted into foster mothers to produce chimeric mice, whose cells are partially derived from the targeted ES-cells and partially from the host blastocyst. Germline incorporation of the mutated ES cells can be determined by breeding the chimeras, generally by using a strain of mice containing a different coat color gene, and by testing their chimeric progeny for the mutated allele. Heterozygotes are then mated to produce a line of mice deficient for the adhesion molecule of choice (54, 158). This procedure including the breeding program typically takes at least 12 months or more. If the targeted gene is essential at some stage of embryonal development, the respective mutations will be embryonically lethal. Null mutants of all three selectins have been produced and are viable, fertile and grossly normal (12, 53, 131, 171).

P-selectin

Gene-targeted mice deficient for expression of P-selectin show a complete absence of leukocyte rolling immediately following tissue exteriorization (159, 171). Neutrophil recruitment into inflammatory lesions elicited by intraperitoneal injection of thioglycollate, a pro-inflammatory irritant, was greatly reduced at early time points (2–4 hrs) (171), but returned to or exceeded that seen in wild type mice 24–48 hrs after the inflammatory challenge (109). Elevated circulating granulocyte counts in P-selectin deficient mice suggest that P-selectin participates in sub-inflammatory neutrophil trafficking through tissues (109). The infiltration of inflammatory cells into oxazolone-induced delayed-type hypersensitivity (DTH) sites in skin was found to be reduced by about 50% in P-selectin deficient mice compared to wild type controls, but tissue edema in response to DTH was unaffected (239).

In P-selectin deficient mice, leukocyte rolling is completely absent after exteriorization of the mesentery. Even after stimulation with the calcium ionophore A23187 or with H_2O_2, both known stimulators of endothelial cell degranulation, leukocyte rolling was almost undetectable, although these treatments doubled the numbers of rolling cells per unit of time in wild type mice (171). These findings confirm that leukocyte rolling following surgical exposure of the mesentery is largely P-selectin dependent. In venules of P-selectin deficient mice, leukocyte rolling is detectable at later time points (>90 min after exteriorization) (128, 159). Interestingly, this residual rolling in P-selectin deficient mice appears to be completely L-selectin dependent (159). Local pro-inflammatory pretreatment by intrascrotal injection of TNF-α restored leukocyte rolling fluxes to control values in P-selectin deficient mice (159). In P-selectin deficient mice treated with TNF-α, an E-selectin antibody (mAb 10E9.6) known to block myeloid cell adhesion *in vitro* and neutrophil recruitment *in vivo* (44) had no effect on leukocyte rolling (159), although a different E-selectin antibody (mAb 9A9) (199) did show complete inhibition (128), showing that TNF-α induced rolling in P-selectin deficient mice is E-selectin dependent. Platelets were also observed to occasionally roll in venules of the exteriorized mouse mesentery, and this rolling was absent in P-selectin deficient mice, suggesting that platelets require endothelial P-selectin to roll *in vivo* (81).

L-selectin

In L-selectin deficient mice, short term neutrophil recruitment into thioglycollate-induced peritonitis was reduced to a similar extent (12) as reported in P-selectin deficient mice. However, the deficiency of neutrophil recruitment into inflamed peritoneum was found to be sustained, amounting to a 50%–70% reduction of neutrophil, macrophage and lymphocyte recruitment at 24 hrs. (247). Also, the edematous inflammatory response to subcutaneous injection of sheep erythrocytes into the foot pad was reduced by 75% in L-selectin deficient mice compared to wild type controls. Ear swelling resulting from oxazolone-induced DTH was reduced by 70% (247). Most remarkably, L-selectin deficient mice appear to survive endotoxin shock induced by a dose of lipopolysaccharide which killed 90% of the wild type mice within 24 hours (247). Peripheral leukocyte and neutrophil counts in these mice were found to be normal, suggesting that the absence of L-selectin does not interfere with basal neutrophil trafficking. L-selectin deficient mice have small lymph nodes, because lymphocyte homing to peripheral lymph nodes and, to a lesser extent, Peyer's patches, is impaired (12).

Leukocyte rolling in exposed mesenteric venules of L-selectin deficient mice was found to be normal initially, but within 90 min declined to less than 20% of rolling seen in wild type controls (12). This phenotype could be reproduced by treating wild type mice with intravenous L-selectin antibody. These findings confirm that leukocyte rolling at 0.5–2 hours after exteriorization of the mesentery is L-selectin dependent. Direct comparison of leukocyte rolling in cremaster venules of L- and P-selectin deficient mice revealed that P-selectin deficient mice lacked leukocyte rolling immediately (<1 hour) after surgical preparation and L-selectin deficient mice developed a marked rolling deficit over time, reaching a minimum at 90–120 min after exteriorization (159) (Figure 3).

E-selectin

E-selectin deficient mice have been found to be normal in all investigated aspects including the inflammatory response to intraperitoneal thioglycollate injection (131). However, in these mice, leukocyte emigration into the peritoneum was completely suppressed by a function-blocking monoclonal antibody to P-selectin, which had no effect in wild type mice. Similar effects of P-selectin antibodies were reported for oxazolone-induced DTH in E-selectin deficient, but not wild type mice (131). These findings suggest that, although E-selectin is not strictly required for neutrophil emigration in inflammation, the absence of E-selectin renders neutrophil recruitment P-selectin dependent. Apparently, E- and P-selectin have partially overlapping functions in inflammatory cell migration. Similar findings have been obtained by using combinations of function-blocking monoclonal antibodies to P- and E-selectin ((44), and B.A. Wolitzky, unpublished results). Conversely, recent experiments have shown that E-selectin antibodies block cytokine-induced leukocyte rolling in P-selectin deficient mice, but have little effect in wild type controls (128).

These findings show that the functions of P-selectin and E-selectin are at least partially redundant and can replace each other in the process of recruiting neutrophils to inflammatory sites *in vivo*. Leukocyte rolling in skin venules of E-selectin deficient mice both with and without TNF-α treatment is indistinguishable from rolling in wild type controls (M.A. Gimbrone, D. Milstone and R. Jain, data presented at Experimental Biology, Atlanta 1995), again suggesting that at least in the mouse system P-selectin is sufficient to mediate rolling in the absence of E-selectin. However, these experiments also showed that firm leukocyte adhesion in TNF-treated venules of E-selectin deficient mice was reduced by 50% in E-selectin mutant mice compared to wild type controls. These findings suggest that E-selectin *in vivo* may play a role beyond mediating leukocyte rolling, including the possibility that E-selectin may promote firm leukocyte attachment.

Rolling Velocities in Selectin-deficient Mice

Post-traumatic leukocyte rolling measured at 1–2 hours after surgical preparation of the cremaster muscle was found to occur at a much higher velocities in P-selectin deficient compared to control mice (290). Since trauma-induced leukocyte rolling in P-selectin deficient mice is known to be L-selectin dependent (159), these findings suggest that L-selectin may preferentially support "fast" rolling (>100 μm/s). This concept is consistent with higher velocities of L-selectin mediated neutrophil rolling observed *in vitro* (136) compared to E- and P-selectin mediated rolling (137, 138), with experimental observations of increased leukocyte rolling velocity after application of P-selectin monoclonal

Table 2 Leukocyte rolling in gene-targeted mice

targeted molecule	*trauma-induced <1 hour*	*trauma-induced 1–2 hours*	*rolling velocity*	*TNF-α induced 2–3 hours*	*rolling velocity*
none (wild type)	+++	+++	20–40 μm/s	+++	3–7 μm/s
P-selectin	0	+	>100 μm/s	+++	3–7 μm/s
L-selectin	+++	+	20–40 μm/s	++	N.D.
E-selectin	+++	+++	20–40 μm/s	+++	20 μm/s
P/E-selectin	0	0	no rolling	0	no rolling
ICAM-1	+++	+++	20–40 μm/s	+++	3–7 μm/s
ICAM-1/P-sel.	0	0	no rolling	++	3–7 μm/s

This table is based on findings reported in (52, 128, 159, 171, 290. 291) and unpublished data.

antibodies *in vivo* (20, 40), and with elevated rolling velocities of L-selectin transfected cells compared to neutrophils *in vivo* (156). Leukocyte rolling in TNF-α treated wild type or P-selectin deficient mice occurs at very slow velocities (~5 μm/s) (291), suggesting that E-selectin supports the slowest type of leukocyte rolling (Table 2). Consistent with this finding, leukocyte rolling velocities in P-selectin deficient mice at 4 hrs after intraperitoneal thioglycollate instillation have been found to be reduced to 7 μm/s, compared to 24 μm/s measured in wild type controls (109). This slow rolling may be ascribed to induction of E-selectin expression in thioglycollate-induced inflammation. The impact of the different selectins on leukocyte rolling velocities *in vivo* has not been fully defined and awaits further investigation.

Combined Deficiency of Selectins and Other Adhesion Molecules

Mice deficient for both P-selectin and Intercellular Adhesion Molecule-1 (ICAM-1) show a complete absence of neutrophil recruitment into a peritoneal inflammatory lesion at 4 hours (53). These mice have no detectable leukocyte rolling in cremaster venules throughout the entire 2 hour observation period (128). After TNF-α-treatment, rolling is induced in P-selectin/ICAM-1 double deficient mice, but significantly less than in mice deficient for ICAM-1 only. Interestingly, leukocyte rolling in ICAM-1 deficient mice is not different from wild type controls (128), whereas the absence of both ICAM-1 and P-selectin produces a phenotype that is clearly different from that seen in mice lacking P-selectin alone (159). These findings suggest an unexpected synergism between P-selectin and ICAM-1, the physiological mechanism of which is not understood at this time.

The most striking deficiency of leukocyte rolling and recruitment is seen in double mutant mice deficient for both E-selectin and P-selectin (52). These mice show no trauma-induced leukocyte rolling, and even after pre-treatment with TNF-α for 2 hours, no rolling or firm adhesion is observed. These mice have very high circulating neutrophil counts (up to 60, 000 μl^{-1}) and suffer from spontaneous infections of the oral cavity (52). In these lesions, neutrophil infiltrates are seen, suggesting that some neutrophil recruitment is still possible, albeit retarded, in the absence of both P- and E-selectin.

OTHER RECEPT7ORS MEDIATING LEUKOCYTE ROLLING

Recently, the dogma that only selectins can mediate leukocyte rolling has been challenged by findings that certain integrins also can support rolling of lymphocytes in flow chamber systems. Lymphocytic cell lines have been shown to be able to use $\alpha_4\beta_1$ integrin to roll on its endothelial ligand vascular cell adhesion molecule-1 (VCAM-1) (8, 111) and $\alpha_4\beta_7$ integrin to roll on MAdCAM-1 (35). A monoclonal antibody blocking α_4 integrin function can partially inhibit eosinophil rolling *in vivo* (233).

In one report, a monoclonal antibody to InterCellular Adhesion Molecule-1 (ICAM-1) has been shown to partially inhibit leukocyte rolling in rat liver microvessels (260). However, ICAM-1 alone cannot support leukocyte rolling as shown in reconstitution experiments *in vitro* (137). Consistent with this notion, rolling in venules of the cremaster muscle is normal in ICAM-1 deficient mice (128), even though these mice have marked deficiencies in various inflammatory functions (226, 281). At low flow velocities, β_2

integrin-dependent leukocyte rolling has been observed *in vivo* (84). Two of the β_2 integrins, Mac-1 and LFA-1, are physiological counter-receptors for ICAM-1 (232). Taken together, these findings suggest that ICAM-1 and β_2 integrins may serve auxiliary functions in leukocyte rolling. The conditions under which integrin-mediated leukocyte rolling is physiologically important and promotes leukocyte recruitment in the presence of shear stress remain to be determined.

MOLECULAR PROPERTIES OF ROLLING RECEPTORS

Affinity and Kinetics of Binding of Soluble Selectins to Cells

Binding of soluble selectin constructs to cells bearing appropriate ligands has been studied in various systems (176, 180, 255, 271). However, equilibrium binding affinity has only been reported for the interaction of P-selectin with myeloid cells (180, 255). In an initial study, binding of detergent-solubilized P-selectin purified from platelets to human neutrophils was reported to occur at an apparent affinity of about 1.5 nM. This result was later revised, because detergent-solubilized P-selectin tends to form oligomers which bind with higher apparent affinity. Truncated P-selectin (lacking the transmembrane and cytoplasmic domains) as well as alternatively spliced P-selectin (lacking the transmembrane domain only) occur as monomers in solution and bind to about 25, 000 sites per neutrophil with an apparent affinity of 70 nM (255). This binding affinity is of the same order of magnitude as values reported for other adhesion molecules including integrins (9, 105, 270).

In contrast to integrin adhesion receptors, however, selectins are supposed to bind to their ligands much more rapidly than integrins. Obtaining estimates of the forward rate of reaction of selectins with their ligands has proved to be extremely difficult. In one study, half-maximal binding of P-selectin to neutrophils was shown to be reached within less than 2 seconds, and reversal of binding was even faster after addition of a 100-fold excess of unlabeled P-selectin (180). From these data, the rate of bond formation has been estimated to range between 0.06 and 0.6 $\mu m^2/s$ on the cell surface, and the rate of detachment of unstressed bonds has been estimated to be about 0.05 to $0.5 \cdot 10^{-3}\ s^{-1}$ (254). More recent estimates based on the behavior of cells on a substrate containing P-selectin at very low site density yielded values for the detachment of unstressed P-selectin bonds of around 1 s^{-1} (7). Even though these estimates differ by three orders of magnitude, it is generally accepted that high on-rates (bond formation) and off-rates (detachment) are unique properties which enable the selectin molecules to support leukocyte rolling in the presence of flow.

Kinetics and Tensile Strength of Selectin Bonds Measured in Flow Chambers

When selectins are coated to glass coverslips or incorporated into planar lipid bilayers at high site densities (50–400 μm^{-2}), which are probably comparable to those found on endothelial cells *in vivo*, cells bearing the appropriate selectin ligands show continuous rolling behavior in flow chamber assays (137, 138). At much lower site densities (1–15 μm^{-2}), the cells stop temporarily, release and stop again, causing a discontinuous movement. Analysis of the duration of these periods of attachment is compatible with a unimolecular interaction between cells and substrate. The estimated off-rate for P-selectin mediated binding was of the order of 1 s^{-1} under resting conditions, and 3.5 s^{-1} at a force of about 100 pN acting on the tethered cell (7). This moderate increase of the off-rate with bond stress suggests a high tensile strength of the selectin-mediated bond. Indeed, at higher selectin densities in a similar flow chamber assay, E-selectin mediated binding has been reported to resist shear stress values of up to 35 dyn/cm^2 (135), which corresponds to the highest values observed in venules *in vivo*. However, integrin-mediated binding of leukocytes to endothelial cells shows similar shear resistance, suggesting that high tensile strength is not unique to the selectins. Further experimentation is required to fully understand why selectins are uniquely suited to mediate leukocyte rolling.

Biomechanical Models of Leukocyte Rolling

Modeling of leukocyte rolling with available estimates of rate constants and binding kinetics is consistent with the rapid association necessary for attachment, but the off-rates determined for binding of soluble selectins to cells are too low to account for effective bond breakage at the trailing end of a rolling cell necessary to accommodate the rolling velocities observed both *in vitro* and *in vivo* (96, 254). This has led to the suggestion that the rate of selectin detachment may be increased by mechanical stress (254), and very recent estimates of the dissociation rate constant of P-selectin bonds in the presence of external fluid forces appear to qualitatively support this concept (7), although quantitative differences remain.

Existing biomechanical models of leukocyte rolling treat the cells as rigid, non-deformable bodies, which is clearly an inaccuracy (see above) that is likely to influence the conclusions drawn from modeling studies. As better kinetic and affinity data on selectin-mediated binding become available, development of more accurate biomechanical models of leukocyte rolling becomes increasingly important. To be meaningful, future modeling attempts will have to account for deformation, cytoskeletal structure and anchorage of adhesion molecules.

ROLLING AND THE INFLAMMATORY PROCESS

Synergistic Interactions Between Different Selectins

Functional experiments suggest that the selectins act synergistically on leukocyte binding. The sum of leukocyte rolling flux fractions in P-selectin and L-selectin deficient mice is smaller at almost any time point than the leukocyte rolling flux fraction observed in wild type controls, which is consistent with a synergistic interaction of L-selectin and P-selectin in mediating leukocyte rolling (159). There is ample evidence that L-selectin is required for inflammatory cell recruitment (152, 179, 186, 261, 272), but it is unclear whether L-selectin is sufficient to ensure effective recruitment in the absence of E- and P-selectin. Although L-selectin can mediate rolling of transfected cell lines which cannot bind to E-selectin, rolling of these transfectants is not as effective as neutrophil rolling in that these cells are retarded by adhesive forces to a lesser extent and consequently exhibit higher rolling velocities (156). While other factors such as cell size, expression levels and distribution of L-selectin on the cell surface may contribute to increased rolling velocities of these cells, there is evidence that P-selectin supports leukocyte rolling at slower velocities than L-selectin, and E-selectin mediated interaction is characterized by still lower rolling velocities (128, 135–137, 156, 290, 291). The functional roles of the selectins, while clearly distinguishable in some assays, are likely to overlap *in vivo*. Although individual selectins can be shown to be sufficient to mediate leukocyte rolling, their temporal pattern of expression *in vivo* suggests that the selectins often cooperate to produce effective leukocyte rolling and adhesion.

A Special Role For L-selectin in Initiating Leukocyte Rolling?

Current data suggest that the function of L-selectin in neutrophil adhesion may be partially different from that of P- and E-selectin. L-selectin appears to promote the attachment of leukocytes from the streaming blood (capture) without being able to sustain an effective rolling interaction. This is also supported by recent findings that L-selectin is needed to initiate interaction of naive lymphocytes with high endothelial venules of Peyer's patches *in vivo* (29). The concept of L-selectin dependent neutrophil capturing preceding E- and P-selectin mediated rolling can explain the more dramatic defect of inflammatory cell recruitment in L-selectin (12, 159, 247) than in P-selectin deficient mice (159, 171, 239). At high levels of P-selectin expression such as immediately following tissue trauma or histamine treatment, P-selectin appears to be able to function both as a capturing and a rolling receptor. This is suggested by the absence of any defect of leukocyte rolling in L-selectin deficient mice during this initial phase (159). By contrast, E-selectin does not appear to be sufficient to capture leukocytes in the presence of shear stress: L-selectin is required for neutrophils to attach to purified E-selectin in a flow chamber assay, but L-selectin function is no longer required once E-selectin mediated rolling has been initiated (135). The concept of a special role of L-selectin in capturing is reinforced by the recent finding that localization of L-selectin to the tips of micro-folds on the neutrophil surface is required for effective attachment in a flow chamber assay *in vitro* (265), but the topographic position of L-selectin does not influence rolling velocity or detachment rate once rolling is established.

Leukocyte Rolling is a Rate-limiting Step in Inflammation

The ability of circulating leukocytes to recognize and adhere to the endothelium of inflamed venules in the presence of shear force is an essential feature of leukocyte function. It has now been widely accepted that recruitment of leukocytes involves several sequential steps which are characterized by distinct molecular mechanisms (55, 149, 162, 231). The first observable phenomenon in this cascade of events is selectin-mediated rolling which allows leukocytes to slow down in the venular blood stream. Rolling neutrophils can become activated by soluble or surface-bound chemoattractants and respond by engaging leukocyte or β_2 integrins.

Interference with selectin mediated adhesion events substantially reduces neutrophil accumulation at inflammatory sites in various animal models (12, 44, 112, 113, 147, 185, 186, 189, 212, 247, 272). Many studies show that L-selectin function is required for neutrophil adhesion *in vivo*. Removal of L-selectin and PSGL-1 by chymotrypsin treatment rendered human neutrophils unable to roll (262) and to respond to a chemotactic stimulus despite the expression of fully functional β_2 integrins (262). When intravascular shear force was removed by vessel ligation, chymotrypsin-treated neutrophils were again able to adhere and emigrate *in vivo*. Therefore, rolling is a prerequisite for neutrophil adhesion only under conditions of physiologic flow.

Numerous studies have shown that blockade of P-selectin function (44, 53, 109, 126, 171, 185, 189, 239) or E-selectin function (131, 183, 185) attenuate the inflammatory response *in vivo*. This is further emphasized by the finding that patients with leukocyte adhesion deficiency (LAD) syndrome type 2 suffer from recurrent bacterial infections and are unable to recruit sufficient numbers of neutrophils to sites of inflammation (76). Neutrophils from these patients do not express sLe^x and related carbohydrates, do not interact with E-selectin *in vitro* (76) and roll poorly in inflamed venules of the rabbit mesentery *in vivo* (263). The clinical syndrome presented by these patients impressively emphasizes the physiological importance of selectin ligands for the inflammatory response.

PHARMACOLOGICAL MODULATION OF LEUKOCYTE ROLLING

Pre-treatment of the rabbit mesentery with the pro-inflammatory cytokine IL-1 has been reported to increase the flux of rolling leukocytes (262). Leukocyte rolling in mouse cremaster venules is very similar after TNF-α treatment compared with rolling seen after surgical tissue exteriorization (128, 159). However, TNF-α treatment greatly increases the number of firmly adherent leukocytes, presumably as a consequence of endothelial activation leading to production of chemoattractant cytokines. Intraperitoneal histamine pretreatment of rats induces leukocyte rolling in mesenteric venules immediately after exteriorization (157) and increases rolling in rats in which the surgery-induced rolling had been suppressed (19, 127). Similar effects were seen with thrombin receptor peptide, an activator of thrombin receptors without pro-coagulant activity (287). Calcium ionophore A23187 can roughly double rolling leukocyte flux in mouse mesenteric venules (171). Histamine, thrombin receptor peptide and ionophore all promote degranulation of endothelial cells and thereby increase expression of P-selectin on the endothelial surface.

Direct stimulation of granulocytes by superfusion of chemoattractants like leukotriene B_4 (68) or formyl peptide fMLP (101, 164) *reduce* rolling leukocyte flux. This phenomenon has been attributed to shedding of L-selectin from the granulocyte surface seen upon chemotactic stimulation (93, 121). In addition, a shift from the pool of rolling leukocytes to firmly adherent leukocytes may add to the apparent reduction of leukocyte rolling.

Nonsteroidal anti-inflammatory drugs have divergent effects on leukocyte rolling. Sulfinpyrazone (277) and high doses of ibuprofen (225) have been reported to reduce leukocyte rolling *in vivo*. This effect may be related to shedding of L-selectin from the neutrophil surface, which has been reported to occur after incubation of neutrophils with indomethacin, diclofenac, ketoprofen and aspirin *in vitro* and upon treatment of volunteers with indomethacin *in vivo* (71). On the other hand, indomethacin increases the prevalence of rolling leukocytes in rat mesenteric venules (18, 225), presumably by shunting arachidonic acid from the prostaglandin to the leukotriene pathway. Corticosteroids can reduce leukocyte rolling (104), possibly by interfering with expression of E-selectin (66).

Compounds which inhibit or limit granulocyte activation, such as prostacyclin PGI_2 (18), adenosine (191), or buflomedil (192), as well as activators of adenylyl cyclase such as β adrenergic agonists, or activators of guanylyl cyclase such as nitric oxide (124) can reduce leukocyte adhesion in animal models. These compounds do not appear to directly modulate leukocyte rolling, but may more generally modulate neutrophil activation and hence activation-dependent firm adhesion. Leukotriene synthesis inhibitors and receptor antagonists as well as platelet activating factor (PAF) receptor antagonists have clearly been shown to attenuate leukocyte accumulation after various experimental challenges *in vivo* (18, 83, 140, 141, 143). Also, vitamin C has been shown to reduce cigarette smoke-induced leukocyte rolling in arterioles and venules (144). It is not known whether the effects of these interventions are related to altered selectin expression and/or function.

Carbohydrate components as competitive inhibitors of selectin function are being explored for their therapeutic potential (203, 278). In one study, moderate inhibitory effects of sialyl Lewis^x, an E-selectin binding tetrasaccharide, on thrombin receptor peptide-induced rolling have been reported (287). Sulfated polysaccharides like dextran sulfate and fucoidin block leukocyte rolling *in vivo* (126, 151, 155, 164, 246) much more effectively than sialyl Lewis^x and can attenuate inflammatory responses to various challenges (11, 92). A variety of other inhibitors including antibodies, peptides and small-molecule analogues of selectin-binding carbohydrates are currently explored in a variety of models for their therapeutic potential in inflammatory diseases.

CONCLUSION

Leukocyte rolling is an important physiological phenomenon in surveillance and inflammation. Most of the leukocyte rolling observed in most models is selectin-dependent, although some integrins can mediate attachment in the presence of shear stress. L-selectin appears to be specialized for the initial capturing of leukocytes from the streaming blood, while E-selectin and P-selectin mediate stable leukocyte rolling. The availability of gene-targeted mice deficient for expression of adhesion molecules has greatly enhanced the value of *in vivo* models for research into inflammatory adhesion molecules. The impact of the absence of one or two selectins on chronic models of inflammation and disease, long-term lymphocyte homing, and secondary physiological processes like wound healing demonstrates the key importance of the selectins in the inflammatory process. Detailed knowledge about the importance of individual selectins and their ligands for leukocyte rolling, adhesion and accumulation form the basis for the development of rational pharmacologic interventions ultimately aimed at protecting patients from the consequences of pathologic inflammation.

Acknowledgements

The author's original work reported in this chapter was supported by grants from the National Institutes of Health (HL 54136), Deutsche Forschungsgemeinschaft (Le 573/1, Le 573/2–1, Le 573/2, Le 573/3, and SFB 366), and Mizutani Foundation.

References

1. Abbassi, O., Lane, C.L., Krater, S., Kishimoto, T.K., Anderson, D.C., McIntire, L.V. and Smith, C.W. (1991). Canine neutrophil margination mediated by Lectin Adhesion Molecule-1 *in vitro*. *J. Immunol.*, **147**:2107–2115.

2. Abbassi, O., Kishimoto, T.K., McIntire, L.V., Anderson, D.C. and Smith, C.W. (1993). E-selectin supports neutrophil rolling *in vitro* under conditions of flow. *J. Clin. Invest.*, **92**:2719–2730.
3. Abrahamson, H.A. (1927). The mechanism of the inflammatory process. I. The electrophoresis of the blood cells of the horse and its relation to leukocyte emigration. *J. Exp. Med.*, **46**:987–1002.
4. Allison, F., Smith, M.R. and Wood, W.B. (1955). Studies on the pathogenesis of acute inflammation. I. The inflammatory reaction to thermal injury as observed in the rabbit ear chamber. *J. Exp. Med.*, **102**:655–668.
5. Alon, R., Rossiter, H., Wang, X., Springer, T.A. and Kupper, T.S. (1994). Distinct cell surface ligands mediate T lymphocyte attachment and rolling on P- and E-selectin under physiological flow. *J. Cell Biol.*, **127**:1485–1495.
6. Alon, R., Feizi, T., Yuen, C-T., Fuhlbrigge, R.C. and Springer, T.A. (1995). Glycolipid ligands for selectins support leukocyte tethering and rolling under physiologic flow conditions. *J. Immunol.*, **154**:5356–5366.
7. Alon, R., Hammer, D.A. and Springer, T.A. (1995). Lifetime of the P-selectin-carbohydrate bond and its response to tensile force in hydrodynamic flow. *Nature*, **374**:539–542.
8. Alon, R., Kassner, P.D., Carr, M.W., Finger, E.B., Hemler, M.E. and Springer, T.A. (1995). The integrin VLA-4 supports tethering and rolling in flow on VCAM-1. *J. Cell Biol.*, **128**:1243–1254.
9. Altieri, D.C. (1991). Occupancy of CD11b/CD18 (Mac-1) divalent ion binding site(s) induces leukocyte adhesion. *J. Immunol.*, **147**:1891–1898.
10. Altieri, D.C., Stamnes, S.J. and Gahmberg, C.G. (1992). Regulated Ca^{2+} signalling through leukocyte CD11b/CD18 integrin. *Biochem. J.*, **288**:465–473.
11. Angstwurm, K., Weber, J.R., Segert, A., Bürger, W., Weih, M., Freyer, D., Einhäupl, K.M. and Dirnagl, U. (1995). Fucoidin, a polysaccharide inhibiting leukocyte rolling, attenuates inflammatory responses in experimental pneumococcal meningitis in rats. *Neurosci. Lett.*, **191**:1–4.
12. Arbones, M.L., Ord, D.C., Ley, K., Ratech, H., Maynard-Curry, C., Otten, G., Capon, D.J. and Tedder, T.F. (1994). Lymphocyte homing and leukocyte rolling and migration are impaired in L-selectin (CD62L) deficient mice. *Immunity*, **1**:247–260.
13. Arfors, K-E., Lundberg, C., Lindbom, L., Lundberg, K., Beatty, P.G. and Harlan, J.M. (1987). A monoclonal antibody to the membrane glycoprotein complex CD18 inhibits polymorphonuclear leukocyte accumulation and plasma leakage *in vivo*. *Blood*, **69**:338–340.
14. Arroyo, A.G., Sanchez-Mateos, P., Campanero, M.R., Martin-Padura, I., Dejana, E. and Sanchez-Madrid, F. (1992). Regulation of the VLA integrin ligand interactions through the β_1 subunit. *J. Cell Biol*, **117**:659–670.
15. Aruffo, A., Kolanus, W., Walz, G., Fredman, P. and Seed, B. (1991). CD62/P-selectin recognition of myeloid and tumor cell sulfatides. *Cell*, **67**:35–44.
16. Aruffo, A., Dietsch, M.T., Wan, H., Hellström, K.E. and Hellström, I. (1992). Granule Membrane Protein-140 (GMP140) binds to carcinomas and carcinoma-derived cell lines. *Proc. Natl. Acad. Sci. USA*, **89**:2292–2296.
17. Asa, D., Raycroft, L., Ma, L., Aeed, P.A., Kaytes, P.S., Elhammer, Å.P. and Geng, J-G. (1995). The P-selectin glycoprotein ligand functions as a common human leukocyte ligand for P- and E-selectins. *J. Biol. Chem.*, **270**:11662–11670.
18. Asako, H., Kubes, P., Wallace, J., Gaginella, T., Wolf, R.E. and Granger, D.N. (1992). Indomethacin-induced leukocyte adhesion in mesenteric venules: role of lipoxygenase products. *Am. J. Physiol.*, **262**:G903–G908.
19. Asako, H., Kurose, I., Wolf, R., DeFrees, S., Zheng, Z.L., Phillips, M.L., Paulson, J.C. and Granger, D.N. (1994). Role of H_1-receptors and P-selectin in histamine-induced leukocyte rolling and adhesion in postcapillary venules. *J. Clin. Invest.*, **93**:1508–1515.
20. Asako, H., Kurose, I., Wolf, R.E. and Granger, D.N. (1994). Mechanisms of lactoferrin-induced leukocyte-endothelial cell adhesion in postcapillary venules. *Microcirculation*, **1**:27–34.
21. Atherton, A. and Born, G.V.R. (1972). Quantitative investigations of the adhesiveness of circulating polymorphonuclear leukocytes to blood vessels. *J. Physiol. (London)*, **222**:447–474.
22. Atherton, A. and Born, G.V.R. (1973). Effects of neuraminidase and N-acetyl neuraminic acid on the adhesion of circulating granulocytes and platelets in venules. *J. Physiol. (London)*, **234**:66P–67P.
23. Atherton, A. and Born, G.V.R. (1973). Relationship between the velocity of rolling granulocytes and that of the blood flow in venules. *J. Physiol. (London)*, **233**:157–165.
24. Baez, S. (1973). An open cremaster muscle preparation for the study of blood vessels by *in vivo* microscopy. *Microvasc. Res.*, **5**:384–394.
25. Bagge, U., Amundson, B. and Lauritzen, C. (1980). White blood cell deformability and plugging of skeletal muscle capillaries in hemorrhagic shock. *Acta Physiol. Scand.*, **180**:159–163.
26. Baggiolini, M. and Kernen, P. (1992). Neutrophil activation: control of shape change, exocytosis, and respiratory burst. *News Physiol. Sci.*, **7**:215–219.

27. Baker, M. and Wayland, H. (1974). On-line volume flow rate and velocity profile measurement for blood in microvessels. *Microvasc. Res.*, **7**:131–143.
28. Bargatze, R.F., Kurk, S., Butcher, E.C. and Jutila, M.A. (1994). Neutrophils roll on adherent neutrophils bound to cytokine-induced endothelial cells via L-selectin on the rolling cells. *J. Exp. Med.*, **180**: 1785–1792.
29. Bargatze, R.F., Jutila, M.A. and Butcher, E.C. (1995). Distinct roles of L-selectin and integrins $\alpha_4\beta_7$ and LFA-1 in lymphocyte homing to Peyer's patch-HEV in situ: the multistep model confirmed and refined. *Immunity*, **3**:99–108.
30. Baumhueter, S., Singer, M.S., Henzel, W., Hemmerich, S., Renz, M., Rosen, S.D. and Lasky, L.A. (1993). Binding of L-selectin to the vascular sialomucin CD34. *Science*, **262**:436–438.
31. Becker, M., Menger, M.D. and Lehr, H-A. (1994). Heparin-released superoxide dismutase inhibits postischemic leukocyte adhesion to venular endothelium. *Am. J. Physiol.*, **267**:H925–H930.
32. Berg, E.L., Robinson, M.K., Mansson, O., Butcher, E.C. and Magnani, J.L. (1991). A carbohydrate domain common to both sialyl Le^a and sialyl Le^x is recognized by the endothelial cell leukocyte adhesion molecule ELAM-1. *J. Biol. Chem.*, **266**:14869–14872.
32a. Berg, E.L., Robinson, M.K., Warnock, R.A. and Butcher, E.C. (1991). The human peripheral lymph node vascular address is a ligand for LECAM-1, the peripheral lymph node homing receptor. *J. Cell Biol.*, **114**:343-349.
33. Berg, E.L., Yoshino, T., Rott, L.S., Robinson, M.K., Warnock, R.A., Kishimoto, T.K., Picker, L.J. and Butcher, E.C. (1991). The Cutaneous Lymphocyte Antigen is a skin lymphocyte homing receptor for the vascular lectin Endothelial Cell-Leukocyte Adhesion Molecule-1. *J. Exp. Med.*, **174**:1461–1466.
34. Berg, E.L., McEvoy, L.M., Berlin, C., Bargatze, R.F. and Butcher, E.C. (1993). L-selectin-mediated lymphocyte rolling on MAdCAM-1. *Nature*, **366**:695–698.
35. Berlin, C., Bargatze, R.F., Campbell, J.J., von Andrian, U.H., Szabo, M.C., Hasslen, S.R., Nelson, R.D., Berg, E.L., Erlandsen, S.L. and Butcher, E.C. (1995). α_4 integrins mediate lymphocyte attachment and rolling under physiologic flow. *Cell*, **80**:413–422.
36. Bevilacqua, M.P., Pober, J.S., Mendrick, D.L., Cotran, R.S. and Gimbrone, M.A.J. (1987). Identification of an inducible endothelial-leukocyte adhesion molecule. *Proc. Natl. Acad. Sci. USA*, **84**:9238–9242.
37. Bevilacqua, M.P., Stengelin, S., Gimbrone, M.A.J. and Seed, B. (1989). Endothelial leukocyte adhesion molecule-1: An inducible receptor for neutrophils related to complement regulatory proteins and lectins. *Science*, **243**:1160–1165.
38. Bevilacqua, M.P., Butcher, E.C., Furie, B., Furie, B.C., Gallatin, W.M., Gimbrone, M.A.J., Harlan, J.M., Kishimoto, T.K., Lasky, L.A., McEver, R.P., Paulson, J.C., Rosen, S.D., Seed, B., Siegelman, M.H., Springer, T.A., Stoolman, L.M., Tedder, T.F., Varki, A., Wagner, D.D., Weissman, I.L. and Zimmerman, G.A. (1991). Selectins — a family of adhesion receptors. *Cell*, **67**:233.
39. Bevilacqua, M.P. and Nelson, R.M. (1993). Selectins. *J. Clin. Invest.*, **91**:379–387.
40. Bienvenu, K. and Granger, D.N. (1993). Molecular determinants of shear rate-dependent leukocyte adhesion in postcapillary venules. *Am. J. Physiol.*, **264**:H1504–H1508.
41. Bierman, H.R., Kelly, K.H., Cordes, F.L., Byron, R.L., Polhemus, J.A. and Rappoport, S. (1952). The release of leukocytes and platelets from the pulmonary circulation by epinephrine. *Blood*, **7**:683–692.
42. Binnerts, M.E., Van Kooyk, Y., Simmons, D.L. and Figdor, C.G. (1994). Distinct binding of T lymphocytes to ICAM-1, -2 or -3 upon activation of LFA-1. *Eur. J. Immunol.*, **24**:2155–2160.
43. Bochner, B.S., Sterbinsky, S.A., Bickel, C.A., Werfel, S., Wein, M. and Newman, W. (1994). Differences between human eosinophils and neutrophils in the function and expression of sialic acid-containing counterligands for E-selectin. *J. Immunol.*, **152**:774–782.
44. Bosse, R. and Vestweber, D. (1994). Only simultaneous blocking of the L- and P-selectin completely inhibits neutrophil migration into mouse peritoneum. *Eur. J. Immunol.*, **24**:3019–3024.
45. Brady, H.R., Spertini, O., Jimenez, W., Brenner, B.M., Marsden, P.A. and Tedder, T.F. (1992). Neutrophils, monocytes, and lymphocytes bind to cytokine-activated kidney glomerular endothelial cells through L-selectin (LAM-1) *in vitro*. *J. Immunol.*, **149**:2437–2444.
46. Brandley, B.K., Ross, T.S. and Schnaar, R.L. (1987). Multiple carbohydrate receptors on lymphocytes revealed by adhesion to immobilized polysaccharides. *J. Cell Biol.*, **105**:991–997.
47. Brånemark, P-I. and Eriksson, E. (1972). Method for studying qualitative and quantitative changes of blood flow in skeletal muscle. *Acta Physiol. Scand.*, **84**:284–288.
48. Breit, G.A. (1988). *Quantitative dynamic analysis of post-junctional leukocyte accumulation*, San Diego: Master's thesis: Bioengineering: University of California.
49. Briskin, M.J., McEvoy, L.M. and Butcher, E.C. (1993). MAdCAM-1 has homology to immunoglobulin and mucin-like adhesion receptors and to IgA1. *Nature*, **363**:461–464.
50. Brustein, M., Kraal, G., Mebius, R.E. and Watson, S.R. (1992). Identification of a soluble form of a ligand for the lymphocyte homing receptor. *J. Exp. Med.*, **176**:1415–1419.
51. Buerke, M., Weyrich, A.S., Murohara, T., Queen, C., Klingbeil, C.K., Co, M.S. and Lefer, A.M. (1994). Humanized monoclonal antibody DREG-200 directed against L-selectin protects in feline myocardial reperfusion injury. *J. Pharmacol. Exp. Ther.*, **271**:134–142.

52. Bullard, D.C., Kunkel, E.J., Kubo, H., Hicks,M.J., Lorenzo, I., Doyle, N.A., Doerschuk, C.M., Ley, K. and Beaudet, A.L. (1996). Infectious susceptibility and severe deficiency of leukocyte rolling and recruitment in E-selectin and P-selectin double mutant mice. *J. Exp. Med.*, **183**:2329–2336.
53. Bullard, D.C., Qin, L., Lorenzo, I., Quinlin, W.M., Doyle, N.A., Bosse, R., Vestweber, D., Doerschuk, C.M. and Beaudet, A.L. (1995). P-selectin/ICAM-1 double mutant mice: Acute emigration of neutrophils into the peritoneum is completely absent but is normal in pulmonary alveoli. *J. Clin. Invest.*, **95**:1782–1788.
54. Bullard, D.C., Sandberg, E.T., Scharffetter-Kochanek, K. and Beaudet, A.L. (1995). Gene targeting for inflammatory cell adhesion molecules. *Agents & Actions Suppl.*, **47**:143–154.
55. Butcher, E.C. (1991). Leukocyte-endothelial cell recognition — Three (or more) steps to specificity and diversity. *Cell*, **67**:1033–1036.
56. Buttrum, S.M., Hatton, R. and Nash, G.B. (1993). Selectin-mediated rolling of neutrophils on immoblized platelets. *Blood*, **82**:1165–1174.
57. Capecchi, M.R. (1989). Altering the genome by homologous recombination. *Science*, **244**:1288–1292.
58. Chien, S., Usami, S. and Skalak, R. (1984). Blood flow in small tubes. In: Renkin, E.M. and Michel, C.C. (Eds.) *Handbook of Physiology. The Cardiovascular System. Microcirculation.* pp. 217–249. Bethesda, MD: American Physiological Society.
59. Chien, S., Tvetenstrand, C.D., Farrell Epstein, M.A. and Schmid-Schönbein, G.W. (1985). Model studies on distributions of blood cells at microvascular bifurcations. *Am. J. Physiol.*, **248**:H568–H576.
60. Clark, E.A. and Brugge, J.S. (1995). Integrins and signal transduction pathways: The road taken. *Science*, **268**:233–239.
61. Cohnheim, J. (1873). *Neue Untersuchungen über die Entzündung*, Berlin: August Hirschwald Verlag.
62. Cohnheim, J. (1889). *Lectures on General Pathology: A Handbook for Practitioners and Students*, London: The New Sydenham Society.
63. Cooke, B.M., Usami, S., Perry, I. and Nash, G.B. (1993). A simplified method for culture of endothelial cells and analysis of adhesion of blood cells under conditions of flow. *Microvasc. Res.*, **45**:33–45.
64. Cotran, R.S., Gimbrone, M.A.J., Bevilacqua, M.P., Mendrick, D.L. and Pober, J.S. (1986). Induction and detection of a human endothelial activation antigen *in vivo*. *J. Exp. Med.*, **164**:661–666.
65. Cozens-Roberts, C., Quinn, J.A. and Lauffenburger, D.A. (1990). Receptor-mediated cell attachment and detachment kinetics. II. Experimental model studies with the radial-flow detachment assay. *Biophys. J.*, **58**:857–872.
66. Cronstein, B.N., Kimmel, S.C., Levin, R.I., Martiniuk, F. and Weissmann, G. (1992). A mechanism for the antiinflammatory effects of corticosteroids: The glucocorticoid receptor regulates leukocyte adhesion to endothelial cells and expression of Endothelial-Leukocyte Adhesion Molecule-1 and Intercellular Adhesion Molecule-1. *Proc. Natl. Acad. Sci. USA*, **89**:9991–9995.
67. Cyster, J.G. and Williams, A.F. (1992). The importance of cross-linking in the homotypic aggregation of lymphocytes induced by antileukosialin (CD43) antibodies. *Eur. J. Immunol.*, **22**:2565–2572.
68. Dahlen, S-E., Björk, J., Hedqvist, P., Arfors, K-E., Hammarström, S., Lindgren, J-Å. and Samuelsson, B. (1981). Leukotrienes promote plasma leakage and leukocyte adhesion in postcapillary venules: *In vivo* effects with relevance to the acute inflammatory response. *Proc. Natl. Acad. Sci. USA*, **78**:3887–3891.
69. Damiano, E.R., Westheider, J., Tözeren, A. and Ley, K. (1995). Variation of velocity, deformation, and adhesion energy density of leukocytes rolling in venules. Submitted.
70. Davenpeck, K.L., Gauthier, T.W., Albertine, K.H. and Lefer, A.M. (1994). Role of P-selectin in microvascular leukocyte-endothelial interaction in splanchnic ischemia-reperfusion. *Am. J. Physiol.*, **267**:H622–H630.
71. Díaz-González, F., González-Alvaro, I., Campanero, M.R., Mollinedo, F., Del Pozo, M.A., Muñoz, C., Pivel, J.P. and Sánchez-Madrid, F. (1995). Prevention of *in vitro* neutrophil-endothelial attachment through shedding of L-selectin by nonsteroidal antiinflammatory drugs. *J. Clin. Invest.*, **95**:1756–1765.
72. Doré, M., Korthuis, R.J., Granger, D.N., Entman, M.L. and Smith, C.W. (1993). P-selectin mediates spontaneous leukocyte rolling *in vivo*. *Blood*, **82**:1308–1316.
73. Duling, B.R. (1973). The preparation and use of the hamster cheek pouch for studies of the microcirculation. *Microvasc. Res.*, **5**:423–429.
74. Endrich, B., Asaishi, K., Goetz, A. and Messmer, K. (1980). Technical report — A new chamber technique for microvascular studies in unanesthetized hamsters. *Res. Exp. Med.*, **177**:125–134.
75. Engler, R.L., Schmid-Schönbein, G.W. and Pavelec, R.S. (1983). Leukocyte capillary plugging in myocardial ischemia and reperfusion in the dog. *Am. J. Pathol.*, **111**:98–111.
76. Etzioni, A., Frydman, M., Pollack, S., Avidor, I., Phillips, M.L., Paulson, J.C. and Gershoni-Baruch, R. (1993). Recurrent severe infections caused by a novel leukocyte adhesion deficiency. *New Engl. J. Med.*, **327**:1789–1792.
77. Fiebig, E., Ley, K. and Arfors, K-E. (1991). Rapid leukocyte accumulation by "spontaneous" rolling and adhesion in the exteriorized rabbit mesentery. *Int. J. Microcirc:Clin. Exp.*, **10**:127–144.
78. Firrell, J.C. and Lipowsky, H.H. (1989). Leukocyte margination and deformation in mesenteric venules of rat. *Am. J. Physiol.*, **256**:H1667–H1674.

79. Foster, N.K., Martyn, J.B., Rangno, R.E., Hogg, J.C. and Pardy, R.I. (1986). Leukocytosis of exercise: Role of cardiac output and catecholamines. *J. Appl. Physiol.*, **61**:2218–2223.
80. Foxall, C., Watson, S.R., Dowbenko, D., Fennie, C., Lasky, L.A., Kiso, M., Hasegawa, A., Asa, D. and Brandley, B.K. (1992). The three members of the selectin receptor family recognize a common carbohydrate epitope, the sialyl Lewis^x oligosaccharide. *J. Cell Biol.*, **117**:895–902.
81. Frenette, P.S., Johnson, R.C., Hynes, M.R. and Wagner, D.D. (1995). Platelets roll on stimulated endothelium *in vivo*: an interaction mediated by endothelial P-selectin. *Proc. Natl. Acad. Sci. USA*, **92**:7450–7454.
82. Fukuda, M., Spooncer, E.S., Oates, J.E., Dell, A. and Klock, J.C. (1984). Structure of sialylated fucosyl lactosaminoglycan isolated from human granulocytes. *J. Biol. Chem.*, **259**:10925–10935.
83. Gaboury, J.P., Anderson, D.C. and Kubes, P. (1994). Molecular mechanisms involved in superoxide-induced leukocyte-endothelial cell interactions *in vivo*. *Am. J. Physiol.*, **266**:H637–H642.
84. Gaboury, J.P. and Kubes, P. (1994). Reductions in physiologic shear rates lead to CD11/CD18-dependent, selectin-independent leukocyte rolling *in vivo*. *Blood*, **83**:345–350.
85. Gaehtgens, P., Ley, K., Pries, A.R. and Müller, R. (1985). Mutual interaction between leukocytes and microvascular blood flow. In: Meßmer, K. and Hammersen, F. (Eds.) *White cell rheology and inflammation*, 7th edn. pp. 15–28. Basel: Karger (Progress in Applied Microcirculation Series).
86. Gallatin, W.M., Weissman, I.L. and Butcher, E.C. (1983). A cell-surface molecule involved in organ-specific homing of lymphocytes. *Nature*, **304**, 30–34.
87. Goldman, A.J., Cox, R.G. and Brenner, H. (1967). Slow viscous motion of a sphere parallel to a plane wall — II. Couette flow. *Chem. Eng. Sci.*, **22**:653–660.
88. Goldsmith, H.L. and Spain, S. (1984). Margination of leukocytes in blood flow through small tubes. *Microvasc. Res.*, **27**:204–222.
89. Gotsch, U., Jäger, U., Dominis, M. and Vestweber, D. (1994). Expression of P-selectin on endothelial cells is upregulated by LPS and TNF-α *in vivo*. *Cell Adhesion and Communication*, **2**:7–14.
90. Görög, P., Kovacs, I.B. and Born, G.V.R. (1980). Suppression of the intravascular adherence of granulocytes by N-acetyl neuraminic (sialic) acid. *Br. J. Exp. Pathol.*, **610**:490–496.
91. Görög, P. and Born, G.V.R. (1982). Increased adhesiveness of granulocytes in rabbit ear-chamber blood vessels perfused with neuraminidase. *Microvasc. Res.*, **23**:380–384.
92. Granert, C., Raud, J., Xie, X., Lindquist, L. and Lindbom, L. (1994). Inhibition of leukocyte rolling with polysaccharide fucoidin prevents pleocytosis in experimental meningitis in the rabbit. *J. Clin. Invest.*, **93**:929–936.
93. Griffin, J.D., Spertini, O., Ernst, T.J., Belvin, M.P., Levine, H.B., Kanakura, Y. and Tedder, T.F. (1990). Granulocyte-macrophage colony-stimulating factor and other cytokines regulate surface expression of the leukocyte adhesion molecule-1 on human neutrophils, monocytes and their precursors. *J. Immunol.*, **145**:576–584.
94. Groves, R.W., Allen, M.H., Ross, E.L., Ahsan, G., Barker, J.N.W.N. and Macdonald, D.M. (1993). Expression of selectin ligands by cutaneous squamous cell carcinoma. *Am. J. Pathol.*, **143**:1220–1225.
95. Hakomori, S. (1992). Le^x and related structures as adhesion molecules. *Histochem. J.*, **24**:771–776.
96. Hammer, D.A. and Apte, S.M. (1992). Simulation of cell rolling and adhesion on surfaces in shear flow: general results and analysis of selectin-mediated neutrophil adhesion. *Biophys. J.*, **63**:35–57.
97. Happel, J. and Byrne, B.J. (1954). Motion of a sphere and fluid in a cylindrical tube. *Ind. Engin. Chem.*, **46**:1181–1186.
98. Hattori, R., Hamilton, K.K., Fugate, R.D., McEver, R.P. and Sims, P.J. (1989). Stimulated secretion of endothelial von Willebrand factor is accompanied by rapid redistribution to the cell surface of the intracellular granule membrane protein GMP-140. *J. Biol. Chem.*, **264**:7768–7771.
99. Hoover, R., Folger, R., Haering, W., Ware, B. and Karnovsky, M. (1980). Adhesion of leukocytes to endothelium: Roles of divalent cations, surface charge, chemotactic agents and substrate. *J. Cell Sci.*, **45**:73–86.
100. House, S.D. and Lipowsky, H.H. (1987). Leukocyte-endothelium adhesion: Microhemodynamics in mesentery of the cat. *Microvasc. Res.*, **34**:363–379.
101. House, S.D. and Lipowsky, H.H. (1988). *In vivo* determination of the force of leukocyte-endothelium adhesion in the mesenteric microvasculature of the cat. *Circ. Res.*, **63**:658–668.
102. Hsu-Lin, S-C., Berman, C.L., Furie, B.C., August, D. and Furie, B. (1984). A platelet membrane protein expressed during activation and secretion: Studies using a monoclonal antibody specific for thrombin-activated platelets. *J. Biol. Chem.*, **259**:9121–9126.
103. Huber, A.R., Kunkel, S.L., Todd, R.F. and Weiss, S.J. (1991). Regulation of transendothelial neutrophil migration by endogenous interleukin-8. *Science*, **254**:99–102.
104. Humphrey, S.A., Rocholl, C.W. and House, S.D. (1993). Chronic exposure to glucocorticoids reduces leukocyte-endothelium interactions during inflammation. *FASEB J.*, **7**:A488 (Abstract).
105. Hynes, R.O. (1992). Integrins: Versatility, modulation, and signaling in cell adhesion. *Cell*, **69**:11–25.
106. Imai, Y., Lasky, L.A. and Rosen, S.D. (1993). Sulphation requirement for GlyCAM-1, an endothelial ligand for L-selectin. *Nature*, **361**:555–557.

107. Iwai, K., Ishikura, H., Kaji, M., Sugiura, H., Ishizu, A., Takahashi, C., Kato, H., Tanabe, T. and Yoshiki, T. (1993). Importance of E-selectin (ELAM-1) and sialyl Lewis[a] in the adhesion of pancreatic carcinoma cells to activated endothelium. *Int. J. Cancer*, **54**:972–977.
108. Janssen, G.H.G.W., Tangelder, G.J., oude Egbrink, M.G.A. and Reneman, R.S. (1994). Spontaneous leukocyte rolling in venules in untraumatized skin of conscious and anesthetized animals. *Am. J. Physiol.*, **267**:H1199–H1204.
109. Johnson, R.C., Mayadas, T.N., Frenette, P.S., Mebius, R.E., Subramaniam, M., Lacasce, A., Hynes, R.O. and Wagner, D.D. (1995). Blood cell dynamics in P-selectin deficient mice. *Blood*, **86**:1106–1114.
110. Jones, D.A., Abbassi, O., McIntire, L.V., McEver, R.P. and Smith, C.W. (1993). P-selectin mediates neutrophil rolling on histamine-stimulated endothelial cells. *Biophys. J.*, **65**:1560–1569.
111. Jones, D.A., McIntire, L.V., Smith, C.W. and Picker, L.J. (1994). A two-step adhesion cascade for T cell/endothelial cell interactions under flow conditions. *J. Clin. Invest.*, **94**:2443–2450.
112. Jutila, M.A., Rott, L., Berg, E.L. and Butcher, E.C. (1989). Function and regulation of the neutrophil MEL-14 antigen *in vivo*: Comparison with LFA-1 and Mac-1. *J. Immunol.*, **143**:3318–3324.
113. Jutila, M.A., Kishimoto, T.K. and Finken, M. (1991). Low-dose chymotrypsin treatment inhibits neutrophil migration into sites of inflammation *in vivo*: Effects on Mac-1 and Mel-14 adhesion protein expression and function. *Cell. Immunol.*, **132**:201–214.
114. Jutila, M.A., Bargatze, R.F., Kurk, S., Warnock, R.A., Ehsani, N., Watson, S.R. and Walcheck, B. (1994). Cell surface P- and E-selectin support shear-dependent rolling of bovine gamma/δ T cells. *J. Immunol.*, **153**:3917–3928.
115. Kansas, G.S., Spertini, O., Stoolman, L.M. and Tedder, T.F. (1991). Molecular mapping of functional domains of the leukocyte receptor for endothelium, LAM-1. *J. Cell Biol.*, **114**:351–358.
116. Kansas, G.S., Ley, K., Munro, J.M. and Tedder, T.F. (1993). Regulation of leukocyte rolling and adhesion to endothelium by the cytoplasmic domain of L-selectin. *J. Exp. Med.*, **177**:833–838.
117. Kansas, G.S., Saunders, K.B., Ley, K., Zakrzewicz, A., Gibson, R.M., Furie, B.C., Furie, B. and Tedder, T.F. (1994). A role for the epidermal growth factor-like domain of P-selectin in ligand recognition and cell adhesion. *J. Cell Biol.*, **124**:609–618.
118. Kanwar, S. and Kubes, P. (1994). Ischemia/reperfusion-induced granulocyte influx is a multistep process mediated by mast cells. *Microcirculation*, **1**:175–182.
119. Keelan, E.T., Licence, S.T., Peters, A.M., Binns, R.M. and Haskard, D.O. (1994). Characterization of E-selectin expression *in vivo* with use of a radiolabeled monoclonal antibody. *Am. J. Physiol.*, **266**:H278–H290.
120. Keizer, G.D., Visser, W., Vliem, M. and Figdor, C.G. (1988). A monoclonal antibody (NK1-L16) directed against a unique epitope on the α-chain of human leukocyte-function associated antigen 1 induces homotypic cell-cell interactions. *J. Immunol.*, **140**:1393–1400.
121. Kishimoto, T.K., Jutila, M.A., Berg, E.L. and Butcher, E.C. (1989). Neutrophil Mac-1 and MEL-14 adhesion proteins inversely regulated by chemotactic factors. *Science*, **245**:1238–1241.
122. Kishimoto, T.K., Warnock, R.A., Jutila, M.A., Butcher, E.C., Lane, C.L., Anderson, D.C. and Smith, C.W. (1991). Antibodies against human neutrophil LECAM-1 (LAM-1/Leu-8/DREG-56 antigen) and endothelial cell ELAM-1 inhibit a common CD18-independent adhesion pathway *in vitro*. *Blood*, **78**:805–811.
123. Klotz, K.F., Pries, A.R., Jepsen, H., Gossrau, R. and Gaehtgens, P. (1991). A new approach to intravital videomicroscopy of rat spinotrapezius muscle. *Int. J. Microcirc:Clin. Exp.*, **10**:205–218.
124. Kubes, P., Suzuki, M. and Granger, D.N. (1991). Nitric oxide — an endogenous modulator of leukocyte adhesion. *Proc. Natl. Acad. Sci. USA*, **88**:4651–4655.
125. Kubes, P., Kurose, I. and Granger, D.N. (1994). NO donors prevent integrin-induced leukocyte adhesion but not P-selectin-dependent rolling in postischemic venules. *Am. J. Physiol.*, **267**:H931–H937.
126. Kubes, P., Jutila, M. and Payne, D. (1995). Therapeutic potential of inhibiting leukocyte rolling in ischemia/ reperfusion. *J. Clin. Invest.*, **95**:2510–2519.
127. Kubes, P. and Kanwar, S. (1994). Histamine induces leukocyte rolling in post-capillary venules: A P-selectin mediated event. *J. Immunol.*, **152**:3570–3577.
128. Kunkel, E.J., Jung, U., Bullard, D.C., Norman, K.E., Wolitzky, B.A., Vestweber, D., Beaudet, A.L. and Ley, K. (1996). Absence of trauma-induced leukocyte rolling in mice deficient in both P-selectin and InterCellular Adhesion Molecule-1 (ICAM-1). *J. Exp. Med.* , **183**: 57–65.
129. Kunzendorf, U., Notter, M., Hock, H., Distler, A., Diamantstein, T. and Walz, G. (1993). T cells bind to the endothelial adhesion molecule GMP-140 (P-selectin). *Transplantation*, **56**:1213–1217.
130. Kunzendorf, U., Kruger-Krasagakes, S., Netter, M., Hock, H., Walz, G. and Diamantstein, T. (1994). A sialyl-Lewis[x] negative melanoma cell line binds to E-selectin but not to P-selectin. *Cancer Res.*, **54**:1109–1112.
131. Labow, M.A., Norton, C.R., Rumberger, J.M., Lombard-Gillooly, K.M., Shuster, D.J., Hubbard, J., Bertko, R., Knaack, P.A., Terry, R.W., Harbison, M.L., Kontgen, F., Stewart, C.L., McIntyre, K.W., Will, P.C., Burns, D.K. and Wolitzky, B.A. (1994). Characterization of E-selectin-deficient mice: demonstration of overlapping function of the endothelial selectins. *Immunity*, **1**:709–720.

132. Lasky, L.A., Singer, M.S., Dowbenko, D., Imai, Y., Henzel, W.J., Grimley, C., Fennie, C., Gillett, N., Watson, S.R. and Rosen, S.D. (1992). An endothelial ligand for L-selectin is a novel mucin-like molecule. *Cell*, **69**:927–938.

133. Laudanna, C., Constantin, G., Baron, P., Scardini, E., Scarlato, G., Cabrini, G., Dechecchi, C., Rossi, F., Cassatella, M.A. and Berton, G. (1994). Sulfatides trigger increase of cytosolic free calcium and enhanced expression of tumor necrosis factor α and interleukin-8 messenger RNA in human neutrophils — evidence for a role of L-selectin as a signaling molecule. *J. Biol. Chem.*, **269**:4021–4026.

134. Lawrence, M.B., McIntire, L.V. and Eskin, S.G. (1987). Effect of flow on polymorphonuclear leukocyte/endothelial cell adhesion. *Blood*, **70**:1284–1290.

135. Lawrence, M.B., Bainton, D.F. and Springer, T.A. (1994). Neutrophil tethering to and rolling on E-selectin are separable by requirement for L-selectin. *Immunity*, **1**:137.

136. Lawrence, M.B., Berg, E.L., Butcher, E.C. and Springer, T.A. (1995). Rolling of lymphocytes and neutrophils on peripheral node addressin and subsequent arrest on ICAM-1 in shear flow. *Eur. J. Immunol.*, **25**:1025–1031.

137. Lawrence, M.B. and Springer, T.A. (1991). Leukocytes roll on a selectin at physiologic flow rates: Distinction from and prerequisite for adhesion through integrins. *Cell*, **65**:859–873.

138. Lawrence, M.B. and Springer, T.A. (1993). Neutrophils roll on E-selectin. *J. Immunol.*, **151**:6338–6346.

139. Lefer, A.M. and Ma, X.L. (1994). PMN adherence to cat ischemic-reperfused mesenteric vascular endothelium under flow: role of P-selectin. *J. Appl. Physiol.*, **76**:33–38.

140. Lehr, H-A., Hübner, C., Finckh, B., Angermüller, S., Nolte, D., Beisiegel, U., Kohlschütter, A. and Messmer, K. (1991). Role of leukotrienes in leukocyte adhesion following systemic administration of oxidatively modified human low density lipoprotein in hamsters. *J. Clin. Invest.*, **88**:9–14.

141. Lehr, H-A., Kress, E. and Menger, M.D. (1993). Involvement of 5-lipoxygenase products in cigarette smoke-induced leukocyte/endothelium interaction in hamsters. *Int. J. Microcirc:Clin. Exp.*, **12**:61–73.

142. Lehr, H-A., Kress, E., Menger, M.D., Friedl, H.P., Hübner, C., Arfors, K-E. and Messmer, K. (1993). Cigarette smoke elicits leukocyte adhesion to endothelium in hamsters: Inhibition by CuZn-SOD. *Free Rad. Biol. Med.*, **14**:573–581.

143. Lehr, H-A., Kröber, M., Hübner, C., Vajkoczy, P., Menger, M.D., Nolte, D., Kohlschütter, A. and Messmer, K. (1993c). Stimulation of leukocyte/endothelium interaction by oxidized low-density lipoprotein in hairless mice: Involvement of CD11b/CD18 adhesion receptor complex. *Lab. Invest.*, **68**:388–395.

144. Lehr, H-A., Frei, B. and Arfors, K-E. (1994). Vitamin C prevents cigarette smoke-induced leukocyte aggregation and adhesion to endothelium *in vivo*. *Proc. Natl. Acad. Sci. USA*, **91**:7688–7692.

145. Lehr, H-A., Olofsson, A.M., Carew, T.E., Vajkoczy, P., von Andrian, U.H., Hübner, C., Berndt, M.C., Steinberg, D., Messmer, K. and Arfors, K-E. (1994). P-selectin mediates the interaction of circulating leukocytes with platelets and microvascular endothelium in response to oxidized lipoprotein *in vivo*. *Lab. Invest.*, **71**:380–386.

146. Lenter, M., Levinovitz, A., Isenmann, S. and Vestweber, D. (1994). Monospecific and common glycoprotein ligands for E- and P-selectin in myeloid cells. *J. Cell Biol.*, **125**:471–481.

147. Lewinsohn, D.M., Bargatze, R.F. and Butcher, E.C. (1987). Leukocyte-endothelial cell recognition: Evidence of a common molecular mechanism shared by neutrophils, lymphocytes and other leukocytes. *J. Immunol.*, **138**:4313–4321.

148. Ley, K., Pries, A.R. and Gaehtgens, P. (1988). Preferential distribution of leukocytes in rat mesentery microvessel networks. *Pfluegers Arch.*, **412**:93–100.

149. Ley, K. (1989). Granulocyte adhesion to microvascular and cultured endothelium. *Studia Biophys.*, **134**:179–184.

150. Ley, K., Lundgren, E., Berger, E.M. and Arfors, K-E. (1989). Shear-dependent inhibition of granulocyte adhesion to cultured endothelium by dextran sulfate. *Blood*, **73**:1324–1330.

151. Ley, K., Cerrito, M. and Arfors, K-E. (1991). Sulfated polysaccharides inhibit leukocyte rolling in rabbit mesentery venules. *Am. J. Physiol.*, **260**:H1667–H1673.

152. Ley, K., Gaehtgens, P., Fennie, C., Singer, M.S., Lasky, L.A. and Rosen, S.D. (1991). Lectin-like Cell Adhesion Molecule-1 mediates leukocyte rolling in mesenteric venules *in vivo*. *Blood*, **77**:2553–2555.

153. Ley, K., Gaehtgens, P. and Spanel-Borowski, K. (1992). Differential adhesion of granulocytes to five distinct phenotypes of cultured microvascular endothelial cells. *Microvasc. Res.*, **43**:119–133.

154. Ley, K., Baker, J.B., Cybulsky, M.I., Gimbrone, M.A.J. and Luscinskas, F.W. (1993). Intravenous interleukin-8 inhibits granulocyte emigration from rabbit mesenteric venules without altering L-selectin expression or leukocyte rolling. *J. Immunol.*, **151**:6347–6357.

155. Ley, K., Linnemann, G., Meinen, M., Stoolman, L.M. and Gaehtgens, P. (1993). Fucoidin, but not yeast polyphosphomannan PPME inhibits leukocyte rolling in venules of the rat mesentery. *Blood*, **81**:177–185.

156. Ley, K., Tedder, T.F. and Kansas, G.S. (1993). L-selectin can mediate leukocyte rolling in untreated mesenteric venules *in vivo* independent of E- or P-selectin. *Blood*, **82**:1632–1638.

157. Ley, K. (1994). Histamine can induce leukocyte rolling in rat mesenteric venules. *Am. J. Physiol.*, **267**:H1017–H1023.
158. Ley, K. (1995). Gene-targeted mice in leukocyte adhesion research. *Microcirculation*, **2**:141–150.
159. Ley, K., Bullard, D.C., Arbones, M.L., Bosse, R., Vestweber, D., Tedder, T.F. and Beaudet, A.L. (1995). Sequential contribution of L- and P-selectin to leukocyte rolling *in vivo*. *J. Exp. Med.*, **181**:669–675.
160. Ley, K., Zakrzewicz, A., Hanski, C., Stoolman, L.M. and Kansas, G.S. (1995). Sialylated O-glycans and L-selectin sequentially mediate myeloid cell rolling *in vivo*. *Blood*, **85**:3727–3735.
161. Ley, K. and Gaehtgens, P. (1991). Endothelial, not hemodynamic differences are responsible for preferential leukocyte rolling in venules. *Circ. Res.*, **69**:1034–1041.
162. Ley, K. and Tedder, T.F. (1995). Leukocyte interactions with vascular endothelium: New insights into selectin-mediated attachment and rolling. *J. Immunol.*, **155**:525–528.
163. Lien, D.C., Wagner, W.W., Capen, R.L., Haslett, C., Hanson, W.L., Hofmeister, S.E., Henson, P.M. and Worthen, G.S. (1987). Physiological neutrophil sequestration in the lung: Visual evidence for localization in capillaries. *J. Appl. Physiol.*, **62**:1236–1243.
164. Lindbom, L., Xie, X., Raud, J. and Hedqvist, P. (1992). Chemoattractant-induced firm adhesion of leukocytes to vascular endothelium *in vivo* is critically dependent on initial leukocyte rolling. *Acta Physiol. Scand.*, **146**:415–421.
165. Lipowsky, H.H. and Zweifach, B.W. (1978). Application of the "two-slit" photometric technique to the measurement of microvascular volumetric flow rates. *Microvasc. Res.*, **15**:93–101.
166. Lorant, D.E., Patel, K.D., McIntyre, T.M., McEver, R.P., Prescott, S.M. and Zimmerman, G.A. (1991). Coexpression of GMP-140 and PAF by endothelium stimulated by histamine or thrombin: A juxtacrine system for adhesion and activation of neutrophils. *J. Cell Biol.*, **115**:223–224.
167. Luscinskas, F.W., Kansas, G.S., Ding, H., Pizcueta, P., Schleiffenbaum, B., Tedder, T.F. and Gimbrone, M.A., Jr. (1994). Monocyte rolling, arrest and spreading on IL-4-activated vascular endothelium under flow is mediated via sequential action of L-selectin, β_1-integrins, and β_2-integrins. *J. Cell Biol.*, **125**:1417–1427.
168. Luscinskas, F.W., Ding, H. and Lichtman, A.H. (1995). P-selectin and vascular cell adhesion molecule-1 mediate rolling and arrest, respectively, of $CD4^+$ T lymphocytes on tumor necrosis factor α-activated vascular endothelium under flow. *J. Exp. Med.*, **181**:1179–1186.
169. Majuri, M.L., Mattila, P. and Renkonen, R. (1992). Recombinant E-selectin-protein mediates tumor cell adhesion via sialyl-Le(a) and sialyl-Le(x). *Biochemical & Biophysical Research Communications*, **182**:1376–1382.
170. Marzi, I., Knee, J., Buhren, V., Menger, M.D. and Trentz, O. (1992). Reduction by superoxide dismutase of leukocyte-endothelial adherence after liver transplantation. *Surgery*, **111**:90–97.
171. Mayadas, T.N., Johnson, R.C., Rayburn, H., Hynes, R.O. and Wagner, D.D. (1993). Leukocyte rolling and extravasation are severely compromised in P-selectin-deficient mice. *Cell*, **74**:541–554.
172. Mayrovitz, H.N., Tuma, R.F. and Wiedeman, M.P. (1980). Leukocyte adherence in arterioles following extravascular tissue trauma. *Microvasc. Res.*, **20**:264–274.
173. Mayrovitz, H.N. (1992). Leukocyte rolling: A prominent feature of venules in intact skin of anesthetized hairless mice. *Am. J. Physiol.*, **262**:H157–H161.
174. McEver, R.P., Moore, K.L. and Cummings, R.D. (1995). Leukocyte trafficking mediated by selectin-carbohydrate interactions. *J. Biol. Chem.*, **270**:11025–11028.
175. McEver, R.P. and Martin, M.N. (1984). A monoclonal antibody to a membrane glycoprotein binds only to activated platelets. *J. Biol. Chem.*, **259**:9799–9804.
176. Mebius, R.E. and Watson, S.R. (1993). L- and E-selectin can recognize the same naturally occurring ligands on high endothelial venules. *J. Immunol.*, **151**:3252–3260.
177. Merwin, J.R., Madri, J.A. and Lynch, M.J. (1992). Cancer cell binding to E-selectin transfected human endothelia. *Biochem. Biophys. Res. Commun.*, **189**:315–323.
178. Metschnikoff, M.E. (1887). Sur la lutte des cellules de l'organisme contre l'invasion des microbes. *Ann. Inst. Pasteur*, **1**:321–336.
179. Mihelcic, D., Schleiffenbaum, B., Tedder, T.F., Sharar, S.R., Harlan, J.M. and Winn, R.K. (1994). Inhibition of leukocyte L-selectin function with a monoclonal antibody attenuates reperfusion injury to the rabbit ear. *Blood*, **84**:2322–2328.
180. Moore, K.L., Varki, A. and McEver, R.P. (1991). GMP-140 binds to a glycoprotein receptor on human neutrophils: Evidence for a lectin-like interaction. *J. Cell Biol.*, **112**:491–499.
181. Moore, K.L., Patel, K.D., Breuhl, R.E., Fugang, L., Johnson, D.A., Lichenstein, H.S., Cummings, R.D., Bainton, D.F. and McEver, R.P. (1995). P-selectin glycoprotein ligand-1 mediates rolling of human neutrophils on P-selectin. *J. Cell Biol.*, **128**:661–671.
182. Muir, A.L., Cruz, M., Martin, B.A., Thommasen, H., Belzberg, A. and Hogg, J.C. (1984). Leukocyte kinetics in the human lung: Role of exercise and catecholamines. *J. Appl. Physiol.*, **57**:711–719.
183. Mulligan, M.S., Varani, J., Dame, M.K., Lane, C., Smith, C.W., Anderson, D.C. and Ward, P.A. (1991). Role of Endothelial-Leukocyte Adhesion Molecule-1 (ELAM-1) in neutrophil-mediated lung injury in rats. *J. Clin. Invest.*, **88**:1396–1406.

184. Mulligan, M.S., Polley, M.J., Bayer, R.J., Nunn, M.F., Paulson, J.C. and Ward, P.A. (1992). Neutrophil-dependent acute lung injury. Requirement for P-selectin (GMP-140). *J. Clin. Invest.*, **90**:1600–1607.
185. Mulligan, M.S., Watson, S.R., Fennie, C. and Ward, P.A. (1993). Protective effects of selectin chimeras in neutrophil-mediated lung injury. *J. Immunol.*, **151**:6410–6417.
186. Mulligan, M.S., Miyasaka, M., Tamatani, T., Jones, M.L. and Ward, P.A. (1994). Requirements for L-selectin in neutrophil-mediated lung injury in rats. *J. Immunol.*, **152**:832–840.
187. Nazziola, E. and House, S.D. (1992). Effects of hydrodynamics and leukocyte-endothelium specificity on leukocyte-endothelium interactions. *Microvasc. Res.*, **44**:127–142.
188. Needham, L.K. and Schnaar, R.L. (1993). The HNK-1 reactive sulfoglucuronyl glycolipids are ligands for L-selectin and P-selectin but not E-selectin. *Proceedings of the National Academy of Sciences of the United States of America*, **90**:1359–1363.
189. Nelson, R.M., Cecconi, O., Roberts, W.G., Aruffo, A., Linhardt, R.J. and Bevilacqua, M.P. (1993). Heparin oligosaccharides bind L- and P-selectin and inhibit acute inflammation. *Blood*, **82**:3253–3258.
190. Nobis, U., Pries, A.R., Cokelet, G.R. and Gaehtgens, P. (1985). Radial distribution of white cells during blood flow in small tubes. *Microvasc. Res.*, **29**:295–304.
191. Nolte, D., Lehr, H-A. and Messmer, K. (1991). Adenosine inhibits postischemic leukocyte-endothelium interaction in postcapillary venules of the hamster. *Am. J. Physiol.*, **261**:H651–H655.
192. Nolte, D., Lehr, H-A., Sack, F.U. and Messmer, K. (1991). Reduction of postischemic reperfusion injury by the vasoactive drug buflomedil. *Blood Vessels*, **28**:8–14.
193. Nolte, D., Schmid, P., Jäger, U., Botzlar, A., Roesken, F., Hecht, R., Uhl, E., Messmer, K. and Vestweber, D. (1994). Leukocyte rolling in venules of striated muscle and skin is mediated by P-selectin, not by L-selectin. *Am. J. Physiol.*, **267**:H1637–H1642.
194. Norgard, K.E., Moore, K.L., Diaz, S., Stults, N.L., Ushiyama, S., McEver, R.P., Cummings, R.D. and Varki, A. (1993). Characterization of a specific ligand for P-selectin on myeloid cells. A minor glycoprotein with sialylated *O*-linked oligosaccharides. *J. Biol. Chem.*, **268**:12764–12774.
195. Norgard-Sumnicht, K.E., Varki, N.M. and Varki, A. (1993). Calcium-dependent heparin-like ligands for L-selectin in nonlymphoid endothelial cells. *Science*, **261**:480–483.
196. Norgard-Sumnicht, K. and Varki, A. (1995). Endothelial heparan sulfate proteoglycans that bind to L-selectin have glucosamine residues with unsubstituted amino groups. *J. Biol. Chem.*, **270**:12012–12024.
197. Norman, K.E., Moore, K.L., McEver, R.P. and Ley, K. (1995). Leukocyte rolling *in vivo* is mediated by P-Selectin Glycoprotein Ligand-1. *Blood*, **86**:4417–4421.
198. Norman, K.E., Scheding, C., Kunkel, E.J., Heavner, G.A. and Ley, K. (1996). Peptides derived from the lectin domain of selectin adhesion molecules inhibit leukocyte rolling *in vivo*. *Microcirculation*, **3**:29–38.
199. Norton, C.R., Rumberger, J.M., Burns, D.K. and Wolitzky, B.A. (1993). Characterization of murine E-selectin expression *in vitro* using novel anti-mouse E-selectin monoclonal antibodies. *Biochem. Biophys. Res. Commun.*, **195**:250–258.
200. Ohta, K., Gotoh, F., Tomita, M., Tanahashi, N., Kobari, M., Shinohara, T.T.Y., Mihara, B. and Takeda, H. (1992). Animal species differences in erythrocyte aggregability. *Am. J. Physiol.*, **262**:H1009–12.
201. Olofsson, A.M., Arfors, K-E., Ramezani, L., Wolitzky, B.A., Butcher, E.C. and von Andrian, U.H. (1994). E-selectin mediates leukocyte rolling in interleukin-1-treated rabbit mesentery venules. *Blood*, **84**:2749–2758.
202. Packman, C.H. and Lichtman, M.A. (1990). Activation of neutrophils: Measurement of actin conformational changes by flow cytometry. *Blood Cells*, **16**:193–207.
203. Parekh, R.B. and Patel, T. (1992). Carbohydrate ligands of the LECAM family as candidates for the development of anti-inflammatory compounds. *J. Pharm. Pharmacol.*, **44(Suppl.1)**:168–171.
204. Pavalko, F.M., Otey, C.A., Simon, K.O. and Burridge, K. (1991). α-Actinin — A direct link between actin and integrins. *Biochem. Soc. Trans.*, **19**:1065–1069.
205. Pavalko, F.M., Walker, D.M., Graham, L., Goheen, M., Doerschuk, C.M. and Kansas, G.S. (1995). The cytoplasmic domain of L-selectin interacts with cytoskeletal proteins via α-actinin: Receptor positioning in microvilli does not require interaction with α-actinin. *J. Cell Biol.*, **129**:1155–1164.
206. Perry, M.A. and Granger, D.N. (1991). Role of CD11/CD18 in shear rate-dependent leukocyte-endothelial cell interactions in cat mesenteric venules. *J. Clin. Invest.*, **87**:1798–1804.
207. Phat, D., Ley, K., Arfors, K-E. and Intaglietta, M. (1991). *In vitro* determination of the force required for detachment of human polymorphonuclear granulocytes from cultured endothelium and immobilized albumin. *Clin. Hemorheol.*, **11**:417–428.
208. Phillips, M.L., Nudelman, E.D., Gaeta, F.C.A., Perez, M., Singhal, A.K., Hakomori, S. and Paulson, J.C. (1990). ELAM-1 mediates cell adhesion by recognition of a carbohydrate ligand, sialyl-Le^x. *Science*, **250**:1130–1132.
209. Picker, L.J., Kishimoto, T.K., Smith, C.W., Warnock, R.A. and Butcher, E.C. (1991). ELAM-1 is an adhesion molecule for skin-homing T cells. *Nature*, **349**:796–799.

210. Picker, L.J., Warnock, R.A., Burns, A.R., Doerschuk, C.M., Berg, E.L. and Butcher, E.C. (1991). The neutrophil selectin LECAM-1 presents carbohydrate ligands to the vascular selectins ELAM-1 and GMP-140. *Cell*, **66**:921–933.
211. Pittman, R.N. and Ellsworth, M.L. (1986). Estimation of red cell flow in microvessels: Consequences of the Baker-Wayland spatial averaging model. *Microvasc. Res.*, **32**:371–388.
212. Pizcueta, P. and Luscinskas, F.W. (1994). Monoclonal antibody blockade of L-selectin inhibits mononuclear leukocyte recruitment to inflammatory sites *in vivo*. *Am. J. Pathol.*, **145**:461–469.
213. Polley, M.J., Phillips, M.L., Wayner, E., Nudelman, E.D., Singhal, A.K., Hakomori, S. and Paulson, J.C. (1991). CD62 and endothelial cell leukocyte adhesion molecule-1 (ELAM-1) recognize the same carbohydrate ligand, sialyl-Lewis-X. *Proc. Natl. Acad. Sci. USA*, **88**:6224–6228.
214. Postigo, A.A., Marazuela, M., Sánchez-Madrid, F. and De Landázuri, M.O. (1994). B lymphocyte binding to E- and P-selectins is mediated through the de novo expression of carbohydrates on *in vitro* and *in vivo* activated human B cells. *J. Clin. Invest.*, **94**:1585–1596.
215. Reneman, R.S., Woldhuis, B., oude Egbrink, M.G.A., Slaaf, D.W. and Tangelder, G.J. (1992). Concentration and velocity profiles of blood cells in the microcirculation. In: Hwang, N.H.C., Turitto, V.T. and Yen, M.R.T. (Eds.) *Advances in cardiovascular engineering*, pp. 25–40. New York: Plenum Press.
216. Robinson, M.K., Andrew, D., Rosen, H., Brown, D., Ortlepp, S., Stephens, P. and Butcher, E.C. (1992). Antibody against the Leu-CAM β-chain (CD18) promotes both LFA-1-dependent and CR3-dependent adhesion events. *J. Immunol.*, **148**:1080–1085.
217. Rosen, S.D. (1993). L-selectin and its biological ligands. *Histochemistry*, **100**:185–191.
218. Sako, D., Chang, X-J., Barone, K.M., Vachino, G., White, H.M., Shaw, G., Veldman, G.M., Bean, K.M., Ahern, T.J., Furie, B., Cumming, D.A. and Larsen, G.R. (1993). Expression cloning of a functional glycoprotein ligand for P-selectin. *Cell*, **75**:1179–1186.
219. Sanders, W.E., Wilson, R.W., Ballantyne, C.M. and Beaudet, A.L. (1992). Molecular cloning and analysis of *in vivo* expression of murine P-selectin. *Blood*, **80**:795–800.
220. Sawada, R., Tsuboi, S. and Fukuda, M. (1994). Differential E-selectin dependent adhesion efficiency in sublines of a human colon cancer exhibiting distinct metastatic potentials. *J. Biol. Chem.*, **269**:1425– 1431.
221. Schmid-Schönbein, G.W., Fung, Y.C. and Zweifach, B.W. (1975). Vascular endothelium-leukocyte interaction. Sticking shear force in venules. *Circ. Res.*, **36**:173–184.
222. Schmid-Schönbein, G.W., Usami, S., Skalak, R. and Chien, S. (1980). The interaction of leukocytes and erythrocytes in capillary and postcapillary vessels. *Microvasc. Res.*, **19**:45–70.
223. Schmidt, E.E., MacDonald, I.C. and Groom, A.C. (1990). Interactions of leukocytes with vessel walls and with other blood cells, studied by high-resolution intravital videomicroscopy of spleen. *Microvasc. Res.*, **40**:99–117.
224. Silber, A., Newman, W., Reimann, K.A., Hendricks, E., Walsh, D. and Ringler, D.J. (1994). Kinetic expression of endothelial adhesion molecules and relationship to leukocyte recruitment in two cutaneous models of inflammation. *Lab. Invest.*, **70**:163–175.
225. Slater, C. and House, S.D. (1993). Effects of nonsteroidal anti-inflammatory drugs on microvascular dynamics. *Microvasc. Res.*, **45**:166–179.
226. Sligh, J.E., Jr., Ballantyne, C.M., Rich, S.S., Hawkins, H.K., Smith, C.W., Bradley, A. and Beaudet, A.L. (1993). Inflammatory and immune responses are impaired in mice deficient in intercellular adhesion molecule-1. *Proc. Natl. Acad. Sci. USA*, **90**:8529–8533.
227. Smith, C.W., Kishimoto, T.K., Abbassi, O., Hughes, B.J., Rothlein, R., McIntire, L.V., Butcher, E.C. and Anderson, D.C. (1991). Chemotactic factors regulate lectin adhesion molecule 1 (LECAM-1)-dependent neutrophil adhesion to cytokine-stimulated endothelial cells *in vitro*. *J. Clin. Invest.*, **87**:609–618.
228. Spertini, O., Kansas, G.S., Reimann, K.A., MacKay, C.R. and Tedder, T.F. (1991). Function and evolutionary conservation of distinct epitopes on the Leukocyte Adhesion Molecule-1 (TQ-1, Leu-8) that regulate leukocyte migration. *J. Immunol.*, **147**:942–949.
229. Spertini, O., Luscinskas, F.W., Kansas, G.S., Munro, J.M., Griffin, J.D., Gimbrone, M.A.J. and Tedder, T.F. (1991). Leukocyte adhesion molecule-1 (LAM-1) interacts with an inducible endothelial cell ligand to support leukocyte adhesion. *J. Immunol.*, **147**:2565–2573.
230. Spertini, O., Luscinskas, F.W., Gimbrone, M.A.J. and Tedder, T.F. (1992). Monocyte attachment to activated human vascular endothelium *in vitro* is mediated by Leukocyte Adhesion Molecule-1 (L-selectin) under nonstatic conditions. *J. Exp. Med.*, **175**:1789–1792.
231. Springer, T.A. (1994). Traffic signals for lymphocyte recirculation and leukocyte emigration: the multistep paradigm. *Cell*, **76**:301–314.
232. Springer, T.A. (1995). Traffic signals on endothelium for lymphocyte recirculation and leukocyte emigration. *Annu. Rev. Physiol.*, **57**:827–872.
233. Sriramarao, P., von Andrian, U.H., Butcher, E.C., Bourdon, M.A. and Broide, D.H. (1994). L-selectin and very late antigen-4 integrin promote eosinophil rolling at physiological shear rates *in vivo*. *J. Immunol.*, **153**:4238–4246.

234. Steegmaier, M., Levinovitz, A., Isenmann, S., Borges, E., Lenter, M., Kocher, H.P., Kleuser, B. and Vestweber, D. (1995). The E-selectin-ligand ESL-1 is a variant of a receptor for fibroblast growth factor. *Nature*, **373**:615–620.
235. Steininger, C.N., Eddy, C.A., Leimgruber, R.M., Mellors, A. and Welply, J.K. (1992). The glycoprotease of Pasteurella Haemolytica A1 eliminates binding of myeloid cells to P-selectin but not to E-selectin. *Biochem. Biophys. Res. Commun.*, **188**:760–766.
236. Stone, J.P. and Wagner, D.D. (1993). P-selectin mediates adhesion of platelets to neuroblastoma and small cell lung cancer. *J. Clin. Invest.*, **92**:804–813.
237. Stoolman, L.M., Yednock, T.A. and Rosen, S.D. (1987). Homing receptors of human and rodent lymphocytes — Evidence for a conserved carbohydrate-binding specificity. *Blood*, **70**:1842–1850.
238. Stoolman, L.M. and Rosen, S.D. (1983). Possible role for cell-surface carbohydrate-binding molecules in lymphocyte recirculation. *J. Cell Biol.*, **96**:722–729.
239. Subramaniam, M., Saffaripour, S., Watson, S.R., Mayadas, T.N., Hynes, R.O. and Wagner, D.D. (1995). Reduced recruitment of inflammatory cells in a contact hypersensitivity response in P-selectin deficient mice. *J. Exp. Med.*, **181**:2277–2282.
240. Symington, F.W., Hedges, D.L. and Hakomori, S. (1985). Glycolipid antigens of human polymorphonuclear neutrophils and the inducible HL-60 myeloid leukemia line. *J. Immunol.*, **134**:2498–2506.
241. Tangelder, G.J., Slaaf, D.W. and Reneman, R.S. (1982). Fluorescent labeling of blood platelets *in vivo*. *Thromb. Res.*, **28**:803–820.
242. Tangelder, G.J., Slaaf, D.W., Muijtjens, A.M.M., Arts, T., oude Egbrink, M.G.A. and Reneman, R.S. (1986). Velocity profiles of blood platelets and red blood cells flowing in arterioles of the rabbit mesentery. *Circ. Res.*, **59**:505–514.
243. Tangelder, G.J., Neumann, C. and Arfors, K-E. (1988). Reduction of leukocyte sticking by dextran sulfate-induced inhibition of leukocyte rolling. *FASEB J.*, **2**:A1881 (Abstract).
244. Tangelder, G.J., Janssens, C.J.J.G., Slaaf, D.W., oude Egbrink, M.G.A. and Reneman, R.S. (1995). *In vivo* differentiation of leukocytes rolling in mesenteric postcapillary venules. *Am. J. Physiol.*, **268**:H909–H915.
245. Tangelder, G.J. and Arfors, K-E. (1987). Inhibition of leukocyte rolling in venules by sulfated polysaccharides. *Fed. Proc.*, **46**:1542 (Abstract).
246. Tangelder, G.J. and Arfors, K-E. (1991). Inhibition of leukocyte rolling in venules by protamin and sulfated polysaccharides. *Blood,* **77**:1565–1571.
247. Tedder, T.F., Steeber, D.A. and Pizcueta, P. (1995). L-selectin deficient mice have impaired leukocyte recruitment into inflammatory sites. *J. Exp. Med.*, **181**:2259–2264.
248. Tedder, T.F. and Engel, P. (1995). The selectins: Vascular adhesion molecules. *FASEB J.,* **9**: 866–873.
249. Thompson, P.L., Papadimitriou, J.M. and Walters, M.N.I. (1967). Suppression of leukocyte sticking and emigration by chelation of calcium. *J. Path. Bact.*, **94**:389–396.
250. Thorlacius, H., Raud, J., Rosengren-Beezley, S., Forrest, M.J., Hedqvist, P. and Lindbom, L. (1994). Mast cell activation induces P-selectin-dependent leukocyte rolling and adhesion in postcapillary venules *in vivo*. *Biochem. Biophys. Res. Commun.*, **203**:1043–1049.
251. Tiemeyer, M., Swiedler, S.J., Ishihara, M., Moreland, M., Schweingruber, H., Hirtzer, P. and Brandley, B.K. (1991). Carbohydrate ligands for endothelial-leukocyte adhesion. *Proc. Natl. Acad. Sci. USA*, **88**:1138–1142.
252. Tözeren, A., Kleinman, H.K., Wu, S., Mercurio, A.M. and Byers, S.W. (1994). Integrin $\alpha_6\beta_4$ mediates dynamic interactions with laminin. *J. Cell Sci.*, **107**:3153–3163.
253. Tözeren, A., Kleinman, H.K., Grant, D.S., Morales, D., Mercurio, A.M. and Byers, S.W. (1995). E-selectin-mediated dynamic interactions of breast- and colon-cancer cells with endothelial-cell monolayers. *Int. J. Cancer*, **60**:426–431.
254. Tözeren, A. and Ley, K. (1992). How do selectins mediate leukocyte rolling in venules? *Biophys. J.*, **63**:700–709.
255. Ushiyama, S., Laue, T.M., Moore, K.L., Erickson, H.P. and McEver, R.P. (1993). Structural and functional characterization of monomeric soluble P-selectin and comparison with membrane P-selectin. *J. Biol. Chem.*, **268**:15229–15237.
256. van Oss, C.J. (1986). Phagocytosis: An overview. *Methods Enzymol.*, **132**:3–15.
257. Varki, A. (1994). Selectin ligands. *Proc. Natl. Acad. Sci. USA*, **91**:7390–7397.
258. Vejlens, G. (1938). The distribution of leukocytes in the vascular system. *Acta Pathol. Microbiol. Scand.*, **Suppl.33**:1–239.
259. Virchow, R. (1871). *Die Cellularpathologie in ihrer Begründung auf physiologische und pathologische Gewebelehre*, 4th edn. Berlin: August Hirschwald Verlag.
260. Vollmar, B., Glasz, J., Menger, M.D. and Messmer, K. (1995). Leukocytes contribute to hepatic ischemia/reperfusion injury via intercellular adhesion molecule-1-mediated venular adherence. *Surgery*, **117**:195–200.

261. von Andrian, U.H., Chambers, J.D., McEvoy, L.M., Bargatze, R.F., Arfors, K-E. and Butcher, E.C. (1991). Two step model of leukocyte-endothelial cell interaction in inflammation: Distinct roles for LECAM-1 and the leukocyte β_2 integrins *in vivo*. *Proc. Natl. Acad. Sci. USA*, **88**:7538–7542.
262. von Andrian, U.H., Hansell, P., Chambers, J.D., Berger, E.M., Torres Filho, I., Butcher, E.C. and Arfors, K.-E. (1992). L-selectin function is required for β_2 integrin-mediated neutrophil adhesion at physiological shear rates *in vivo*. *Am. J. Physiol.*, **263**:H1034–H1044.
263. von Andrian, U.H., Berger, E.M., Ramezani, L., Chambers, J.D., Ochs, H.D., Harlan, J.M., Paulson, J.C., Etzioni, A. and Arfors, K-E. (1993). *In vivo* behavior of neutrophils from two patients with distinct inherited leukocyte adhesion deficiency syndromes. *J. Clin. Invest.*, **91**:2893–2897.
264. von Andrian, U.H., Chambers, J.D., Berg, E.L., Michie, S.A., Brown, D.A., Karolak, D., Ramezani, L., Berger, E.M., Arfors, K-E. and Butcher, E.C. (1993). L-selectin mediates neutrophil rolling in inflamed venules through sialyl Lewis^x-dependent and -independent recognition pathways. *Blood*, **82**:182–191.
265. von Andrian, U.H., Hasslen, S.R., Nelson, R.D., Erlandsen, S.L. and Butcher, E.C. (1995). Essential role for microvillous receptor presentation in leukocyte adhesion under flow. *Cell*, **82**:989–999.
266. Waddell, T.K., Fialkow, L., Chan, C.K., Kishimoto, T.K. and Downey, G.P. (1994). Potentiation of the oxidative burst of human neutrophils. A signaling role for L-selectin. *J. Biol. Chem.*, **269**:18485–18491.
267. Wagner, R. (1839). *Erläuterungstafeln zur Physiologie und Entwicklungsgeschichte*, Leipzig: Leopold Voss.
268. Walz, G., Aruffo, A., Kolanus, W., Bevilacqua, M.P. and Seed, B. (1990). Recognition by ELAM-1 of the sialyl-Le^x determinant on myeloid and tumor cells. *Science*, **250**:1132–1135.
269. Walzog, B., Seifert, R., Zakrzewicz, A., Gaehtgens, P. and Ley, K. (1994). Cross-linking of CD18 in human neutrophils induces an increase of intracellular free Ca^{2+}, exocytosis of azurophilic granules, quantitative up-regulation of CD18, shedding of L-selectin, and actin polymerization. *J. Leukocyte Biol.*, **56**:625–635.
270. Walzog, B., Schuppan, D., Heimpel, C., Hafezi-Moghadam, A., Gaehtgens, P. and Ley, K. (1995). The leukocyte integrin Mac-1 (CD11b/CD18) contributes to binding of human granulocytes to collagen. *Exp. Cell Res.*, **218**:28–38.
271. Watson, S.R., Imai, Y., Fennie, C., Geoffroy, J.S., Rosen, S.D. and Lasky, L.A. (1990). A homing receptor-IgG chimera as a probe for adhesive ligands of lymph node high endothelial venules. *J. Cell Biol.*, **110**:2221–2229.
272. Watson, S.R., Fennie, C. and Lasky, L.A. (1991). Neutrophil influx into an inflammatory site inhibited by soluble homing receptor-IgG chimaera. *Nature*, **349**:164–167.
273. Wayland, H. and Johnson, P.C. (1967). Erythrocyte velocity measurement in microvessels by two-slit photometric method. *J. Appl. Physiol.*, **22**:333–337.
274. Weller, A., Isenmann, S. and Vestweber, D. (1992). Cloning of the mouse endothelial selectins: expression of both E- and P-selectin is inducible by tumor necrosis factor. *J. Biol. Chem.*, **267**:15176–15183.
275. Whitmore, R.L. (1967). A theory of blood flow in small vessels. *J. Appl. Physiol.*, **22**:333–337.
276. Wiedeman, M.P. (1973). Preparation of the bat wing for *in vivo* microscopy. *Microvasc. Res.*, **5**:417–422.
277. Wiedeman, M.P. (1982). Movement of leukocytes in flowing blood. In: Bagge, U., Born, G.V.R. and Gaehtgens, P. (Eds.) *White blood cells*, pp. 78–81. The Hague: Martinus Nijhoff.
278. Winkelhake, J.L. (1991). Will complex carbohydrate ligands of vascular selectins be the next generation of non-steroidal anti-inflammatory drugs? *Glycoconjugate J.*, **8**:381–386.
279. Worthen, G.S., Smedley, L.A., Tonnesen, M.G., Ellis, D., Voelkel, N.F., Reeves, J.T. and Henson, P.M. (1987). Effects of shear stress on adhesive interaction between neutrophils and cultured endothelial cells. *J. Appl. Physiol.*, **63**:2031–2041.
280. Worthen, G.S., Schwab, B., III., Elson, E.L. and Downey, G.P. (1989). Mechanics of stimulated neutrophils: Cell stiffening induces retention in capillaries. *Science*, **245**:183–186.
281. Xu, H., Gonzalo, J.A., St. Pierre, Y., Williams, I.R., Kupper, T.S., Cotran, R.S., Springer, T.A. and Gutierrez-Ramos, J-C. (1994). Leukocytosis and resistance to septic shock in intercellular adhesion molecule 1-deficient mice. *J. Exp. Med.*, **180**:95–109.
282. Yednock, T.A., Butcher, E.C., Stoolman, L.M. and Rosen, S.D. (1987). Receptors involved in lymphocyte homing: Relationship between a carbohydrate-binding receptor and the Mel-14 antigen. *J. Cell Biol.*, **104**:725–731.
283. Yednock, T.A., Stoolman, L.M. and Rosen, S.D. (1987). Phosphomannosyl-derivatized beads detect a receptor involved in lymphocyte homing. *J. Cell Biol.*, **104**:713–723.
284. Yeo, E.L., Sheppard, J-A.I. and Feuerstein, I.A. (1994). Role of P-selectin and leukocyte activation in polymorphonuclear cell adhesion to surface adherent activated platelets under physiologic shear conditions (an injury vessel wall model). *Blood*, **83**:2498–2507.
285. Yoshida, M., Westlin, W.F., Wang, N., Ingber, D.E., Rosenzweig, A., Resnick, N. and Gimbrone, Jr. , M.A. (1996). Leukocyte adhesion to vascular endothelium induces E-selectin linkage to the actin cytoskeleton. *J. Cell Biol.*, **133**:445–455.

286. Zakrzewicz, A., Graefe, M., Terbeek, D., Bongrazio, M., Auch-Schwelk, W., Walzog, B., Graf, K., Fleck, E., Ley, K. and Gaehtgens, P. (1995). L-selectin dependent leukocyte adhesion to micro- but not-macrovascular endothelial cells of the human coronary system. *Blood*, in press.
287. Zimmerman, B.J., Paulson, J.C., Arrhenius, T.S., Gaeta, F.C.A. and Granger, D.N. (1994). Thrombin receptor peptide-mediated leukocyte rolling in rat mesenteric venules: Roles of P-selectin and sialyl Lewis X. *Am. J. Physiol.*, **267**:H1049–H1053.
288. Zweifach, B.W. (1973). The microcirculation in the intestinal mesentery. *Microvasc. Res.*, **5**:363–367.
289. Zweifach, B.W. and Lipowsky, H.H. (1984). Pressure-flow relations in blood and lymph microcirculation. In: Renkin, E.M. and Michel, C.C. (Eds.) *Handbook of Physiology. The Cardiovascular System. Microcirculation*, pp. 251–307. Bethesda, MD: American Physiological Society.
290. Jung, U., Bullard, D.C., Tedder, T.F. and Ley, K. (1996). Velocity difference between L-selectin and P-selectin dependent neutrophil rolling in venules of the mouse cremaster muscle *in vivo*. *Am. J. Physiol.* in press.
291. Kunkel, E.J. and Ley, K. (1996). Distinct phenotype of E-selectin deficient mice: E-selectin is required for slow leukocyte rolling *in vivo* submitted.

5 The Selectins in Lymphocyte Homing

Alf Hamann and Petra Jonas

Dept. Immunology. Medical Clinic, University Hospital Eppendorf, 20246 Hamburg, FRG.

The suggestion that distinct lymphocyte receptors ("homing receptors") might recognize tissue-specific endothelial ligands and thereby regulate organ specific trafficking dates back to the pioneering work of Gowans (Gowans and Knight, 1964). Two decades later Gallatin *et al.* (1983) identified the first homing receptor by mAbs. Molecular cloning unraveled that this leukocytic antigen, later on named L-selectin, together with the endothelial molecules E- and P-selectin, belongs to a subfamily of carbohydrate binding molecules that are all involved in the extravasation of leukocytes (Bevilacqua *et al.*, 1989; Johnston *et al.*, 1989; Lasky *et al.*, 1989). More recently, it has became clear that the predominant — albeit not exclusive — role of selectins is to provide transient contacts mediating "tethering" or "rolling" of the circulating cells as the initial step in the adhesion cascade, ultimately leading to their extravasation at sites within the body expressing the appropriate "addressins".

Whereas L-selectin controls, by these means, the recirculation of naive lymphocytes through lymph nodes and, to some extent, mucosal tissues in the healthy organism, E- and P-selectin are important initiators of polymorphonuclear leukocyte (PMNL) recruitment into inflamed tissue. However, these functions are not exclusive: L-selectin has been found to be involved in PMNL and monocyte extravasation as well; furthermore some evidence points to a role of all three selectins in lymphocyte trafficking into inflamed tissues.

In this chapter we want to focus on the role of selectins for the migration of lymphocytes, which have the unique capacity to recirculate continuously between blood and tissues, in contrast to most other leukocytes (except, probably, dendritic cells).

HOW LYMPHOCYTES TRAFFIC WITHIN THE BODY

An average lymphocyte enters at least once a day the spleen, lymph nodes, or Peyer's patches from the blood; crosses these organs within 4–20 hours, and re-enters the blood circulation either directly (spleen) or via the efferent lymph (Smith and Ford, 1983). It is this pathway which is mainly followed by naive, resting lymphocytes; only to a smaller extent does recirculation occur through non-lymphoid tissues such as lung, liver or gut.

Upon activation and eventual differentiation into memory cells, the trafficking patterns of lymphocyte populations change dramatically. Their extravasation now takes place predominantly within non-lymphoid tissues (Smith *et al.*, 1980; Hamann and Rebstock, 1993; Mackay, 1993). Inflamed sites in skin or other tissues seem to attract lymphoblasts and memory cells, with some preference (Chin and Hay, 1980; Parrott and Wilkinson, 1981; Picker *et al.*, 1990a; Mackay *et al.*, 1992a), whereas in the intact animal, cells within this differentiation stage target lung, liver and gut from the circulation. As discussed below, some knowledge exists as to the mechanisms regulating lymphocyte extravasation into the intestine and into (inflamed) skin, but the mechanisms and receptors regulating blast and memory cell accumulation within lung and liver do not include selectins and are as yet largely unknown (Hamann and Rebstock, 1993, and unpublished data).

A variety of studies has indicated that, among activated and memory lymphocytes, distinct subpopulations or differentiation stages exhibit organ-specific migration patterns, e.g. mucosal homing or a preferential recirculation through inflamed sites in the skin (Guy-Grand *et al.*, 1974; Smith *et al.*, 1980; Parrott and Wilkinson, 1981; Mackay *et al.*, 1992b). Differentiation processes ensuring that cells primed at a given site will home back to the same or related sites have been postulated, although the regulatory events leading to an acquisition of specific homing properties have still to be defined.

Clear evidence that L-selectin is a key receptor required for entry of lymphocytes into peripheral lymph nodes was provided by Gallatin *et al.* (1983). These authors raised an mAb against mouse L-selectin, MEL-14, which blocked lymphocyte adhesion to lymph node high endothelial venules (HEV) *in vitro* and suppressed completely their localization in peripheral lymph nodes *in vivo*. As effects on binding to HEV of Peyer's patches or homing into this tissue were not noticed in the initial study, L-selectin was assumed to be a peripheral lymph node-specific homing receptor. The almost absolute requirement

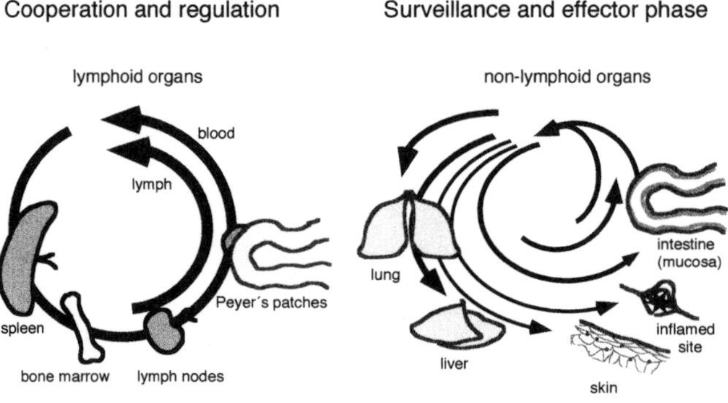

Figure 1 Schematic view of the major recirculation pathways and compartments L-selectin is required for the immigration of naive lymphocytes into peripheral lymph nodes and contributes to immigration into Peyer's patches and the intestine.

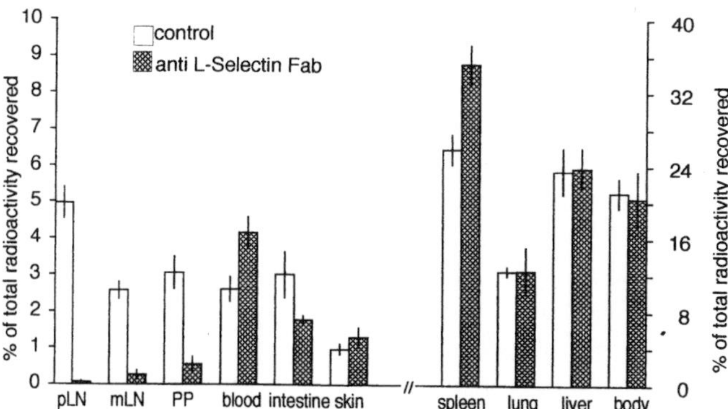

Figure 2 Representative experiment demonstrating the role of L-selectin for the entry of lymphocytes into various compartments of the body. ^{51}Chromium labeled lymph node lymphocytes were injected intravenously without (blank columns) or together with MEL-14 Ab (200 μg Fab/mouse, hatched columns). 1 h later, radioactivity in the organs was counted. Note the different scale for the large organs on the right side of the figure.

for L-selectin to enter peripheral lymph nodes is unique; it is also reflected in the strong reduction of cell numbers present in peripheral, but not mesenterial lymph nodes upon genetic deletion of L-selectin (Arbonès *et al.*, 1994) or long-term application of anti L-selectin mAbs (Bradley *et al.*, 1994; Lepault *et al.*, 1994). It appears that the capacity of L-selectin and its ligands to mediate the initial steps in the adhesion cascade, the transient interactions ("tethering" and "rolling"), is indispensable in the extravasation into lymph nodes, in contrast to other tissues.

Interestingly, recent data indicate that L-selectin may also function as a signalling molecule in lymphocytes: notably, that crosslinking L-selectin or the interaction with the secreted ligand GlyCAM-1 is able to trigger integrin adhesion (Simon *et al.*, 1995; S. Rosen, personal communication), similar to the envisaged function of chemokines within the multistep model. As GlyCAM-1 is abundant in lymph nodes, this contribution to the later steps of the adhesion cascade could contribute to the outstanding role of L-selectin in this type of tissue.

Re-examination of the role of L-selectin *in vivo* using anti-L-selectin Fab fragments or L-selectin chimera disclosed that L-selectin also contributes significantly to the recirculation of resting lymphocytes through Peyer's patches and the gut wall, although the blockage of immigration is not as complete as seen for peripheral lymph nodes (Figure 2) (Hamann *et al.*, 1991; Mebius and Watson, 1993). The role of L-selectin in migration into mucosal sites has long been neglected, not only because of misleading results of *in vitro* assays, but also because the ligands for L-selectin were present in these tissues at a much lower level (Bargatze *et al.*, 1990) and were initially not detected on the endothelial surface by mAbs (Streeter *et al.*, 1988) or L-selectin chimera (Watson *et al.*, 1990).

The exact role of L-selectin for lymphocyte extravasation into mucosal sites has recently been specified by Bargatze *et al.* (1995), dissecting the contribution of L-selectin and integrins to the sequential interactions of lymphocytes with endothelium using intravital microscopy. The authors found that L-selectin but also $\alpha_4\beta_7$ mediate first contact and rolling of naive lymphocytes to HEV of Peyer's patches. These data confirmed not only that L-selectin contributes to mucosal homing, but additionally that, as previously shown for neutrophils, the predominant role of L-selectin for lymphocyte migration is to mediate the initial, transient interaction of the adhesion cascade. The authors also showed that L-selectin becomes dispensable with increasing expression levels or the preactivated stage of the $\alpha_4\beta_7$ integrin, as for example is seen on some memory cells or activated lymphocytes, which may account for the smaller effect of anti L-selectin antibodies on lymphoblast migration to mucosal sites (Hamann *et al.*, 1994).

A role for L-selectin cannot be detected in the other tissues studied beside lymph nodes, Peyer's patches and intestine, at least in the absence of inflammation (see below). An increase in lymphocyte numbers within the spleen upon blockage of entry into lymph nodes and Peyer's patches by anti L-selectin mAb in short term (Hamann *et al.*, 1991), as well as in long term experiments (Bradley *et al.*, 1994; Lepault *et al.*, 1994), is most likely due to redistribution of the cells to this organ; an effect of anti L-selectin mAb on the localization of lymphocytes in the spleen is not observed in *aly*-mice, lacking lymph nodes and Peyer's patches, but not a spleen (Hamann, unpublished data).

In summary, the dominant physiological role of L-selectin in lymphocyte trafficking is to act as homing receptor for naive lymphocytes in their recirculation through HEV-bearing lymphoid tissue. It is absolutely essential for entry into peripheral lymph nodes, of decreasing importance for entry into mesenteric nodes and Peyer's patches, and additionally contributes slightly to entry into the gut wall as the only normal non-lymphoid tissue.

L-SELECTIN AND THE TRAFFICKING OF ACTIVATED AND MEMORY LYMPHOCYTES

Whereas L-selectin is uniformly expressed on naive lymphocytes, the expression pattern on blasts and memory cells is much more complex. L-selectin expression was found to be downregulated on the majority of lymphocytes upon prolonged activation and on T-cell clones (Dailey *et al.*, 1985), which has often led to the view that lymphocyte activation is inadvertently linked with loss of L-selectin. However, even on long term-activated T-cells a substantial minority of cells remains L-selectin$^+$, and transiently a fraction of blasts may even upregulate L-selectin expression *in vitro* and *in vivo* (Dailey *et al.*, 1983; Hamann *et al.*, 1988; Picker *et al.*, 1993a). In addition differentiated effector stages of $CD4^+$ lymphocytes contain L-selectin$^+$ and$^-$ cells (Swain *et al.*, 1991). Whether these L-selectin$^+$ and$^-$ subpopulations display different functions, or relate to populations of different tissue origin is not yet clear.

The loss of L-selectin on part of the activated cells or memory cells has frequently been taken as an explanation for the altered migration behavior of these cells. However, as mentioned in the preceding section, lack of L-selectin function by mAb blockage or upon genetic deletion results in the complete or partial failure to enter lymph nodes and Peyer's patches, yet simultaneously leads to increased levels in spleen and blood. The latter is not observed with blasts; in contrast, activated cells show a reduction in blood levels as well as spleen localization. Rather these cells accumulate preferentially in lung and other non-lymphoid tissues when injected i.v. This indicates that the cells have acquired or activated further adhesion mechanisms extracting them efficiently into non-lymphoid tissues, which may be of greater importance for their trafficking behavior than the loss of L-selectin. This view is supported by the migration behavior found for lymphoma cells which resemble activated lymphocytes in many features and have a homing phenotype similar to that seen for lymphoblasts. Notably, lines with high L-selectin or low L-selectin expression are both similarly unable to localize within lymph nodes (Hamann *et al.*, 1990, and unpublished data).

The biological significance of this redistribution of activated cells into non-lymphoid compartments is yet to be established, although the targeting of effector cells into the major entry ports for microorganisms may increase the efficiency of the immune system. It must also be considered, however, that it is likely that the majority of activated cells does not leave the tissue during proliferation; the increased accumulation of i.v. injected cells in lung and liver may thus rather reflect a "sessile phenotype" characterized by activated adhesion mechanisms which still await characterization.

On memory cells, L-selectin expression is even more complex than on activated lymphocytes and pronounced differences between species are found. In mice, memory cells of the $CD4^+$ subtype seem to be exclusively L-selectin$^-$ (Bradley *et al.*, 1992). However, within the $CD8^+$ subset, memory cells are predominantly found in the L-selectin+ fraction (Lee and Vitetta, 1991; Mobley *et al.*, 1994). Also long term murine B-memory cells seem to be L-selectin$^+$ (Kraal *et al.*, 1988). In the human, $CD4^+$ memory cells can be L-selectin$^+$ or$^-$ (Picker *et al.*, 1990b; Tedder *et al.*, 1990), and this seems to be correlated with the tissue origin (Picker *et al.*, 1993a).

The migratory property of memory cells remains a controversial issue. Whereas studies in the sheep indicated an inability of cells of memory phenotype to recirculate via HEV (Mackay *et al.*, 1990), more recent investigations in the rat suggested a normal recirculatory behavior of CD45 RC- defined memory cells (Westermann *et al.*, 1994). In mice, most $CD4^+$ cells of memory phenotype (L-selectin$^-$, CD45RB low) do not recirculate through lymph nodes and only poorly enter the spleen. Instead, a high preference of the great majority of these cells to localize in lung and liver is observed, similar to what is found for activated cells (Tietz and Hamann, submitted). Again, this could indicate that these cells are either mainly sessile, or recirculate predominantly through non-lymphoid tissues. Whether the observed expression of L-selectin on murine B- and $CD8^+$ or human $CD4^+$ memory cells is sufficient to provide the capacity to recirculate as naive cells do is not yet clear.

L-SELECTIN MAY ALSO PLAY A ROLE IN THE IMMIGRATION OF LYMPHOCYTES INTO INFLAMED SITES

Several findings indicate that L-selectin might have a role for the extravasation of lymphocytes into a variety of inflamed non-lymphoid tissues, where it normally plays no role. Ligands for L-selectin, defined by the mAb MECA 79, have been detected in many chronically inflamed tissues including skin, thyroid (Michie *et al.*, 1993) and pancreatic islets (Hänninen *et al.*, 1993; Faveeuw *et al.*, 1994) whereas they are for example not found in inflamed lung (Michie *et al.*, 1993). Furthermore, in the human, L-selectin was found to be expressed selectively by memory T cells localized within chronically inflamed skin lesions, whereas the majority of memory cells derived from other sites were L-selectin$^-$ (Picker *et al.*, 1994).

Partial blocking effects of anti L-selectin antibodies have been detected in adhesion assays using sections of inflamed synovia (Fischer *et al.*, 1993) and kidney (Turunen *et al.*, 1994) or cytokine-activated human umbilical vein endothelium (Spertini *et al.*, 1991). Moreover, treatment with L-selectin antibodies had a protective effect in certain T-cell mediated autoimmune disease models such as diabetes in the NOD mice (Yang *et al.*, 1993; 1994). These data could suggest that L-selectin may have a significant role for the recruitment not only of neutrophils, as discussed elsewhere in this book, but also of lymphocytes into sites of inflammation. However, only in the case of a cutaneous DTH-like reaction has it to date been shown directly that anti L-selectin antibodies partially inhibit lymphocyte immigration (Dawson *et al.*, 1992).

In this study, however, as well as in the therapeutic settings, indirect effects of the anti L-selectin Abs on lymphocyte trafficking are not formally excluded. As discussed in a previous chapter, L-selectin has a major effect on the accumulation and activation of neutrophils and monocytes at inflamed sites. It has to be taken in consideration that prevention neutrophil immigration or activation by L-selectin blockade may affect the subsequent recruitment of lymphocytes. Therefore, unequivocal data showing that blocking or lack of L-selectin has a major, direct effect on the infiltration of lymphocytes into different inflamed sites are still missing.

ENDOTHELIAL SELECTINS AND LYMPHOCYTE TRAFFIC

The expression of the two endothelial selectins is induced by a variety of inflammatory stimuli and it is well established that E- and P-selectin contribute substantially to the infiltration of neutrophils into inflamed tissue. Less well documented is the role these two selectins may play for lymphocyte migration to sites of active immunoreactivity. In 1990 Graber *et al.* had already demonstrated that anti-E-selectin antibodies lead to a reduced binding of human T cells to cytokine-activated endothelium *in vitro*. Later it was shown that the T cell population binding to E-selectin is composed almost exclusively of memory cells (Picker *et al.*, 1991; Shimizu *et al.*, 1991a), and that γ/δ T cells (Walcheck *et al.*, 1993), subpopulations of activated B cells (Postigo *et al.*, 1994) and NK cells (Pinola *et al.*, 1994) exhibit E-selectin binding capacity, too. More recently it became clear that subpopulations of memory T cells, γ/δ T cells, NK cells and activated B cells can also interact with P-selectin (Moore *et al.*, 1992; Damle *et al.*, 1992; Jutila *et al.*, 1994; Postigo *et al.*, 1994). The subpopulations that bind to either P- or E-selectin do partially overlap, but they seem not to be identical in most cases (Jutila *et al.*, 1994; Rossiter *et al.*, 1994). From these data it was suggested that the endothelial selectins may play a role for the selective recruitment of distinct effector/memory lymphocyte populations at specific phases of an inflammatory response.

Indirect evidence for a role of E-selectin in determining tissue-specific and inflammation-specific recruitment of lymphocyte subpopulations came from studies on the expression of carbohydrate groups serving as ligand for E-selectin, which were detected by MAb HECA 452 in humans. Human memory T cells found in chronically inflamed skin express this epitope, now called CLA, cutaneous lymphocyte antigen (Picker *et al.*, 1991; 1990b). CLA is a modified sialyl-Lewis x carbohydrate epitope (Berg *et al.*, 1991), which binds to E-selectin (Berg *et al.*, 1991). A variety of studies has shown that human T-cells adhere via CLA / E-selectin interactions to endothelium (Picker *et al.*, 1991; Shimizu *et al.*, 1991b; de

Boer *et al.*, 1994; Santamaria-Babi *et al.*, 1995a). Besides in chronically inflamed cutaneous sites (Picker *et al.*, 1990b), it is also found on the few memory cells present in normal human skin (Bos *et al.*, 1993), skin-derived T-cell clones (Rossiter *et al.*, 1994), and cutaneous lymphoma (Noorduyn *et al.*, 1992; Heald *et al.*, 1993). In contrast, in the peripheral blood only a minor subset of memory T cells carries this epitope and only very few T cells in extracutaneous inflammatory sites are CLA^+ (Picker *et al.*, 1990b). This specific expression pattern has given rise to the hypothesis that CLA may mediate skin-specific homing of a subpopulation of memory cells via E-selectin, which is persistently upregulated in skin inflammatory sites containing lymphocyte infiltrates (Silber *et al.*, 1994a). Indeed, in a primate model of delayed type hypersensitivity, antibodies against E-selectin were able to reduce lymphocyte recruitment (Silber *et al.*, 1994b). In mice, E- as well as P-selectin were found to contribute to effector T-cell immigration into the inflamed skin, demonstrating an overlapping function of both endothelial selectins for T-cell recruitment into sites of inflammation (Subramaniam, 1995; Austrup *et al.*, submitted).

Supporting evidence for a tissue-specific differentiation of memory cells presents the finding that circulating memory T-cells in atopic patients, reactive to a sensitizing antigen, were found to be CLA+ in those with dermatitis, but not in those with a broncho-pulmonary or gastrointestional manifestation (Abernathy-Carver *et al.*, 1995; Santamaria-Babi *et al.*, 1995b), and that CLA becomes expressed on a majority of T cells undergoing virgin to memory transition within subcutanous lymph nodes, but not within the appendix (Picker *et al.*, 1993b). These findings have been related to the paradigm of organ-specific homing. This property of specific subsets of lymphocytes is assumed to be shaped by tissue-specific factors, inducing the expression of appropriate homing receptors upon priming within a given tissue (Guy-Grand *et al.*, 1974; Butcher, 1986; Butcher and Picker, 1996).

Doubt as to a skin-specific role of CLA / E-selectin interactions arises from a widespread expression of E-selectin in a variety of other inflamed tissues, and a proven role for neutrophil recruitment in these sites. In mice, we demonstrated a clear function of E- (and P-) selectin for the immigration of effector T-cells into the inflamed skin, but also into the synovia of arthritic mice (Austrup *et al.*, submitted)

Moreover, we found that the expression of carbohydrate epitopes serving as E- and P-selectin ligands is rather dependent on the differentiation into distinct functional subsets than related to the origin of cells: Ligands for P- and E-selectin were selectively expressed on TH1-, but not TH2-cells (Austrup *et al.*, submitted). The differentiation of TH1 cells is driven by Il-12. Interestingly, this cytokine has also been found to induce the CLA epitope in the human system (Picker *et al.*, 1993b; Leung *et al.*, 1995). These data indicate that the development of tissue-specific trafficking patterns under the influence of a local microenvironment is a complex matter and may go in parallel with the development of a functional specialization.

SUMMARY

The physiological role of the selectin family is tightly associated with mechanisms regulating leucocyte trafficking. In the case of lymphocytes, L-selectin and the endothelial selectins fulfill divergent functions: whereas L-selectin guarantees the regular recirculation of naive lymphocytes through HEV-bearing lymphoid tissues and, to a smaller extent, through the normal gut wall, E- and P-selectin rather serve to recruit memory or activated and effector cell populations into sites of inflammation.

The expression of ligands for endothelial selectins is especially obvious in T-cells residing in the skin, and it seems to distinguish specific functional subsets of T-cells. Especially TH1-cells are endowed with the respective carbohydrate epitopes; by recruiting only these pro-inflammatory cells into sites of acute immune reaction, the endothelial selectins contribute to shape direction and character of a local immune reaction.

References

Abernathy-Carver, K.J., Sampson, H.A., Picker, L.J. and Leung, D.Y.M. (1995). Milk-induced eczema is associated with the expansion of T cells expressing cutaneous lymphocyte antigen. *J. Clin. Invest.*, **95**:913–918.

Arbonès, M.L., Ord, D.C., Ley, K., Ratech, H., Maynard-Curry, C., Otten, G., Capon, D.J. and Tedder, T.F. (1994). Lymphocyte homing and leukocyte rolling and migration are impaired in L-selectin-deficient mice. *Immunity*, **1**:247–260.

Austrup, F., Vestweber, D., Borges, E., Bräuer, R., Hallmann, R. and Hamann, A. P- and E-selectin mediate recruitment of T helper 1 but not T helper 2 cells into inflamed tissues. Submitted.

Bargatze, R.F., Streeter, P.R. and Butcher, E.C. (1990). Expression of low levels of peripheral lymph node-associated vascular addressin in mucosal lymphoid tissues: possible relevance to the dissemination of passaged AKR lymphomas. *J. Cell Biochem.*, **42**:219–227.

Bargatze, R.F., Jutila, M.A. and Butcher, E.C. (1995). Distinct roles of L-selectin and integrin alpha4β7 and LFA-1 in lymphocyte homing to Peyer's patch-HEV In situ: the multistep model confirmed and refined. *Immunity*, **3**:99–108.

Berg, E.L., Yoshino, T., Rott, L.S., Robinson, M.K., Warnock, R.A., Kishimoto, T.K., Picker, L.J. and Butcher, E.C. (1991). The cutaneous lymphocyte antigen is a skin lymphocyte homing receptor for the vascular lectin endothelial cell-leukocyte adhesion molecule 1. *J. Exp. Med.*, **174**:1461–1466.

Bevilacqua, M.P., Stengelin, S., Gimbrone, M.A. and Seed, B. (1989). Endothelial leukocyte adhesion molecule 1: an inducible receptor for neutrophils related to complement regulatory proteins and lectins. *Science*, **243**:1160–1165.

Bos, J.D., de, B.O., Tibosch, E., Das, P.K. and Pals, S.T. (1993). Skin-homing T lymphocytes: detection of cutaneous lymphocyte-associated antigen (CLA) by HECA-452 in normal human skin. *Arch. Dermatol. Res.*, **285**:179–183.

Bradley, L.M., Atkins, G.G. and Swain, S.L. (1992). Long-term CD4$^+$ memory T cells from the spleen lack MEL-14, the lymph node homing receptor. *J. Immunol.*, **148**:324–331.

Bradley, L.M., Watson, S.R. and Swain, S.L. (1994). Entry of naive CD4 T cells into peripheral lymph nodes requires L-selectin. *J. Exp. Med.*, **180**:2401–2406.

Butcher, E.C. (1986). The regulation of lymphocyte traffic. *Curr. Top Microbiol. Immunol.*, **128**:85–122.

Butcher, E. C., and Picker, L.J. (1996). Lymphocyte homing and homeostasis. *Science*, **272**:60–66.

Chin, W. and Hay, J.B. (1980). A comparision of lymphocyte migration through intestinal lymph nodes, subcutaneous lymph nodes, and chronic inflammatory sites of sheep. *Gastroenterology*, **79**:1231–1242.

Dailey, M.O., Gallatin, W.M., Weissman, I.L. and Butcher, E.C. (1983). Surface phenotype and migration properties of activated lymphoctes and T cell clones. In "*Intercellular communication in leucoyte function*", edited by Parker, J.W. and O'Brien, R.L., pp. 641–644.

Dailey, M.O., Gallatin, W.M. and Weissman, I.L. (1985). The *in vivo* behavior of T cell clones: altered migration due to loss of the lymphocyte surface homing receptor. *J. Mol. Cell Immunol.*, **2**:27–36.

Damle, N.K., Klussman, K., Dietsch, M.T., Mohagheghpour, N. and Aruffo, A. (1992). GMP-140 (P-selectin/CD62) binds to chronically stimulated but not resting CD4$^+$ T lymphocytes and regulates their production of proinflammatory cytokines. *Eur. J. Immunol.*, **22**:1789–1793.

Dawson, J., Sedgwick, A.D., Edwards, J.C. and Lees, P. (1992). The monoclonal antibody MEL-14 can block lymphocyte migration into a site of chronic inflammation. *Eur. J. Immunol.*, **22**:1647–1650.

de Boer, O.J., Horst, E., Pals, S.T., Bos, J.D. and Das, P.K. (1994). Functional evidence that the HECA-452 antigen is involved in the adhesion of human neutrophils and lymphocytes to tumour necrosis factor-alpha-stimulated endothelial cells. *Immunology*, **81**:359–365.

Faveeuw, C., Gagnerault, M.C. and Lepault, F. (1994). Expression of homing and adhesion molecules in infiltrated islets of Langerhans and salivary glands of nonobese diabetic mice. *J. Immunol.*, **152**:5969–5978.

Fischer, C., Thiele, H-G. and Hamann, A. (1993). Lymphocyte-endothelial interactions in inflamed synovia: involvement of several adhesion molecules and integrin epitopes. *Scand. J. Immunol.*, **38**:158–166.

Gallatin, W.M., Weissmann, I.L. and Butcher, E.C. (1983). A cell-surface molecule involved in organ-specific homing of lymphocytes. *Nature*, **303**:30–34.

Gowans, J.L. and Knight, E.J. (1964). The route of recirculation of lymphocytes in the rat. *Proceed. Roy. Soc. London, B.*, **159**:257–282.

Graber, N., Gopal, T.V., Wilson, D., Beall, L.D., Polte, T. and Newman, W. (1990). T cells bind to cytokine-activated endothelial cells via a novel, inducible sialoglycoprotein and endothelial leukocyte adhesion molecule-1. *J. Immunol.*, **145**:819–830.

Guy–Grand, D., Griscelli, C. and Vassalli, P. (1974). The gut associated lymphoid system: nature and properties of the large dividing cells. *Eur. J. Immunol.*, **4**:435.

Hamann, A., Jablonski-Westrich, D., Scholz, K-U., Duijvestijn, A., Butcher, E.C. and Thiele, H-G. (1988). Regulation of lymphocyte homing. I. Alterations in homing receptor expression and organ-specific high endothelial venule binding of lymphocytes upon activation. *J. Immunol.*, **140**:737–743.

Hamann, A., Berlin, C., Bührer, C., Butcher, E.C., Jablonski-Westrich, D., Jonas, P. and Thiele, H-G. (1990). Homing mechanisms of lymphoma cells and activated lymphocytes. *J. Cancer Res. Clin. Oncol.*, **116 (Suppl. II)**:1210.

Hamann, A., Jablonski-Westrich, D., Jonas, P. and Thiele, H-G. (1991). Homing receptors reexamined: mouse LECAM-1 (MEL-14 antigen) is involved in lymphocyte migration into gut-associated lymphoid tissue. *Eur. J. Imm.*, **21**:2925–2929.

Hamann, A. and Rebstock, S. (1993). Migration of activated lymphocytes. In *Adhesion in leucocyte homing and differentiation*, edited by C.R.M.D. Dunon and B. Imhof, pp. 109–124. Berlin-Heidelberg-New York: Springer Verlag.

Hamann, A., Andrew, D.P., Jablonski-Westrich, D., Holzmann, B. and Butcher, E.C. (1994). Role of a4-Integrins in lymphocyte homing to mucosal tissues *in vivo*. *J. Immunol.*, **152**:3282–3293.

Hänninen, A., Taylor, C., Streeter, P.R., Stark, L.S., Sarte, J.M., Shizuru, J.A., Simell, O. and Michie, S.A. (1993). Vascular addressins are induced on islet vessels during insulitis in nonobese diabetic mice and are involved in lymphoid cell binding to islet endothelium. *J. Clin. Invest.*, **92**:2509–2515.

Heald, P.W., Yan, S.L., Edelson, R.L., Tigelaar, R. and Picker, L.J. (1993). Skin-selective lymphocyte homing mechanisms in the pathogenesis of leukemic cutaneous T-cell lymphoma. *J. Invest. Dermatol.*, **101**:222–226.

Johnston, G.I., Cook, R.G. and McEver, R.P. (1989). Cloning of GMP-140, a granula membrane protein of platelates and endothelium: sequence similarity to proteins involved in cell adhesion and inflammation. *Cell*, **56**.

Jutila, M.A., Bargatze, R.F., Kurk, S., Warnock, R.A., Ehsani, N., Watson, S.R. and Walcheck, B. (1994). Cell surface P- and E-selectin support shear-dependent rolling of bovine gamma/delta T cells. *J. Immunol.*, **153**:3917–3928.

Kraal, G., Weissman, I.L. and Butcher, E.C. (1988). Memory B cells express a phenotype consistent with migratory competence after secondary but not short-term primary immunization. *Cell Immunol.*, **115**:78–87.

Lasky, L.A., Singer, M.S., Yednock, T.A., Dowbenko, D., Fennie, C., Rodriguez, H., Nguyen, T., Stachel, S. and Rose, S.D. (1989). Cloning of a lymphocyte homing receptor reveals a lectin domain. *Cell*, **56**:1045–1055.

Lee, W.T. and Vitetta, E.S. (1991). The differential expression of homing and adhesion molecules on virgin and memory T cells in the mouse. *Cell Immunol.*, **132**:215–222.

Lepault, F., Gagnerault, M.C., Faveeuw, C. and Boitard, C. (1994). Recirculation, phenotype and functions of lymphocytes in mice treated with monoclonal antibody MEL-14. *Eur. J. Immunol.*, **24**:3106–3112.

Leung, D.Y.M., Gately, M., Trumble, A., Ferguson-Darnell, B., Schievert, P.M. and Picker, L.J. (1995). Bacterial superanigens induce T cell expreeion of the skin-selective homing receptor, the cutanous lymphocyte-associated antigen, via stimulation of interleukin 12 production. *J. Exp. Med.*, **181**:747–753.

Mackay, C.R., Marston, W.L. and Dudler, L. (1990). Naive and memory T cells show distinct pathways of lymphocyte recirculation. *J. Exp. Med.*, **171**:801–817.

Mackay, C.R., Marston, W. and Dudler, L. (1992a). Altered patterns of T cell migration through lymph nodes and skin following antigen challenge. *Eur. J. Immunol.*, **22**:2205–2210.

Mackay, C.R., Marston, W.L., Dudler, L., Spertini, O., Tedder, T.F. and Hein, W.R. (1992b). Tissue-specific migration pathways by phenotypically distinct subpopulations of memory T cells. *Eur. J. Immunol.*, **22**:887–895.

Mackay, C.R. (1993). Homing of naive, memory and effector lymphocytes. *Curr. Opin. Immunol.*, **5**:423–427.

Mebius, R.E. and Watson, S.R. (1993). L- and E-selectin can recognize the same naturally occurring ligands on high endothelial venules. *J. Immunol.*, **151**:3252–3260.

Michie, S.A., Streeter, P.R., Bolt, P.A., Butcher, E.C. and Picker, L.J. (1993). The human peripheral lymph node vascular addressin. An inducible endothelial antigen involved in lymphocyte homing. *Am. J. Pathol.*, **143**:1688–1698.

Mobley, J.L., Rigby, S.M. and Dailey, M.O. (1994). Regulation of adhesion molecule expression by CD8 T cells *in vivo*. II. Expression of L-selectin (CD62L) by memory cytolytic T cells responding to minor histocompatibility antigens. *J. Immunol.*, **153**:5443–5452.

Moore, K.L., Stults, N.L., Diaz, S., Smith, D.F., Cummings, R.D., Varki, A. and McEver, R.P. (1992). Identification of a specific glycoprotein ligand for P-selectin (CD62) on myeloid cells. *J. Cell Biol.*, **118**:445–456.

Noorduyn, L.A., Beljaards, R.C., Pals, S.T., van, H.P., Radaszkiewicz, T., Willemze, R. and Meijer, C.J. (1992). Differential expression of the HECA-452 antigen (cutaneous lymphocyte associated antigen, CLA) in cutaneous and non-cutaneous T-cell lymphomas. *Histopathology*, **21**:59–64.

Parrott, D.M.V. and Wilkinson, P.C. (1981). Lymphocyte locomotion and migration. *Prog. Allergy*, **28**:193–284.

Picker, L.J., Michie, S.A., Rott, L.S. and Butcher, E.C. (1990a). A unique phenotype of skin-associated lymphocytes in humans. Preferential expression of the HECA-452 epitope by benign and malignant T cells at cutaneous sites. *Am. J. Pathol.*, **136**:1053–1068.

Picker, L.J., Terstappen, L.W., Rott, L.S., Streeter, P.R., Stein, H. and Butcher, E.C. (1990b). Differential expression of homing-associated adhesion molecules by T cell subsets in man. *J. Immunol.*, **145**:3247–3255.

Picker, L.J., Kishimoto, T.K., Smith, C.W. and Butcher, E.C. (1991). ELAM-1 is an adhesion molecule for skin-homing T-cells. *Nature*, **349**:796–799.

Picker, L.J. and Butcher, E.C. (1992). Physiological and molecular mechanisms of lymphocyte homing. *Annu. Rev. Immunol.*, **10**:561–591.

Picker, L.J., Treer, J.R., Ferguson, D.B., Collins, P.A., Buck, D. and Terstappen, L.W. (1993a). Control of lymphocyte recirculation in man. I. Differential regulation of the peripheral lymph node homing receptor L-selectin on T cells during the virgin to memory cell transition. *J. Immunol.*, **150**:1105–1121.

Picker, L.J., Treer, J.R., Ferguson, D.B., Collins, P.A., Bergstresser, P.R., and Terstappen, L.W. (1993b). Control of lymphocyte recirculation in man. II. Differential regulation of the cutaneous lymphocyte-associated antigen, a tissue-selective homing receptor for skin-homing T cells. *J. Immunol.*, **150**:1122–1136.

Picker, L.J., Martin, R.J., Trumble, A., Newman, L.S., Collins, P.A., Bergstresser, P.R., and Leung, D.Y. (1994). Differential expression of lymphocyte homing receptors by human memory/effector T cells in pulmonary versus cutaneous immune effector sites. *Eur. J. Immunol.*, **24**:1269–1277.

Pinola, M., Renkonen, R., Majuri, M.L., Tiisala, S. and Saksela, E. (1994). Characterization of the E-selectin ligand on NK cells. *J. Immunol.*, **152**:3586–3594.

Postigo, A.A., Marazuela, M., Sanchez Madrid, F., and de Landazuri, M.O. (1994). B lymphocyte binding to E- and P-selectins is mediated through the de novo expression of carbohydrates on *in vitro* and *in vivo* activated human B cells. *J. Clin. Invest.*, **94**:1585–1596.

Rossiter, H., van-Reijsen, F., Mudde, G.C., Kalthoff, F., Bruijnzeel-Koomen, C.A., Picker, L.J. and Kupper, T.S. (1994). Skin disease-related T cells bind to endothelial selectins: expression of cutaneous lymphocyte antigen (CLA) predicts E-selectin but not P-selectin binding. *Eur. J. Immunol.*, **24**:205–210.

Santamaria-Babi, L.F., Moser, R., Soler, M.T.P., Picker, L.J., Blaser, K., and Hauser, C. (1995a). Migration of skin-homing T cells across cytokine-activated human endothelial cell layers involves interaction of the cutanous lymphocyte-associated antigene (CLA), the very late antigene-4 (VLA-4), and the lymphocyte associated antigen-1 (LFA-1). *J. Immunol.*, **154**:1543–1550.

Santamaria-Babi, L.F.S., Picker, L.J., Soler, M.T.P., Drzimalla, K., Flohr, P., Blaser, K. and Hauser, C. (1995b). Circulating allergen-reactive T cells from patients with atopic dermatitis express the skin-selective homing receptor, the cutaneous lymphocyte-associated antigen. *J. Exp. Med.*, **181**:1935–1940.

Shimizu, Y., Shaw, S., Graber, N., Gopal, T.V., Horgan, K.J., van Seventer, G.A. and Newman, W. (1991a). Activation-independent binding of human memory T cells to adhesion molecule ELAM-1. *Nature*, **349**:799–802.

Shimizu, Y., Newman, W., Gopal, T.V., Horgan, K.J., Graber, N., Beall, L.D., van Seventer, G.A. and Shaw, S. (1991b). Four molecular pathways of T cell adhesion to endothelial cells: roles of LFA-1, VCAM-1, and ELAM-1 and changes in pathway hierarchy under different activation conditions. *J. Cell Biol.*, **113**:1203–1212.

Silber, A., Newman, W., Reimann, E., Hendricks, E., Walsh, D. and Ringler, D.J. (1994a). Kinetic expression of endothelial adhesion molecules and relationship to leucocyte recruitment in two cutaneous models of inflammation. *Laboratory Invest.*, **70**:63–175.

Silber, A., Newman, W., Sasseville, V.G., Pauley, D., Beall, D., Walsh, D.G. and Ringler, D.J. (1994b). Recruitment of lymphocytes during cutaneous delayed hypersensitivity in nonhuman primates is dependent on E-selectin and vascular cell adhesion molecule 1. *J. Clin. Invest.*, **93**:1554–1563.

Simon, S. I., A. R. Burns, A. D. Taylor, P. K. Gopalan, E. B. Lynam, L. A. Sklar, and C. W. Smith. 1995. L-selectin (CD62L) cross-linking signals neutrophil adhesive functions via the Mac-1 (CD11b/CD18) beta 2-integrin. *J. Immunol.*, **155**:1502–14.

Smith, M.E., Martin, A.F. and Ford, W.L. (1980). Migration of lymphoblasts in the rat. *Monogr. Allergy*, **16**:203–232.

Smith, M.E. and Ford, W.L. (1983). The recirculating lymphocyte pool of the rat: a systematic description of the migratory behaviour of recirculating lymphocytes. *Immunology.*, **49**:83–94.

Spertini, O., Luscinskas, F.W., Kansas, G.S., Munro, J.M., Griffin, J.D., Gimbrone, M.J. and Tedder, T.F. (1991). Leukocyte adhesion molecule-1 (LAM-1, L-selectin) interacts with an inducible endothelial cell ligand to support leukocyte adhesion. *J. Immunol.*, **147**:2565–2573.

Streeter, P.R., Rouse, B.T. and Butcher, E.C. (1988). Immunohistologic and functional characterization of a vascular addressin involved in lymphocyte homing into peripheral lymph nodes. *J. Cell Biol.*, **107**:1853–1862.

Subramaniam, M., S. Saffaripour, S. R. Watson, T. N. Mayadas, R. O. Hynes, and D. D. Wagner. 1995. Reduced recruitment of inflammatory cells in a contact hypersensitivity response in P-selectin-deficient mice. *J. Exp. Med.*, **181**:2277–82.

Swain, S.L., Bradley, L.M., Croft, M., Tonkonogy, S., Atkins, G., Weinberg, A.D., Duncan, D.D., Hedrick, S.M., Dutton, R.W. and Huston, G. (1991). Helper T-cell subsets: phenotype, function and the role of lymphokines in regulating their development. *Immunol. Rev.*, **123**:115–144.

Tedder, T.F., Matsuyama, T., Rothstein, D., Schlossman, S.F. and Morimoto, C. (1990). Human antigen-specific memory T cells express the homing receptor (LAM-1) necessary for lymphocyte recirculation. *Eur. J. Immunol.*, **20**:1351–1355.

Turunen, J.P., Paavonen, T., Majuri, M.L., Tiisala, S., Mattila, P., Mennander, A., Gahmberg, C.G., Hayry, P., Tamatani, T., Miyasaka, M. and Renkonen, R. (1994). Sialyl Lewis(x)- and L-selectin-dependent site-specific lymphocyte extravasation into renal transplants during acute rejection. *Eur. J. Immunol.*, **24**:1130–1136.

Walcheck, B., Watts, G. and Jutila, M.A. (1993). Bovine gamma/delta T cells bind E-selectin via a novel glycoprotein receptor: first characterization of a lymphocyte/E-selectin interaction in an animal model. *J. Exp. Med.*, **178**:853–863.

Watson, S.R., Imai, Y., Fennie, C., Geoffroy, J.S., Rosen, S.D. and Lasky, L.A. (1990). A homing receptor-IgG chimera as a probe for adhesive ligands of lymph node high endothelial venules. *J. Cell Biol.*, **110**:2221–2229.

Westermann, J., Persin, S., Matyas, J., van-der-Meide, P. and Pabst, R. (1994). Migration of so-called naive and memory T lymphocytes from blood to lymph in the rat. The influence of IFN-gamma on the circulation pattern. *J. Immunol.*, **152**:1744–1750.

Yang, X.D., Karin, N., Tisch, R., Steinman, L. and McDevitt, H.O. (1993). Inhibition of insulitis and prevention of diabetes in nonobese diabetic mice by blocking L-selectin and very late antigen 4 adhesion receptors. *Proc. Natl. Acad. Sci. USA*, **90**:10494–10498.

Yang, X.D., Michie, S.A., Tisch, R., Karin, N., Steinman, L. and McDevitt, H.O. (1994). A predominant role of integrin alpha 4 in the spontaneous development of autoimmune diabetes in nonobese diabetic mice. *Proc. Natl. Acad. Sci. USA*, **91**:12604–12608.

6 The *in vivo* Role of the Selectins — Lessons from Leukocyte Adhesion Deficiency (LAD) II

A. Etzioni and J.M. Harlan

Dept. Pediatrics, Rambam Medical Center, B. Rappaport Medical School, Haifa, Israel and Div. Hematol. Univ. of Washington, Seattle, Washington, USA

The ability of leukocytes to leave the circulation and enter specific tissues is critical for the local control of infections and for the repair of injured tissue (Etzioni and Douglas, 1993). The process by which leukocytes migrate out of the circulation is a dynamic one and involves at least three distinct steps (Springer, 1994). The orchestration of these steps must be precisely regulated to ensure a rapid response to isolate and destroy the invading pathogen. Neutrophil interactions with vascular endothelial cells are of central importance in guiding the acute inflammatory response and are mediated by three adhesion molecule families: the integrins, the immunoglobulin superfamily, and the selectins (Carlos and Harlan, 1994; Etzioni, 1994). The initial interaction appears to be transient and of low affinity, resulting in the rolling of leukocyte along the blood vessel wall under conditions of flow. The rolling phase is prerequisite for the next phase in which the leukocyte becomes activated, resulting in the arrest and firm adhesion of the cell to the endothelium. Finally, the leukocyte tranverses the endothelium (diapedesis) and migrates through the subendothelial matrix to the site of infection. The initial rolling of cells depends primarily upon the interaction of the selectin family of proteins with their carbohydrate counter-structures (Varki, 1994). Firm adhesion and diapedesis are mediated by the interaction of immunoglobulin superfamily protein expressed on endothelial cells with integrin receptors expressed on leukocytes (Bevilacqua, 1993).

The use of specific monoclonal antibodies (mAbs) directed against the various adhesion molecules has undoubtedly contributed much to our understanding of the adhesion cascade. However, perhaps the most powerful and elegant approach to define the role of a specific adhesion molecule(s) is provided by genetic conditions in which one of the molecules is deficient or dysfunctional. Indeed, gene targeting techniques have been used to generate mutant mice that are deficient in various adhesion molecules, including E-, P- and L- selectins (Mayadas, 1993; Labow, 1995; Ley, 1995). Although these animal models are quite informative, they do not mimic exactly the human situation, and, therefore, those very rare congenital disorders of leukocyte adhesion deficiency are crucial for revealing the *in vivo* role of these molecules in man (Etzioni, 1994a). Two leukocyte adhesion deficiency (LAD) syndromes have been described. LAD I is caused by deficiency of the $\beta 2$ integrin receptor (Harlan, 1993; Anderson and Springer, 1987), while in LAD II the fucosylated ligand for the selectin family, Sialyl Lewis X (SLeX), is absent (Etzioni, 1992).

This review will focus on the current knowledge of LAD II and will compare it to LAD I.

CLINICAL PICTURE

LAD II was described in two unrelated boys, each the offspring of consanguineous parents. They were born after uneventful pregnancies with normal height and weight. No delay in the separation of the umbilical cord, as is reported in LAD I, was observed. Both children have severe mental retardation, short stature, a distinctive facial appearance (Figure 1) and the rare Bombay (hh) blood phenotype (Frydman *et al.*, 1992; Etzioni *et al.*, 1993a). It should be noticed that both children were born normally and no intrauterine retardation was observed. LAD II is transmitted in an autosomal recessive mode, and, recently, prenatal diagnosis was made by revealing the Bombay blood phenotype in a 21 week female fetus in one of the families (Frydman *et al.*, 1995 submitted). From early life both children have suffered from recurrent episodes of bacterial infection, mainly pneumonias proven by X-ray, periodontitis, otitis media, and localized cellulitis without obvious pus formation. These are non life-threatening events and have usually been treated in the outpatient clinic.

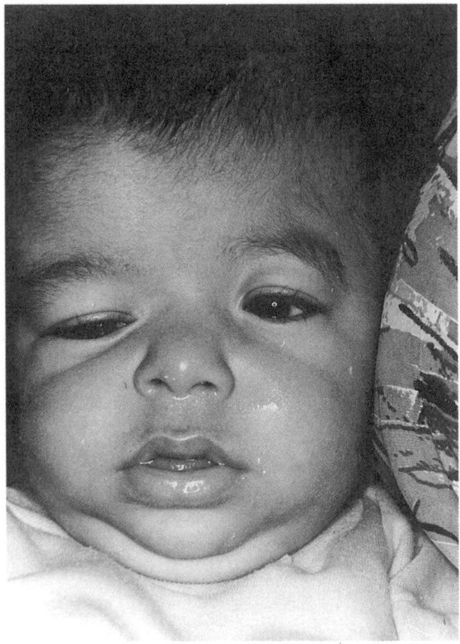

Figure 1 Typical facial appearance of a patient with LAD II. note.

Table 1 Neutrophil function in two patients with leukocyte adhesion deficiency type 2

Source of Neutrophils	*Chemoattractant or Activator**	*Patient 1*	*Patient 2*
		percent of control	
Chemiluminescence			
Normal subject	SOZ (P)	95 ± 26	109 ± 24
Patient	SOZ (N)	74 ± 13	91 ± 18
Patient	SOZ (P)	81 ± 17	89 ± 14
Patient	PMA	140 ± 23	108 ± 21
Chemotaxis			
Normal subject	ZAS (P)	109 ± 14	104 ± 13
Patient	ZAS (N)	9 ± 1	11 ± 3
Patient	ZAS (P)	17 ± 3	13 ± 1
Patient	Bacterial filtrate	18 ± 1.8	9 ± 4
Patient	None	14 ± 3	7 ± 2

*SOZ denotes serum-opsonized zymosan, PMA phorbol myristate acetate, and ZAS zymosan-activated serum. P indicates that patient serum was used, and N that serum from a normal subject was used.

In the last three years the frequence of infections has decreased, and they no longer receive prophylactic antibiotics. Currently, their main infectious problem is severe periodontitis as is also observed in patients with LAD I. From the first days of life extreme neutrophilia was noted. During times of infections the neutrophil count is as high as 150,000/ul and ranges between 20–35,000/ul even when free of any infections. The severity of the clinical symptoms is less than that observed in the classical form of LAD I, in which a high mortality rate is observed in early childhood. The infectious episodes in LAD II are more compatible with the moderate type of LAD I (Table 1). The molecular defect in LAD II is a generalized abnormality in fucose metabolism, resulting in the absence of the H antigen on the erythrocyte (Bombay phenotype), the Sialyl Lewis X (SLeX) on neutrophil, the Cutaneous Lymphocyte Antigen (CLA) on lymphocytes as well as other fucosylated structures. Binding of fucose specific lectins to the patients' B-lymphocytes transformed with Epstein-Bar virus was reduced, whereas the binding of mannose-specific lectins was normal, providing further evidence for a general fucose deficiency as the primary defect (Etzioni, 1995).

NEUTROPHIL STUDIES

The clinical picture (skin infections without evidence for pus formation, pneumonia and periodontitis) associated with very high neutrophil count were suggestive of a neutrophil defect as the cause of this syndrome. It was found that although the opsonophagocytic activity of the patients' neutrophil was normal, there was a marked defect in their motility ability (Etzioni *et al.*, 1992). Both random migration as well as directed migration towards chemotactic factors were markedly decreased (10% of normal). The patient' serum did not have any inhibitory effect on neutrophil chemotaxis (Table 2). Furthermore the homotypic aggregation of their neutrophils was impaired (Etzioni *et al.*, 1994b). These initial studies identified the problem as a defect in neutrophil capacity to adhere to surfaces, and therefore, adhesion molecules were examined. Indeed, while patients' (and their parents') neutrophils exhibited normal levels of the $\beta2$ integrin subunits, LAD II

Table 2 Leukocyte adhesion deficiency syndromes

	LAD I	*LAD II*
Clinical Manifestation		
Recurrent severe infections	+++	+
Neutrophilia – basal	+	+++
– with infection	+++	+++
Gingivitis	++	++
Skin infection	++	+
Delay separation of the umbilical abnormalities	+++	–
Developmental abnormalities	–	+++
Laboratory Findings		
CD18 expression	↓↓↓	NL
SLeX expression	NL	Absent
Neutrophil motility	↓↓↓	↓↓↓
Neutrophil rolling	NL	↓↓↓
Neutrophil adherence	↓↓↓	↓
Opsonophagocytic activity	↓	NL
T + B cell function	↓	NL

neutrophils were found to be deficient in the expression of SLeX antigen (Figure 2) (Etzioni *et al.*, 1992).

Interestingly the poor migration toward chemoattractants in an under-agarose assay and the defective homotypic neutrophil adhesion cannot be readily explained by the bio-

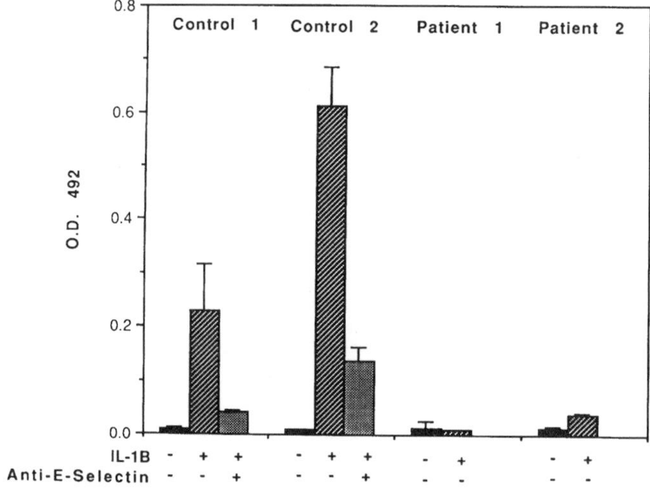

Figure 2 Expression of Sialyl-Lewis X and CD18 on Neutrophils from patients and control. (Reprinted by permission of *The New England Journal of Medicine*, Etzioni *et al.*, **327**, 1789–92, 1992. Copyright (1992) Massachusettes Medical Society).

chemical deficiency of SLeX. Whether these phenomena are directly or indirectly related to the SLeX defect is still under investigation.

Marked and persistent neutrophilia even in the absence of clinical infections was observed in the two patients. Interestingly chronic neutrophilia was also observed in the P-selectin deficient mice (Wagner, 1995). Marked neutrophilia is also observed in LAD I, although the levels of neutrophils are not necessarily grossly elevated in the absence of infection. In CD18-deficiency a prolonged intravascular neutrophil survival was observed (Davis *et al.*, 1991), which in part may be explained by the inability of the cells to migrate from the blood vessels to the site of infection. Suprisingly, in LAD II the neutrophil intravascular half disappearance time was much shorter than normal (3.2 hours versus 6–9 hours in control). The neutrophil turnover rate was markedly elevated with a calculated marrow neutrophil production rate of approximately eight times normal (Price *et al.*, 1994). The mechanism of increased marrow production with shortened neutrophil half-life time is unclear. It is possible that the underlying defect in fucose metabolism, and resulting abnormality in fucosylation of other glycoproteins and glycolipids alters the neutrophil surface membrane producing rapid clearance from the circulation.

In this regard other fucosylating antigens, such as LeX or CLA were found to be lacking on patients' cells. (unpublished observation).

In order to determine the significance of this defect we performed both *in vitro* as well as *in vivo* experiments. Neutrophils isolated from peripheral blood of LAD II patients were examined *in vitro* for their ability to adhere to human umbilical cord endothelial cells before and after the induction of the expression of E-selectin with interleukin-1β. As shown in Figure 3, the neutrophils from normal controls bound to endothelial cells after activation with interleukin-1β, and the binding was largely (80%) inhibited by a blocking monoclanal antibody to the E-selectin. In contrast, the patients' neutrophils were unable to bind to the interleukin-1β activated andothelial cells.

Rolling, the first step in neutrophil recruitment to sites of inflammation, is thought to be mediated in large part by the selectins and their ligand, SLeX. To determine the relative ability of LAD II neutrophils to roll, we studied the *in vivo* behavior of fluorescein-labeled neutrophils during their passage through inflamed venules in rabbit mesenteries, using the intravital microscopy (von Andrian *et al.*, 1991). The rolling fraction of control neutrophils in this assay is around 30%. Similar results were observed when LAD I neutrophils were studied. In contrast to control and LAD I cells, LAD II neutrophils rolled poorly. The rolling fraction was only 5% (Figure 4), and most of the cells that did interact with endothelial cells had a higher rolling velocity, rolled only over short distances, and frequently detached from the vessel wall (von Andrian *et al.*, 1993). This remarkable inability to react to an inflammatory stimulus appears to be dependent on the presence of intravascular shear force. In an additional work, the mesentric blood flow was stopped and LAD II cells were injected into the unperfused microvasculature. When blood flow was than restored, numerous LAD II cells had adhered firmly and were not detached by the flowing blood. These experiments indicate that the defect in neutrophil emigration in LAD II is in fact due to an inability of neutrophils to roll and slow down in inflamed venules under condition of shear forces, and not due to a dysfunction of later steps in the adhesion cascade process that are independent of shear forces. The ability of LAD II neutrophils to adhere and emigrate with reduced flow, may account for the milder phenotype of these patients. With marked inflammation there may be stasis, allowing selectin independent neutrophil emigration.

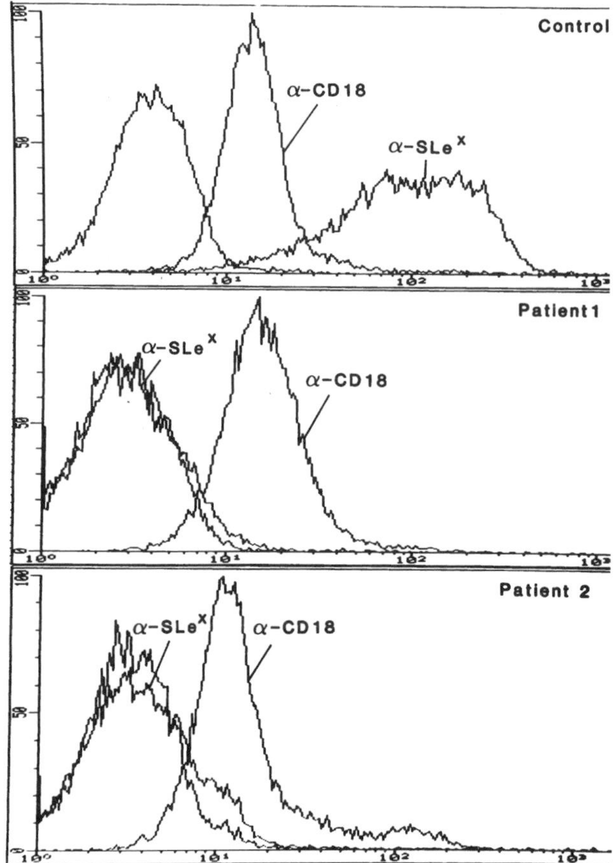

Figure 3 Binding of Neutrophils from the patients and controls to activated endothelial cells. (Reprinted by permission of *The New England Journal of Medicine*, Etzioni *et al.*, **327**, 1789–92, 1992. Copyright (1992) Massachusettes Medical Society).

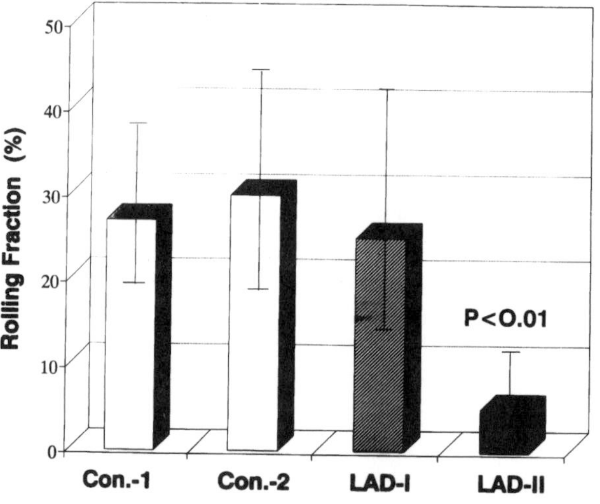

Figure 4 Neutrophil rolling from control and patients with LAD I and II, as measured by in intravital microscopy. (Adapted from JCI 1993 with permission).

We also examined the *in vivo* chemotaxis of patient neutrophils. The response to mild cutaneous inflammation was assessed by both skin chamber and skin window techniques. Neutrophil emigration was markedly diminished in both tests, the values being approximately 1.5% and 6% of normal in the skin window and skin chamber tests, respectively. (Price *et al.*, 1994). Monocyte migration to the skin window site was reduced to a similar degree as neutrophil.

Neutrophils from patients with LAD I showed the same magnitude of defect in these tests (Price *et al.*, 1994).

LYMPHOCYTE STUDIES

Adhesion molecules also participate in the immune response at the level of T lymphocyte function (Springer, 1990). Indeed several lymphocyte functions are defective in LAD I (Fisher *et al.*, 1988). In assessing the immune function in LAD II, a normal proportion and number of subpopulation of T lymphocytes were detected. Their proliferative response to various mitogens and the natural killing (NK) activity, as well as immunoglobulin levels were all within the normal range (Price *et al.*, 1994). Furthermore, antibody production using the bacteriophage 0X174, was found to be normal with normal switch from IgM to IgG. This finding is in contrast to patients with LAD I. Still, the patient failed to react to various antigens injected intradermally. This apperent anergy might have been due to the absence of the skin homing T cell antigen, CLA, which contains fucose in its structure (Picker, 1994). Recently we found out that although no delayed type hypersensitivity reaction occurs after an antigen challenge is induced in the patients' skin, accumulation of lymphocytes still exists at the site of the antigen injection (Kuijpers *et al.*, unpublished).

CONCLUSION

Leukocyte adhesion deficiency syndrome type II is a new and recently described syndrome that has provided new insights into the role of adhesion molecules in host defense mechanisms.

The general defect in fucose metabolism, the underlying abnormality in this syndrome, is responsible for the absence of various fucosylated molecules including SLeX, the ligand for the selectins. The association with short stature and mental retardation, is intriguing. It is not clear whether these abnormalities are related to defect in adhesion or to an unknown role of fucose in development.

It has recently been proposed that adhesive interactions between leukocytes and endothelial cells involve multiple sequential steps (Butcher, 1991; Springer, 1994). The importance of each of these molecules can best be provided by those rare experiments of nature in which a single molecular defect is found that can be directly related to the observed clinical picture.

The crucial role of the $\beta 2$ integrin receptor in neutrophil emigration was established when LAD I was described.

The newly described LAD II clarifies the role of the selectin receptors and their fucosylated carbohydrate ligands on leukocytes, SLeX. *In vitro* as well as *in vivo* studies of

LAD II confirm that selectins are essential for the first step in neutrophil emigration through the blood vessel, the rolling phase. Clinically, patients with LAD II suffer from an apparently less severe type of disease than LAD I patients with complete deficiency, and resemble the moderate phenotype of LAD I, where there is only partial deficiency. This may be due in part to the ability of neutrophils to emigrate in a selectin independent manner with reduced flow, and to the observed normal T and B lymphocyte function in LAD II, as opposed to LAD I.

References

Anderson, D.C. and Springer, T.A. (1987). Leukocyte adhesion deficiency: an inherited defect in Mac-1, LFA-1, and p150, 95 glycoprotien. *Annu. Rev. Med.*, **38**:175–192.

Bevilacqua, M.P. (1993). Endothelial-leukocyte adhesion molecules. *Ann. Rev. Immunol.*, **11**:767–804.

Butcher, E.C. (1991). Leukocyte-endothelial cell recognition: three (or more) steps to specificity and diversity. *Cell*, **67**:1033–1036.

Etzioni, A., Frydman, M., Pollack, S. *et al.* (1992). Severe recurrent infections due to a novel adhesion molecule defect. *N. Eng. J. Med.*, **327**:1789–1792.

Etzioni, A. and Douglas, S.D. (1993). *Microbial phagocytosis and killing in host defense.* In Pediatric Immunology. Edited by Spirer, Z., Roifman, C.M., pp. 17–27. Basel, Karger.

Etzioni, A., Harlan, J.M., Pollack, S., Phillips, L.M., Gershoni-Baruch, R. and Paulson, J.C. (1993a). Leukocyte adhesion deficiency (LAD) II: A new adhesion defect due to absence of sialyl lewis x, the ligand for selectins. *Immunodeficiency*, **4**:307–308.

Etzioni, A. (1994). Adhesion molecules in host defense *Clin. Diag. Lab. Immunol.*, **1**:1–4.

Etzioni, A. (1994a). Adhesion molecule deficiencies and their clinical significance. *Cell Adhes. Comm.*, **2**:257–260.

Etzioni, A., Gershoni-Baruch, R. and Harlan, J.M. (1994b). Sialyl lewis x and neutrophil aggregation. *Blood*, **83**:876–877.

Etzioni, A., Phillips, L.M., Paulson, J.C. and Harlan, J.M. (1995). Leukocyte adhesion deficiency (LAD) II. in *Cell adhesion and human disease*, edited by R.O. Hynes pp. 51–62. New York: J. Wiley and sons.

Fisher, A., Lisowska-Grospierre, B., Anderson, D.C. and Springer, T.A. (1988). Leukocyte adhesion deficiency: molecular basis and functional consequences. *Immunodefic. Rev.*, **1**:39–54.

Frydman, M., Etzioni, A., Eidlitz-Markus, T. *et al.* (1992). Rambam-Hasharon syndrome of psychomotor retardation, short stature, defective neutrophil motility, and bombay phenotype. *Amer. J. Med. Genet.*, **44**:297–302.

Harlan, J.M. (1993). Leukocyte adhesion deficiency syndrome: insight into the molecular basis of leukocyte emigration. *Clin. Immunol. Immunopathol.*, **69**:S16–S24.

Labow, M.A., Norton, C.R., Rumberger, J.M., *et al.* (1994). Characterization of E-selectin deficienct mice: demonstration of overlapping function of the endothelial selectins. *Immunity*, **1**:709–720.

Ley, K., Bullard, D.C., Arbones, M.L., Bosse, R., Vestweber, D., Tedder, T.F. and Beaudet, A.L. (1995). Sequential contribution of L- and P-selectin to leukocyte rolling *in vivo*. *J. Exp. Med.*, **181**:669–675.

Mayadas, T.N., Johnson, R.C., Rayburn, H., Hynes, R.O. and Wagner, D.D. (1993). Leukocyte rolling and extravasation are severely compromised in P-selectin deficient mice. *Cell*, **74**:541–554.

Picker, L.J. (1994). Control of lymphocyte homing. *Curr. Opin. Immunol.*, **6**:394–406.

Phillips, M.L., Schwartz, B.R., Etzioni, A. *et al.* (1995). Neutrophil adhesion in leukocyte adhesion deficiency syndrome type 2 (LAD II). *J. Clin. Invest.*, **96**:2898–2906.

Price, T.H., Harlan, J.M., Ochs, H.D., Gershoni-Baruch, R. and Etzioni, A. (1994). *In vivo* neutrophil and lymphocyte function studies in a patient with leukocyte adhesion deficiency type II. *Blood*, **84**:1635–1639.

Springer, T.A. (1990). Adhesion receptors of the immune system *Nature* (Lond.), **346**:425–434.

Springer, T.A. (1994). Traffic signals for lymphocyte recirculation and leukocyte emigration: the multiple paradigm. *Cell*, **76**:301–314.

Varki, A. (1994). Selectin ligands. *Proc. Natl. Acad. Sci. USA*, **91**:7390–7397.

von Andrian, U.H., Chambers, J.D., McEvoy, L.M., Bargatze, R.E.F., Arfors, K.E. and Buther, E.C. (1991). Two step model of leukocyte endothelial cell interaction in inflammation: distinct roles for LECAM-1 and the leukocyte b2 integrins *in vivo*. *Proc. Natl. Acad. Sci. USA*, **88**:7538–7542.

von Andrian, U.H., Berger, E.M., Ramezani, L., *et al.* (1993). *In vivo* behavior of neutrophils from two patients with distinct inherited leukocyte adhesion deficiency syndromes. *J. Clin. Invest.*, **91**:2893–2897.

Wagner, D.D. (1995). P-selectin knockout: a mouse model for various human diseases. In Cell adhesion and human disease. Edited by Marsh, J., Goode, J.A., pp. 2–16. England, John Wiley & Sons Ltd.

7 Selectins in Animal Models of Acute Lung Inflammatory Injury

Peter A. Ward

Professor and Chairman, Department of Pathology, The University of Michigan Medical School, M5240 Medical Science I, Box 0602, 1301 Catherine Road, Ann Arbor, Michigan 48109-0602, Tel: (313) 763-6384, Fax: (313) 763-4782

Selectins appear to play key roles in facilitating adhesive interactions between the activated endothelium and leukocytes (neutrophils, monocytes, lymphocytes), the result of which is to facilitate the ultimate development of firm adhesiveness and eventual transmigration of leukocytes beyond the vascular confines (reviewed, 1–6). Selectins appear to play important roles in recruitment of leukocytes to extravascular locales, although, because of the apparent low efficiency of selectin-mediated binding interactions, these adhesion-promoting molecules are not sufficient (in the absence of other adhesion-promoting molecules) to allow for leukocyte transmigration. Additional adhesion-promoting molecules such as the $\beta2$ integrins/intercellular cell adhesion molecule-1 (ICAM-1) and platelet endothelial cell adhesion molecule-1 (PECAM-1) are part of the sequence of engagement of adhesion-promoting molecules, the combination of which facilitates transmigration of leukocytes (6). Selectins have been shown to be important *in vivo* in responses related to ischemia-reperfusion events (7–10), in shock following (11) and in a variety of inflammatory reactions (12–20). The emphasis in this review will be the role of selectins in lung inflammatory responses.

DEFINING THE *in vivo* ROLES OF SELECTINS

While *in vitro* adhesion assays have provided important information on the nature of selectin-dependent adhesive interactions between leukocytes and endothelial cells, determining the role of selectins *in vivo* is quite another matter. As outlined in Table 1, several different approaches have been employed to define selectin requirements *in vivo* in a variety of inflammatory models. Perhaps the most direct strategies involve the use of *monoclonal antibodies*, but, as indicated below, their use also creates a series of problems that may affect interpretation of the results. For instance, most monoclonal antibodies to L-selectin cause neutropenia when infused *in vivo*. In order to circumvent this complication, we have used $F(ab')_2$ antibodies which do not affect the levels of blood neutrophils

Table 1 Approaches for definition of *in vivo* role of selectins

1. Monoclonal antibodies
2. Gene knockouts
3. Selectin-Ig chimeras
4. Oligosaccharides
5. Other approaches

(18). However, the half-life of $F(ab')_2$ *in vivo* is very short (circa 15 min). Therefore, their use must be carefully timed to coincide with the early and critical period of time in which endothelial upregulation of adhesion molecules is rapidly occurring. A second approach to defining the role of selectin molecules is the use of *knockout* mice in which expression of a selectin has been prevented. Knockout mice involving each of the three selectins have been developed. In general, P- and E-selectin knockout mice appear not to have an obvious phenotype, although careful evaluation of P-selectin knockout mice has revealed some defects in the time-course for peritoneal recruitment of neutrophils and mononuclear cells. In addition and somewhat surprisingly, defects in the development of cutaneous delayed-type hypersensitivity reactions have been reported in these animals (21). E-selectin knockout mice appear to have neither a distinctive phenotype nor do studies to date indicate that there is a defect in leukocyte recruitment. Only when P-selectin is blocked by the use of an antibody in E-selectin knockout mice are there measurable defects in the inflammatory system (22). Mice with a knockout of L-selectin have splenomegaly, lymphadenopathy and leukocytosis, although the status of their inflammatory response has not yet been determined (23). In general, in contrast to the clear-cut blocking effects of antibodies to selectins, knockout mice seem to have developed an adaptive response which diminishes the effects of the non-expressed protein. The third strategy for defining the role of selectins involves the use of *selectin-Ig chimeric proteins* which can be infused intravenously, resulting in a competitive type of inhibition, since these proteins compete with naturally occurring selectins for binding to their “counter-receptors”. The drawbacks in the use of selectin-Ig proteins are their availability and their relatively low binding affinity. A fourth approach for defining the requirements *in vivo* for selectin engagement is the use of *sialylated and fucosylated oligosaccharides* that represent the sialyl Lewis^x structure. While these oligosaccharides are effective *in vivo* (15, 16, 24), their use does not allow definition of the selectin requirement, since these oligosaccharides bind to varying degrees with all selectins. Finally, other strategies that can be employed *in vivo* involve the use of *selectin-binding agents*, especially the sulfated compounds such as fucoidan and sulfatides. Again, results obtained by the use of compounds are constrained in their interpretation in that these compounds generally interact with all selectins.

SELECTIN REQUIREMENTS IN LUNG INJURY FOLLOWING SYSTEMIC ACTIVATION OF COMPLEMENT

Intravascular infusion into rats of purified cobra venom factor (CVF) leads to acute lung injury that is complement and neutrophil-dependent (27). Both neutrophils as well as the pulmonary vascular endothelium appear to become activated in a manner that results in

Table 2 Role of selectins in lung injury following systemic activation of complement

		*Protection %**					
*Intervention***		*Permeability*		*Hemorrhage*		*MPO*	
Anti-P-selectin	(200 μg)	51	(<.001)	70	(<.001)	41	(<.01)
P-selectin-Ig	(1.0 mg)	52	(<.001)	53	(<.001)	4	(<.001)
E-selectin-Ig	(1.0 mg)	<5		<5		6	(N.S.)
Anti-L-selectin	(200 μg)	51	(<.001)	51	(<.001)	67	(<.001)
L-selectin-Ig	(1.0 mg)	64	(<.001)	68	(<.001)	56	(<.001)

*Protection is defined as reduction (%) in permeability, hemorrhage and MPO content in lung.
**All interventions were given intravenously at time 0. For anti-P-selectin, intact IgG, was used. For anti-L-selectin, F(ab′)2 antibody was infused intravenously at time 0. In this and in subsequent tables, numbers in parenthesis represent p values compared to positive reference controls. Data are from references 18–20, 28.

rapid adhesiveness of leukocytes to the pulmonary vascular endothelium. Generation of toxic oxidants by neutrophils results in focal acute endothelial injury and necrosis, leading to increased vascular permeability and hemorrhage (24). The oxidants appear to be derived from NADPH oxidase present on activated neutrophils. Selectin requirements have been defined according to the interventions described in Table 2. Antibody to P-selectin reduced permeability and hemorrhage parameters by 51–70% and lung neutrophil content (as defined by tissue myeloperoxidase, MPO) by 41% (20). Similar data were obtained following the intravenous infusion of P-selectin-Ig (Table 2) (19). E-selectin-Ig had no significant protective effects, while $F(ab')_2$ anti-L-selectin or L-selectin-Ig reduced parameters of injury and lung content of MPO by 51–68% (18, 19). Accordingly, this model of lung injury appears to be dependent on P- and L-selectin but independent of E-selectin. Immunohistochemical analysis of lung tissue revealed lung capillary and venular staining for P-selectin, which peaked by 15 min after infusion of CVF (20). On the basis of these data, it would appear that neutrophil L-selectin and endothelial P-selectin are the sources for the selectin requirements. In addition to selectins, CD11a/CD18 (LFA-1) and CD11b/CD18 (Mac-1) are required for the full development of injury in this model (25). It is possible that they interact with constitutively expressed ICAM-1 to enhance adhesive interactions since blocking of ICAM-1 by $F(ab')_2$ antibody is also protective (28). It is also possible that *in vivo* activation and oxidant production by intravascular neutrophils requires LFA-1 and Mac-1.

SELECTIN REQUIREMENTS IN LUNG INJURY FOLLOWING INTRA-PULMONARY DEPOSITION OF IMMUNE COMPLEXES

Intrapulmonary deposition of IgG immune complexes have been achieved by airway instillation into rats of rabbit IgG polyclonal antibody to bovine serum albumin (BSA) followed by the intravenous infusion of BSA. This results in intraalveolar deposition of IgG immune complexes, complement activation and injury that is complement and neutrophil-dependent (27). Injury appears to be the result of a complex array of cytokines that are produced by and released from lung macrophages, as well as the release from neutrophils and macrophages of oxidants (derived from NADPH oxidase and inducible

Table 3 Role of selectins in lung injury following intrapulmonary deposition of immune complexes

Class of Immune Complex	*Intervention**	*Protection %*					
		Permeability		*Hemorrhage*		*MPO*	
IgG	Anti-P-selectin	<5	<5	<5			
	P-selectin-Ig	<5	<5	<5			
	Anti-E-selectin	52	(<.003)	94	(<.001)	47	(.02)
	E-selectin-Ig	33	(<.001)	27	(.03)	29	(.006)
	Anti-L-selectin	33	(<.001)	33	(<.001)	43	(<.001)
	L-selectin-Ig	17	(.003)	27	(.03)	29	(.006)
IgA	Anti-P-selectin	<5	<5	<5			
	Anti-E-selectin	<5	<5	<5			
	Anti-L-selectin	<5	<5	<5			

*Selectin-IgG doses were 1.0 mg given intravenously. For antibody interventions, 67 μg F(ab′)2 anti-L-selectin or anti-E-selectin were given intravenously at 2.5, 3.0 and 3.5 hr after induction of lung injury. For anti-P-selectin, 200 μg intact were infused intravenously at time 0. Data are from references 18, 19, 28–31, 33–35.

nitric oxide synthase, iNOS) (28). Selectin requirements in this model of injury are described in Table 3. Blockade of P-selectin, whether by antibody or by P-selectin-Ig, failed to demonstrate any protective effects or any interference in build-up of lung MPO (17). In contrast, a role for E-selectin was clearly shown (17). Blocking of E-selectin by antibody reduced permeability, hemorrhage and MPO parameters by 52%, 94% and 47%, respectively. Immunostaining showed lung capillary and venular staining for E-selectin detectable as early as 2 hr after initiation of reaction and peaking by 3–3.5 hr (17). While not as effective as blocking antibody, E-selectin-Ig nevertheless had significantly protective effects (Table 2) (19). With respect to L-selectin, antibody to this selectin reduced permeability, hemorrhage, and MPO parameters by 33–43%, while L-selectin-Ig reduced these parameters by 17–29% (18,19). Part of the reason for the less effective results with selectin-blocking agents in this model (when compared to the CVF model) may be the longer time-frame (4 hr) for this model and the need to use multiple intravenous infusions when $F(ab')_2$ antibodies (for E- and L-selectins) had to be employed. It is concluded that E- and L-selectins are important in this model of lung injury while P-selectin is not. The protective effects of selectin blockade appear to be attributed to interference with neutrophil recruitment. Neutrophil recruitment in this model vitally depends on TNFα and IL-1, whose main functions appear related to upregulation of lung vascular adhesion molecules (E-selectin and ICAM-1) (30). Recruitment of neutrophils in this model also demonstrates requirements for LFA-1 and Mac-1, as well as ICAM-1 (33).

IgA immune complex-induced lung injury in rats is triggered by airway instillation of IgA myeloma protein that is reactive with intravenously administered dinitrophenol (DNP) linked to carrier protein, BSA. The immune complexes require complement for their tissue damaging effects but the injury is independent of a role for neutrophils. Indeed, neutrophil depletion in this model does not provide protection (34). It would appear that IgA immune complexes together with their complement activation products directly stimulate lung macrophages to produce toxic oxidants (35). Not surprisingly, no selectin requirements have been found in this model of lung injury (Table 3), consistent with the notion that no recruitment of blood leukocytes occurs in events leading to injury (31). This model is remarkable in that, although TNFα and IL-1, play no delectable role, monocyte chemotactic protein-1 (MCP-1) is required in events leading to injury, perhaps because of its role as an autocrine activator of lung macrophages (36).

SELECTIN REQUIREMENTS IN TISSUE INJURY AFTER ISCHEMIA-REPERFUSION EVENTS

Rats were subjected to surgical interruption of blood flow to the lower extremities for 4 hrs, followed by reperfusion (for 4 min). During reperfusion, complement activation occurs (as indicated by a progressive fall in plasma (CH50)), together with the appearance of low levels of TNFα, IL-1 and IL-6, and evidence of neutrophil stimulation (as reflected by increased CD11b/CD18 content of blood neutrophils) (37). Injury of the microvasculature in skeletal muscle of the lower extremities and the lung is both complement and neutrophil-dependent. In lung, the injury does not require P-selectin unless the duration of ischemia and reperfusion is limited to brief periods of time (1 hr ischemia and 1.5 hr reperfusion) (7). Blocking of either L- or E-selectin with antibody reduced lung vascular injury (permeability and hemorrhage) by 51–79%, along with a corresponding reduction in MPO. Lung injury in this model also required TNFα and IL-1, as well as LFA-1, Mac-1 and ICAM-1 (38).

When injury of crural skeletal muscle (which was the direct target of ischemia-reperfusion) was evaluated, a requirement for L-selectin but not for either P- or E-selectin was defined (Table 4). In animals treated with anti-L-selectin, permeability and hemorrhage were reduced 19% and 40%, respectively, while MPO was reduced by 41% (7). When the ischemia and reperfusion times were restricted to 1.0 and 1.5 hr, respectively, there was no reduction in permeability and hemorrhage, although MPO buildup was reduced 20%. These data suggest that neutrophil-mediated lung and skeletal muscle

Table 4 Role of selectins in lung injury after ischemia-reperfusion events

			Protection %		
Organ	*Intervention**	*Time of I/R***	*Permeability*	*Hemorrhage*	*MPO*
Lung	Anti-P-selectin	4/1	12 (N.S.)	<5	<5
	Anti-L-selectin	4/4	51 (<.005)	62 (<.005)	58 (<.005)
	Anti-E-selectin	4/4	55 (<.05)	79 (<.05)	73 (<.05)
	Anti-P-selectin	1/1.5	93 (<.05)	94 (<.05)	87 (<.005)
	Anti-P-selectin	4/4	<5	<5	20 (<.05)
Skeletal Muscle					
	Anti-P-selectin	4/4	<5	<5	<5
	Anti-L-selectin	4/4	19 (<.05)	40 (<.05)	41 (<.05)
	Anti-E-selectin	4/4	<1	13 (N.S.)	<5
	Anti-P-selectin	1/1.5	5	5	20 (<.05)

*Protocols similar to those described in Table 3.
**Duration (hr) of ischemia (I) and reperfusion (R). Data in this table are from references 10, 37–39.

Table 5 Role of selectins in tissue injury (4 hr) after thermal tauma to skin

Site	*Intervention**	*Permeability*		*Hemorrhage*		*MPO*
Skin	Anti-P-selectin	<5		<5		not done
	Anti-L-selectin	41	(<.001)	40	(<.001)	not done
	Anti-E-selectin	59	(<.003)	43	(<.001)	not done
Lung	Anti-P-selectin	30	(N.S.)	38	(.03)	38 (.048)
	Anti-L-selectin	50	(.001)	53	(<.001)	59 (.001)
	Anti-E-selectin	90	(<.001)	58	(.01)	66 (.007)

* Protocols similar to those described in Table 3. Data are contained in references 40 and 41.

injury after ischemia reperfusion events have differing selectin requirements, although with respect to the lung, the pattern of selectin requirement is similar to that found in neutrophil-mediated lung injury following intrapulmonary deposition of IgG immune complexes (Table 3). Accordingly, differences in selectin requirements may be related to organ differences, as well as to the nature of the initiating event in injury (e.g., systemic complement activation versus IgG deposition of IgG immune complexes (Tables 2, 3).

SELECTIN REQUIREMENTS AFTER THERMAL TRAUMA TO SKIN

We have previously shown that thermal trauma to the skin causes at 4 hr massive edema locally, together with some hemorrhage, as well as distal injury in the lung (39). Vascular injury in both areas are neutrophil- and complement-dependent. Using antibodies to define selectin requirements, local (skin) injury was found to be L- and E-selectin-dependent but revealed no requirement for P-selectin (Table 5) (40). In lung, evidence was found for the requirements for all three selectins, except for the effects of anti-P-selectin on permeability changes in which case the effects were not statistically significant. Blocking of any of the three selectins reduced lung injury parameters between 38–90% and reduced MPO content by 38–66%. Thus, as a general rule all selectins appear to participate in this model of neutrophil-dependent lung injury. In the skin, under the conditions employed, L- and E-selectin were required, but not P-selectin.

CONCLUSIONS

Defining the role of selectins in *in vivo* models of inflammatory injury is difficult, although use of blocking monoclonal antibodies appears to yield the most straightforward information. In the cobra venom factor model of lung injury, in which neutrophils (but not cytokines) appear to play a key role, P- and L-selectins (but not E-selectin) appear to be critical for injury. In the IgG immune complex model of lung injury, where TNFα and IL-1 seem to be crucial and linked to upregulation of both ICAM-1 and E-selectin, L- and E-selectins (but not P-selectin) are required. In IgA immune complex induced lung injury, in which residential macrophages seem to play key roles, no selectin requirements appear to exist. In ischemia-reperfusion injury of lung, P-selectin is required only if the times of ischemia and reperfusion are sharply limited. L- and E- selectins appear to play key roles, in keeping with the requirement for recruitment of neutrophils. In ischemia-reperfusion events involving lower extremities of rats, in which case injury is neutrophil-dependent, the only selectin requirement involves L-selectin. Finally, in thermal injury to skin, both L- and E-selectins are required, but not P-selectin. Lung injury that occurs as a secondary event appears to involve all three selectins. In general, neutrophil-dependent inflammatory injury consistently requires L-selectin and usually E- or P-selectin. Injury involving the predominant role of tissue macrophages exhibits no selectin requirements. These data may be relevant to the use of selectin blockade in human inflammatory conditions.

References

1. Butcher, E.C. (1991). Leukocyte-endothelial recognition: Three (or more) steps to specificity and diversity. *Cell*, **67**:1033–1037.
2. Lasky, L.A. (1992). Selectins: Interpreters of cell-specific carbohydrate information during inflammation. *Science*, **258**:964–969.
3. Paulson, J.C. (1991). Selectin/carbohydrate-mediated adhesion of leukocytes. In Adhesion: Its Role in Inflammatory Disease, edited by J.M. Harlan and D.Y. Lui, pp. 19–42. New York: W.H. Freeman and Co.
4. Rosen, S.D. (1990). The LEC-CAMs: An emerging family of cell-cell receptors based on carbohydrate recognition. *Am. J. Resp. Cell Mol. Biol.*, **3**:397–402.
5. Bevilacqua, M.P. and Nelson, R.M. (1993). Selectins *J. Clin. Invest.*, **91**:379–387.
6. Vaporciyan, A.A., DeLisser, H.M., Yan, H., Mendigurer, I.I., Thom, S.R., Jones, M.L., Ward, P.A. and Albelda, S.M. (1993). Involvement of platelet-endothelial cell adhesion molecule-1 in neutrophil recruitment *in vivo*. *Science*, **262**:1580–1582.
7. Stoolman, L.M. (1992). Selectins (LEC-CAMs): lectin-like receptors involved in lymphocyte recirculation and leukocyte recruitment. In *Cell Surface Carbohydrates and Cell Development*. Fukoda, M., ed., CRC Press, 79–98.
8. Weyrich, A.S., Xin-Iiang, M., Lefer, D.J., Albertine, K.H. and Lefer, A.M. (1993). In vivo neutralization of P-selectin protects feline heart and endothelium in myocardial ischemia and reperfusion injury. *J. Clin. Invest.*, **91**:2620–2629.
9. Winn, R.K., Liggitt, D., Vedder, N.B., Paulson, J.C. and Harlan, J.M. (1993). Anti-P-selectin monoclonal antibody attenuates reperfusion injury to the rabbit ear. *J. Clin. Invest.*, **92**:2042–2047.
10. Davenpeck, K.L., Gauthier, T.W., Albertine, K.H. and Lefer, A.M. (1994). Role of P-selectin on microvasculature leucocyte-endothelial interaction in splanchnic ischemia-reperfusion. *Am. J. Physiol.* In Press.
11. Skurk, C., Buerke, M., Guo, J.P., Paulson, J. and Lefer, A.M. (1994). Sialyl Lewis^x-containing oliogosaccharide exerts beneficial effects in murine traumatic shock. *American Journal of Physiology*, **267**:H2124–31.
12. Watson, S.R., Fennie, C. and Lasky, L.A. (1991). Neutrophil influx into an inflammatory site inhibited by a soluble homing receptor-IgG chimera. *Nature*, **349**:164–167.
13. Jutila, M.A., Rott, L., Berg, E.L. and Butcher, E.C. (1989). Function and regulation of the neutrophil MEL-14 antigen *in vivo*: comparison with LFA-1 and MAC-1. *J. Immunol.*, **143**:3318–3324.
14. Ulich, T.R., Howard, S.C., Remick, D.G., Yi, E.S., Collins, T., Guo, K., Yin, S., Keene, J.L., Schmuke, J.J., Steininger, C.N., Welply, J.K. and Williams, J.H. (1994). Intracheal administration of endotoxin and cytokines: VIII. LPS induces E-selectin expression; anti-E-selectin and soluble E-selectin inhibit acute inflammation. *Inflammation*, **18**:4.
15. Mulligan, M.S., Paulson, J.C., DeFrees, S., Zheng, Z.L., Lowe, J.B. and Ward, P.A. (1993). Protective effects of oligosaccharides in P-selectin-dependent lung injury. *Nature*, **364**:149–151.
16. Mulligan, M.S., Lowe, J.B., Larson, R.D., Paulson, J., Zheng, Z.L., DeFrees, S., Maemura, K., Fukuda, M. and Ward, P.A. (1993). Protective effects of sialylated oligosaccharides in immune complex-induced acute lung injury. *J. Exp. Med.*, **178**:623–631.
17. Mulligan, M.S., Varani, J., Dame, M.K., Lane, C.L., Smith, C.W., Anderson, D.C. and Ward, P.A. (1991). Role of endothelial-leukocyte adhesion molecule 1 (ELAM-1) in neutrophil-mediated lung injury in rats. *J. Clin. Invest.*, **88**:1396–1406.
18. Mulligan, M.S., Miyasaka, M., Tamatani, T., Jones, M.L. and Ward, P.A. (1994). Requirements for L-selectin in neutrophil-mediated lung injury in rats. *J. Immunol.*, **152**:832–840.

19. Mulligan, M.S., Watson, S.R., Fennie, C. and Ward, P.A. (1993). Protective effects of selectin chimeras in neutrophil-mediated lung injury. *J. Immunol.*, **151**:6410–6417.
20. Mulligan, M.S., Polley, M.J., Bayer, R.J., Nunn, M.F., Paulson, J.C. and Ward, P.A. (1992). Neutrophil-dependent acute lung injury. Requirement for P-selectin (GMP-140). *J. Clin. Invest.*, **90**:1600–1607.
21. Mayadas, T.N., Johnson, R.C., Rayburn, H., Hynes, R.O. and Wagner, D.D. (1993). Leukocyte rolling and extravasation are severely compromised in P-selectin-deficient mice. *Cell*, **74**:541–54.
22. Lebow, M.A., Norton, C.R., Rumberger, J.M., Lombard-Gillooly, K.M., Shuster, D.J., Hubbard, J., Bertko, R., Knaack, P.A., Terry, R.W., Harbison, M.L., Kontgen, F., Stewart, C. L., McIntyre, K.W., Will, P.C., Burns, D.K. and Wolltky, B.A. (1994). Characterization of E-selectin-deficient mice: Demonstration of overlapping function of the endothelial selectins. *Immunity*, **1**:709–720.
23. Arbones, M.L., Ord, D.C., Ley, K., Ratech, H., Maynard-Curry, C., Otten, G., Capon, D.J. and Tedder, T.F. (1994). Lymphocyte homing and leukocyte rolling and migration are impaired in L-selectin-deficient mice. *Immunity*, **1**:247–260.
24. Buerke, M., Weyrich, A.S., Zheng, Z., Gaeta, F.C.A., Forrest, M.J. and Lefer, A.M. (1994). Sialyl Lewis^x containing oligosaccharide attenuates myocardial reperfusion injury in cats. *J. Clin. Invest.*, **93**:1140–1148.
25. Mulligan, M.S., Varani, J., Warren, J.S., Till, G.O. Smith, C.W., Anderson, D.C., Todd III, R.F. and Ward, P.A. (1992). Roles of $\beta 2$ integrins of rat neutrophils in complement-and oxygen radical-mediated acute inflammatory injury. *J. Immunol.*, **148**:1847.
26. Shappell, S.B., Toman, C., Anderson, D.C., Taylor, A.A., Entman, M.L. and Smith, C.W. (1990). Mac-1 (CD11b/CD18) mediates adherence-dependent hydrogen peroxide production by human and canine neutrophils. *J. Immunol.*, **144**:2702.
27. Johnson, K.J. and Ward, P.A. (1974). Acute immunologic pulmonary alveolitis. *J. Clin. Invest.*, **54**:349–357.
28. Mulligan, M.S., Hevel, J.M., Marletta, M.A. and Ward, P.A. 1991. Tissue injury caused by deposition of immune complexes is L-arginine dependent. *Proc. Natl. Acad. Sci. USA*, **88**:6338–6342.
29. Warren, J.S., Yabroff, K.R., Remick, D.G., Kunkel, S.L., Chensue, S.W., Kunkel, R.G., Johnson, K.J. and Ward, P.A. (1989). Tumor necrosis factor participates in the pathogenesis of acute immune complex alveolitis in the rat *J. Clin. Invest.*, **84**:1873.
30. Warren, J.S. (1991). Intrapulmonary interleukin-1 mediates acute immune complex alveolitis in the rat. *Biochem. Biophys. Res. Commun.*, **175**:604.
31. Mulligan, M.S. and Ward, P.A. (1992). Immune complex-induced lung and dermal vascular injury:differing requirements for tumor necrosis factor-a and IL-1. *J. Immunol.*, **149**:331.
32. Mulligan, M.S., Vaporciyan, A.A., Miyasaka, M., Tamatani, T. and Ward, P.A. (1993). Tumor necrosis factor a regualtes *in vivo* intrapulmonary expression of ICAM-1. *Am J. Pathol.*, **142**:1739.
33. Mulligan, M.S., Wilson, G.P., Todd, R.F., Smith, C.W., Anderson, D.C., Varani, J., Issekutz, T., Miyasaka, M., Tamatani, T., Rusche, J.R., Vaporciyan, A.A. and Ward, P.A. (1993). Role of $\beta 1$, $\beta 2$ integrins and ICAM-1 in lung injury following deposition of IgG and IgA immune complexes. *J. Immunol.*, **150**:2407.
34. Johnson, K.J., Wilson, B.S., Till, G.O. and Ward, P.A. (1984). Acute lung injury in rat caused by immunoglobulin A immune complexes. *J. Clin. Invest.*, **74**:358–369.
35. Mulligan, M.S., Warren, J.S. , Smith, C.W., Anderson, D.C., Yeh, C.G., Rudolph, A.R. and Ward, P.A. (1992). Lung injury after deposition of IgA immune complexes: Requirements for CD18 and L-arginine *J. Immunol.*, **148**:3086–3092.
36. Jones, M.L., Mulligan, M.S., Flory, C.M., Ward, P.A. and Warren, J.S. (1992). Potential role of monocyte chemoattractant protein 1/JE in monocyte/macrophage-dependent IgA immune complex alveolitis in the rat. *J. Immunol.*, **149**:2147–2154.
37. Seekamp, A., Warren, J.S., Remick, D.G., Till, G.O. and Ward, P.A. (1993). Requirements for tumor necrosis factor-a and interleukin-1 in limb ischemia/reperfusion injury and associated lung injury. *Amer. J. Pathol.*, **143**:453–463.
38. Seekamp, A., Mulligan, M.S., Till, G.O., Smith, C.W., Miyasaka, M., Tamatani, T., Todd III, R.F. and Ward, P.A. (1993). Role of $\beta 2$-integrins and ICAM-1 in lung injury following ischemia-reperfusion of rat hind limbs. *Am. J. Pathol.*, **143**:464–472.
39. Mulligan, M.S., Till, G.O., Smith, C.W., Anderson, D.C., Miyasaka, M., Tamatani, T., Todd, R.F., III, Issekutz, J.C. and Ward, P.A. (1994). Role of leukocyte adhesion molecules in lung and dermal vascular injury following thermal trauma of skin. *Amer. J. Pathol.*, **144**:1008–1015.
40. Mulligan, M.S., Till, G.O., Smith, C.W., Anderson, D.C., Miyasaka, M., Tamatani, T., Todd III, R.F., Issekutz, T.B. and Ward, P.A. (1994). Role of leukocyte adhesion molecules in lung and dermal vascular injury after thermal trauma of skin. *American Journal Pathology*, **144**:1008–1015.

8 Analysis of Selectin Deficient Mice

Daniel C. Bullard[1] and Arthur L. Beaudet[1,2]

[1]*Department of Molecular and Human Genetics, Baylor College of Medicine and*
[2]*Howard Hughes Medical Institute, Houston, TX 77030*
Correspondent: Daniel Bullard, Department of Molecular and Human Genetics, Baylor College of Medicine, One Baylor Plaza, Houston, TX 77030

Other chapters in this book describe various aspects of the biology of selectins in great detail, and other reviews of the selectins and related leukocyte and endothelial cell adhesion molecules are available (Lasky, 1992; Springer, 1994; Tedder *et al.*, 1995a). Any discussion of the function of selectins must recognize the importance of the corresponding ligand molecules many of which have mucin-like properties and provide a carbohydrate binding ligand for the selectins. The selectin ligand molecules are generally less well studied than the selectins themselves, but some reviews include discussions of the selectin ligands (Springer, 1994; McEver *et al.*, 1995). There are now extensive data indicating the importance of selectins in leukocyte rolling and emigration. Much of the available information is based on the use of monoclonal antibodies directed against the selectins, particularly antibodies which block adhesion. An alternative has been to use a genetic strategy to further analyze the biology and function of the selectins focusing particularly on gene targeting in the mouse.

L-selectin, P-selectin, and E-selectin map together in a gene cluster on chromosome 1 in the mouse and human (Watson *et al.*, 1990). The fact that these genes are found within a cluster has numerous implications for genetic analysis. In any linkage analysis, the cluster will be inherited as a single unit. Large deletions or other mutations could affect more than one selectin gene, and it will not be possible in the mouse to combine independent selectin mutations through a breeding strategy due to lack of opportunity for crossover between the genes. There are no reported spontaneously occurring deleterious mutations in the selectin genes in humans or other species. A rare form of leukocyte adhesion deficiency is reported in association with a defect in fucose metabolism which is thought to affect the structure of selectin ligand molecules (Etzioni *et al.*, 1992; Frydman *et al.*, 1992; Etzioni, 1994). Amino acid and nucleotide polymorphisms are reported in the human selectin genes (Vora *et al.*, 1994; Wenzel *et al.*, 1994), and it will be of interest to determine if these polymorphisms are associated with any disease susceptibility or variation in inflammatory response.

Introduction of mutations into the germline of the mouse using homologous recombination in embryonic stem (ES) cells is a well established methodology (Capecchi, 1989;

Frohman and Martin, 1989; Sedivy and Joyner, 1992). The most common strategy is to introduce loss of function mutations frequently referred to as "knockout" mutations. These mutations are first introduced into chimeric mice which are bred to obtain animals carrying the mutation in the germline in heterozygous form. These heterozygous knockout animals most often have a normal phenotype. Heterozygote animals are then bred to obtain homozygous mutant animals. In many experiments, it would be expected that absence of gene function due to a homozygous mutation or absence of gene function due to blocking with a monoclonal antibody would have similar or identical effects. In other instances, it might be anticipated that there would be different biological effects, either because the mutant animal develops some compensatory response to the mutation or because a blocking monoclonal antibody does not completely block function or has some effect in addition to the disabling of the target molecule. In many instances blocking monoclonal antibodies were available prior to the development of mutant mice. The mutant mice have a considerable advantage for analysis of the role of selectin molecules in chronic experiments or disease processes. The use of blocking monoclonal antibodies and the availability of mutant mice provide complementary approaches for the study of the biology of selectins.

P-SELECTIN MUTANT MICE

Two different lines of mice with mutations in P-selectin have been reported; in both cases, expression of the normal protein was completely abolished (Mayadas *et al.*, 1993; Bullard *et al.*, 1995b). Homozygous mutant mice were viable, fertile, and without obvious disease phenotype. Examination of peripheral blood leukocyte counts revealed a modest but significant increase in neutrophil numbers when compared to wild type mice of similar genetic background. Cytofluorimetric evaluation of surface expression on leukocytes in blood, lymph nodes, spleen and thymus using antibodies to CD4, CD8, CD3, Mac-1, IgM, Thy 1.2, and natural killer cells did not reveal any significant differences, except for an increased number of Mac-1 positive cells in peripheral blood attributed to the increased circulating neutrophil count (Johnson *et al.*, 1995).

Johnson *et al.* (1995) attempted to determine the mechanism for increased circulating counts in peripheral blood of P-selectin deficient mice. Analyses of erythroid and myeloid precursors from bone marrow of P-selectin mutant mice did not identify any significant differences when compared to marrow from wild type mice suggesting that the mechanism was not due to increased bone marrow production. Epinephrine injections were used to stimulate release of neutrophils from the marginated pool of mutant and wild type mice in order to define whether the increased number of circulating neutrophils was due to a reduction in the marginated pool. The absolute numbers of neutrophils released were found to be comparable using this technique. Finally, neutrophil clearance was investigated by injecting ^{51}chromium-labeled human neutrophils into mutant and wild type mice. The amount of radioactivity recovered from peripheral blood neutrophils at 40, 75, and 135 minutes after injection was significantly higher in mutant mice, suggesting that defective neutrophil clearance may be part of the mechanism leading to neutrophilia in P-selectin mutant mice.

P-selectin was previously demonstrated to mediate leukocyte rolling in artificial flow chambers; however, one of the first demonstrations of P-selectin mediated rolling *in vivo* was accomplished using P-selectin mutant mice (Mayadas *et al.*, 1993). Leukocyte rolling in mesenteric and cremaster venules was shown to be significantly reduced in P-selectin mutant mice when analyzed by intravital microscopy (Mayadas *et al.*, 1993; Ley *et al.*, 1995). Ley *et al.* (1995) showed that initial rolling was completely absent in mutant mice up to 40 minutes after exteriorization of the cremaster muscle, although significant but reduced rolling was detected at later time points up to 2 hours. Johnson *et al.* (1995) found a severe reduction in rolling 4 hours after induction of thioglycolate peritonitis, although the number of adherent cells was normal.

Thioglycolate peritonitis and *Streptococcus pneumoniae*-induced peritonitis have been used to determine the effects of the P-selectin deficiency on leukocyte emigration. In both peritonitis models, significant inhibition of neutrophil emigration into the peritoneal cavity was observed at 4 hours after injection (Mayadas *et al.*, 1993; Bullard *et al.*, 1995b). However, neutrophil emigration was not significantly inhibited in mutant mice at 24 or 48 hours when compared to wild type mice (Johnson *et al.*, 1995). In the *Streptococcus pneumoniae* model, edema formation was significantly reduced at 4, but not 24, hours following intraperitoneal instillation (Bullard *et al.*, 1995b; Bullard *et al.*, 1995a). Bacterial clearance in P-selectin mutant mice was not significantly different from wild type mice at 4 and 24 hours (Bullard *et al.*, 1995a). Johnson *et al.* (1995) reported a significant decrease in macrophage numbers at 48 hours following thioglycolate injection into the peritoneal cavity of P-selectin mutant mice. No inhibition of lymphocyte emigration in either peritonitis model has been described in P-selectin mutant mice.

Two groups have studied oxazolone-induced delayed type hypersensitivity (DTH) in P-selectin mutant mice (Subramaniam *et al.*, 1995; Staite *et al.*, 1995). In this model, mice are first sensitized with oxazolone and then subsequently rechallenged (usually on a single ear) several days later. Edema formation in the challenged ear is determined by measuring ear thickness, ear biopsy weight, or accumulation of iodinated albumin in the ear. Both studies reported no differences in ear swelling or edema formation in P-selectin mutant mice when compared to wild type mice. However, contrasting results were obtained when the number and area of epidermal microabscesses were measured in sensitized ears of mutant animals. Subramanian *et al.* (1995) reported a reduction in both the number and area of focal neutrophil infiltrates in the epidermis of P-selectin mutant animals, while Staite *et al.* (1995) found a significant increase in both of these parameters in mutant mice when compared to wild type. This discrepancy may be due to differences in the methods used in each study. A reduction in infiltrating CD4 T cells and monocytes was also reported in P-selectin mutant animals as well as a reduction in mast cell degranulation (Subramaniam *et al.*, 1995).

P-selectin mutant mice offer an excellent resource to study the role of P-selectin expression on platelets, and several abnormalities in adhesion of platelets to endothelium or neutrophils have been identified. P-selectin was previously shown to mediate activated platelet/neutrophil interactions *in vitro* using blocking monoclonal antibodies against P-selectin (Larsen *et al.*, 1989; Hamburger and McEver, 1990). Platelets from P-selectin mutant mice activated with thrombin failed to rosette with neutrophils from either wild type or P-selectin mutant animals confirming that P-selectin mediates this interaction (Johnson *et al.*, 1995). Frenette *et al.* (1995) used P-selectin mutant mice to define a P-selectin-dependent component which mediates platelet rolling *in vivo*. Fluorescently labeled platelets were injected into mutant or wild type mice, and platelet/endothelial interactions were analyzed by intravital microscopy. Both untreated and activated platelets from mutant or wild type mice were used for these studies. A significant reduction in rolling was observed when platelets from wild type or mutant mice were injected into the tail vein of P-selectin mutant mice, independent of platelet activation. Platelet rolling flux was increased when mutant or wild type platelets were injected into wild type mice following stimulation of the endothelium with the ionophore A23187. Platelet rolling was reduced in P-selectin mutant mice with or without treatment with A23187. Other studies using platelets from P-selectin mutant animals have not yet revealed any obvious defects in platelet activation or aggregation (Mayadas *et al.*, 1993).

L-SELECTIN MUTANT MICE

A single line of L-selectin deficient mice has been reported to date with homozygous mutant animals completely lacking L-selectin (Arbonés *et al.*, 1994). Gross examination of various tissues from L-selectin deficient mice did not reveal any abnormalities except for a slight increase in spleen size and a reduction in size in some of the lymphoid organs. The inguinal, axillary, periaortic, and cervical lymph nodes were substantially smaller when compared to those from wild type animals. The average number of resident lymphocytes was reduced by 70%, and the number of distinct high endothelial venules (HEV) in lymph nodes was also considerably reduced. Peripheral blood neutrophil counts, unlike in P-selectin mutants, were not elevated in L-selectin mutants. Examination of leukocyte subpopulations in bone marrow, thymus, spleen, and other lymphoid organs did not reveal any significant differences in L-selectin mutant animals when compared to wild type.

Using the Stamper-Woodruff assay, attachment of splenocytes from L-selectin deficient animals to HEV was reduced by 95%–99%, confirming the previous observation in which antibodies against L-selectin inhibited lymphocyte adhesion to HEV (Arbonés *et al.*, 1994; Gallatin *et al.*, 1983). Consistent with this observation, a significant reduction of migration of fluorescently labeled splenocytes from L-selectin mutant mice into peripheral lymph nodes, mesenteric lymph nodes, and Peyer's patches was observed one hour following intravenous injection. A significant increase was observed in migration of these cells to the spleen. When analyzed 48 hours later, reductions in lymphocyte migration into the peripheral and mesenteric lymph nodes was once again observed when compared to injection of wild type cells. However, a significant increase in the numbers of mutant lymphocytes was found in the Peyer's patches and spleen.

Significant reductions in acute leukocyte rolling and peritoneal emigration in response to thioglycolate were also observed in L-selectin mutant mice (Arbonés *et al.*, 1994; Ley *et al.*, 1995). Intravital microscopic examination of mesenteric venules from L-selectin mutant animals revealed a significant reduction in the fraction of rolling leukocytes up to 2 hours following the surgical stimulus. In contrast to the phenotype observed in P-selectin mutant animals, initial rolling (0–20 minutes) was not significantly different from wild type mice. Thioglycolate-induced peritonitis using L-selectin mutants revealed a significant reduction in neutrophil emigration (36%–78%) up to 4 hours following intraperitoneal instillation (Arbonés *et al.*, 1994). Significant reductions in emigrated neutrophils (56%–62%) and lymphocytes (70%–75%) were also observed 24 and 48 hours after administration of thioglycolate (Tedder *et al.*, 1995b). Monocyte emigration was inhibited (72%–78%) at these time points (Tedder *et al.*, 1995b).

Oxazolone-induced DTH, sheep red blood cell (SRBC)-induced DTH, and LPS-induced toxic shock models were also used to evaluate the leukocytic response in L-selectin deficient mice (Tedder *et al.*, 1995b). Following oxazolone sensitization and challenge, ear swelling in L-selectin mutant animals was significantly reduced (69%) after 24 hours when compared to wild type mice. SRBC-induced DTH experiments in L-selectin mutant mice also revealed a significant reduction (69–89%) in foot pad swelling 24–72 hours following challenge. A comparison of the draining lymph node weights of SRBC challenged wild type and mutant mice revealed a significant change in weight of lymph nodes from wild type, but not from mutant mice. This suggests that the loss of L-selectin inhibited leukocyte proliferation or homing into the draining lymph nodes following stimulation. Intraperitoneal injection of LPS resulted in increased survival (60%) of L-selectin mutant mice when compared to non-mutant mice. This protection was interpreted to be due to reduced accumulation of leukocytes in tissues, which is a significant step in the pathophysiology of endotoxin shock.

E-SELECTIN MUTANT MICE

Two lines of E-selectin null mutant mice have been described, and unlike P- or L-selectin mutants, no obvious abnormalities of the inflammatory response have been observed (Labow *et al.*, 1994; Bullard *et al.*, 1996). Leukocyte numbers in peripheral blood were similar to that of wild type mice, and no defect in trauma induced leukocyte rolling in cremaster venules was observed (Bullard *et al.*, 1996). Both *Streptococcus pneumoniae* and thioglycolate-induced peritonitis studies failed to show any significant inhibition of neutrophil emigration at 2, 6, or 24 hours after injection when compared to wild type mice (Labow *et al.*, 1994; Bullard *et al.*, 1996). Oxazolone-induced DTH studies also did not reveal any significant differences in edema formation or ear swelling (Labow *et al.*, 1994). However, Staite *et al.* (1996) found a slight increase in the number of neutrophil containing microabscesses in the challenged ears of E-selectin mutant mice.

SELECTIN MUTANTS TREATED WITH SELECTIN MONOCLONAL ANTIBODIES

Initial studies indicated that the absence of a single selectin only partially affected leukocyte rolling and subsequent leukocyte emigration. In the case of E-selectin mutant mice, rolling and emigration were not inhibited. In order to examine possible redundancies of function among the selectins, several studies have used both selectin mutations and anti-selectin monoclonal antibodies in various inflammatory models. Labow *et al.* (1994) found that the addition of a P-selectin antibody to E-selectin mutant mice significantly inhibited neutrophil emigration into the peritoneum 6 hours following thioglycollate instillation, whereas the mutation or antibody treatment alone did not show any effect. This same strategy was used in the oxazolone DTH model and resulted in a significant decrease in edema formation when a P-selectin antibody was used in E-selectin mutant mice (Labow *et al.*, 1994). These findings suggest that E-selectin and P-selectin share some overlapping functions such that the combined absence results in a significant impairment of the inflammatory response under certain stimulation conditions.

Several different combinations of selectin mutations and antibodies have been used to analyze the effects on leukocyte rolling. Ley *et al.* (1995) found that the addition of L-selectin antibodies to P-selectin mutant mice completely blocked residual rolling at >60 minutes following surgical stimulation in cremaster venules. Likewise, addition of a P-selectin antibody to L-selectin mutant mice significantly blocked early rolling (<60 minutes) and late residual rolling (>60 minutes). Thus it appears that surgically induced rolling in this model is mediated by both L-selectin and P-selectin. Pretreatment of wild type or P-selectin mutant mice with TNF-α, a cytokine which upregulates expression of E-selectin and P-selectin, increased the rolling leukocyte flux fraction in venules of the cremaster muscle (Ley *et al.*, 1995). TNF-α induced rolling in P-selectin or P-selectin/ ICAM-1 double mutant mice (see below) was almost completely inhibited by addition of an L-selectin monoclonal antibody (Ley *et al.*, 1995; Kunkel *et al.*, 1996). The first demonstration of E-selectin mediated rolling in the mouse system was observed when TNF-α induced rolling in P-selectin/ICAM-1 mutant mice was blocked with an E-selectin monoclonal antibody; however, this effect was dependent on which E-selectin antibody was used (Kunkel *et al.*, 1996). Monoclonal E-selectin antibody 9A9 completely blocked TNF-α induced rolling (Kunkel *et al.*, 1996) while monoclonal antibody 10E9.6 did not reduce the rolling flux fraction (Ley *et al.*, 1995).

P-SELECTIN/ICAM-1 DOUBLE MUTANTS

P-selectin mutant mice were bred to a line of mice carrying a mutation in ICAM-1 (Sligh *et al.*, 1993), and double homozygotes showed several interesting inflammatory defects (Bullard *et al.*, 1995). In contrast to P-selectin mutants alone, double P-selectin/ICAM-1 mutant mice showed a total absence of surgically-induced rolling in cremaster venules for at least 2 hours (Kunkel *et al.*, 1996). ICAM-1 mutant mice did not show any difference in rolling leukocyte flux compared to wild type mice (Kunkel *et al.*, 1996). Residual rolling in P-selectin mice treated with an ICAM-1 antibody was not inhibited. As mentioned above, L-selectin and E-selectin mediated leukocyte rolling in these double mutants could be induced by pretreatment with TNF-α. The exact mechanism by which ICAM-1 affects leukocyte rolling is not known.

Early neutrophil emigration (2–4 hours) in P-selectin/ICAM-1 double mutant mice was also completely blocked after intraperitoneal instillation of *Streptococcus pneumoniae*, compared to only partial inhibition in each single mutant strain (Bullard *et al.*, 1995). However, neutrophil emigration, edema formation, and bacterial clearance were similar to that in wild type mice after 24 hours (D. Bullard and C. Doerschuk unpublished observations). Neutrophil emigration into the alveolar space following intratracheal instillation of *S. pneumoniae* was not significantly inhibited in P-selectin/ICAM-1 mutant mice (Bullard *et al.*, 1995). This suggests that different adhesive mechanisms (other than P-selectin or ICAM-1) mediate neutrophil emigration in the lung even when the same bacterial stimulus is used.

E-/P-SELECTIN DOUBLE MUTANT MICE

The severe inhibition in both leukocyte rolling and neutrophil emigration following inactivation of multiple selectins suggests that these combined adhesive interactions are crucial for the early leukocytic response to inflammatory stimuli. In order to further address the inflammatory defects involved with chronic loss of selectin function, different lines of mice containing multiple selectin mutations are now being generated. The close linkage of the selectin genes on chromosome 1 (Watson *et al.*, 1990) in the mouse prevents the generation of mice with multiple selectin mutations by simply crossing different mutant strains together. A line of E-/P-selectin double mutant mice has been generated by mutating both genes on the same chromosome in ES cells with subsequent establishment of germline transmission (Bullard *et al.*, 1996).

E-/P-selectin double homozygous mice displayed an increased susceptibility to bacterial infections with the majority of animals developing chronic inflammatory lesions of the oral mucosa and skin (Bullard *et al.*, 1996). Other manifestations in these mice included an extreme plasma cell proliferation in the cervical lymph nodes, spleen, and bone marrow; hypergammaglobulinemia; marked increase in numbers of peripheral blood neutrophils, lymphocytes and monocytes; and an increased cellularity in the alveolocapillary walls in the lung. Electron microscopic studies revealed that the accumulation of leukocytes in the lung is primarily within capillaries (personal communication C. Doerschuk). Decreased levels of L-selectin expression as determined by flow cytometry were also observed on peripheral blood neutrophils and lymphocytes. This reduction was most likely due to increased shedding and activation in peripheral blood, since analysis of L-selectin expression on CD45 positive bone marrow cells revealed only a slight decrease in expression levels.

Examination of leukocyte rolling in cremaster venules in E-/P-selectin double mutants revealed a total absence of both trauma and TNF-α induced leukocyte rolling for up to 2 hours (Bullard *et al.*, 1996). Neutrophil emigration during *Streptococcus pneumoniae*-induced peritonitis was also completely blocked 4 hours following instillation (Bullard *et al.*, 1996). In this model, several interesting features were observed 24 hours after instillation of the bacteria. No significant decrease in neutrophil emigration was observed compared to wild type mice, but an increase in edema formation and bacterial recovery was found at 24 hours. Thus, other adhesive mechanisms, including L-selectin, may mediate neutrophil emigration at later time points into the peritoneum.

OVERVIEW AND RELEVANCE TO DISEASE

Selectin deficient mice have proven to be valuable reagents for the study of selectin function. New insights into the roles of these molecules in inflammatory processes have been gained through analyses of these different mutant lines (see summary table). Unexpected findings include the absence of an obvious phenotype in E-selectin deficient mice, the dramatic disease phenotype in E-/P-selectin double mutant mice, and the complete block in early leukocyte rolling and peritoneal neutrophil emigration in P-selectin/ICAM-1 mutant mice. The studies described above also confirm the importance of the selectins in early leukocyte rolling, an event which is necessary for subsequent leukocyte emigration.

The generation of other combinations of selectin mutations, including a line of mice deficient in all three molecules, will help to further define both the individual functions and the redundancies which exist between each gene. The phenotype of both L-selectin/ICAM-1 and E-selectin/ICAM-1 double mutants should be investigated in order to determine how these combined mutations affect leukocyte rolling and emigration.

New technologies involving gene targeting in mice are rapidly being developed which will allow the determination of the role of selectin expression in specific cell lineages. The *loxP*/Cre recombinase system can be used to produce mutations in specific cell types or to induce mutations postnatally. A premutation containing *loxP* sites is introduced into the germline, and the premutation can be converted to the mutation using transgenes which express the Cre recombinase in a tissue specific or inducible manner (Gu *et al.*, 1994; Kühn *et al.*, 1995). For example, it may be possible to mutate the L-selectin gene in T cells, but not in granulocytes. Likewise, it may be possible to ablate individual selectin gene expression in adult animals using an inducible promoter. These types of experiments would further establish the role of the selectins in inflammatory processes.

Our laboratory has been particularly interested in using mutant mice to test the hypothesis that reduced expression of leukocyte and endothelial cell adhesion molecules will result in decreased susceptibility to a variety of inflammatory disease processes including atherosclerosis, autoimmune disease, arthritis, asthma, inflammatory bowel disease, and other disorders. The effect of heterozygous or homozygous deficiency of inflammatory cell adhesion molecules can be tested using mouse models of disease processes. Relevant models include collagen induced arthritis (Wooley, 1988), autoimmune encephomyelitis (Zamvil and Steinman, 1990), and atherosclerosis (Breslow, 1993; Paigen *et al.*, 1994; Smithies and Maeda, 1995). Mice with increased susceptibility to atherosclerosis can be used in such models including C57BL/6 mice on high fat diet; transgenic mice expressing cholesterol ester transfer protein (CETP), apolipoprotein(a), or other apolipoproteins; or mice with mutations in apolipoprotein E or the LDL receptor.

Another strategy to evaluate the role of selectins in human disease processes is to use the reported amino acid and nucleotide polymorphisms to perform association and linkage studies in patient populations with various diseases. Amino acid polymorphisms such as those identified in E-selectin and ICAM-1 (Vora *et al.*, 1994; Wenzel *et al.*, 1994) are of particular interest for this type of genetic analysis. A possible association of specific alleles of ICAM-1 with various forms of inflammatory bowel disease was found in

Summary of data for mice with selectin mutations

Mutation	*Health Status*	*Leukocyte Counts*	*Neutrophil Emigration in Peritonitis*		*Delayed Type Hypersensitivity*	*Leukocyte Rolling*
			0–4 h	*24 h*		
P-selectin	normal	↑	reduced	normal	normal	↓↓
E-selectin	normal	normal	normal	normal	normal	normal
L-selectin	normal	normal	reduced	reduced	impaired	↓↓
P-selectin /ICAM-1	normal	↑↑	absent	normal*	not determined	↓↓↓
P-/E-selectin	spontaneous infections	↑↑↑↑	absent	normal*	impaired	↓↓↓↓

*May depend on chemical versus bacterial stimulus and does not imply "normal" physiology.

one study, but no association with selectin alleles was observed (Yang *et al.*, 1995). It has been suggested that analysis of polymorphisms in populations with various forms of atherosclerosis would be of interest, and a higher frequency Arg at codon 128 compared to Ser for E-selectin was reported to be associated with atherosclerosis in one study (Wenzel *et al.*, 1994). Genetic studies of the selectins in mice and human populations should provide considerable additional in sights into inflammatory disease processes, and it will be of interest to extend these studies to the selectin ligand molecules.

References

Arbonés, M.L., Ord, D.C., Ley, K., Ratech, H., Maynard-Curry, C., Otten, G., Capon, D.J. and Redder, T.F. (1994). Lymphocyte homing and leukocyte rolling and migration are impaired in L-selectin-deficient mice. *Immunity*, **1**:247–260.

Breslow, J.L. (1993). Transgenic mouse models of lipoprotein metabolism and atherosclerosis. *Proc. Natl. Acad. Sci. USA*, **90**:8314–8318.

Bullard, D.C., Qin, L., Lorenzo, I., Quinlin, W.M., Doyle, N.A., Bosse, R., Vestweber, D., Doerschuk, C.M. and Beaudet, A.L. (1995). P-selectin/ICAM-1 double mutant mice: Acute emigration of neutrophils into the peritoneum is completely absent but is normal into pulmonary alveoli. *J. Clin. Invest.*, **95**:1782–1788.

Bullard, D.C., Kunkel, E.J., Kubo, H., Hicks, M.J., Lorenzo, I., Doyle, N.A., Doerschuk, C.M., Ley, K. and Beaudet, A.L. (1996). Infectious susceptibility and severe deficiency of leukocyte rolling and recruitment in E-selectin and P-selectin double mutant mice. *J. Exp. Med.*, **183**:2329–2336.

Capecchi, M. (1989). Altering the genome by homologous recombination. *Science*, **244**:1288–1292.

Etzioni, A., Frydman, M., Pollack, S., Avidor, I., Phillips, M.L., Paulson, J.C. and Gershoni-Baruch, R. (1992). Brief report: Recurrent severe infections caused by a novel leukocyte adhesion deficiency. *N. Engl. J. Med.*, **327**:1789–1792.

Etzioni, A. (1994). Adhesion molecule deficiencies and their clinical significance. *Cell Adh. Comm.*, **2**:257–260.

Frenette, P.S., Johnson, R.C., Hynes, R.O. and Wagner, D.D. (1995). Platelets roll on stimulated endothelium *in vivo*: An interaction mediated by endothelial P-selectin. *Proc. Natl. Acad. Sci. USA*, **92**:7450–7454.

Frohman, M.A. and Martin, G.R. (1989). Cut, paste, and save: New approaches to altering specific genes in mice. *Cell*, **56**:145–147.

Frydman, M., Etzioni, A., Eiditz-Markus, T., Avidor, I., Varsano, I., Shechter, Y., Orlin, J.B. and Gershoni-Baruch, R. (1992). Rambam-Hasharon syndrome of psychomotor retardation, short stature, defective neutrophil motility, and bombay phenotype. *J. Med. Genet.*, **44**:297–302.

Gallatin, W.M., Weissman, I.L. and Butcher, E.C. (1983). A cell-surface molecule involved in organ-specific homing of lymphocytes. *Nature*, **303**:30–34.

Gu, H., Marth, J.D., Orban, P.C., Mossmann, H. and Rajewsky, K. (1994). Deletion of a DNA polymerase β gene segment in T cells using cell type-specific gene targeting. *Science*, **265**:103–108.

Hamburger, S.A. and McEver, R.P. (1990). GMP-140 mediates adhesion of stimulated platelets of neutrophils. *Blood*, **75**:550–554.

Johnson, R.C., Mayadas, T.N., Frenette, P.S., Mebius, R.E., Subramaniam, M., Lacasce, A., Hynes, R.O. and Wagner, D.D. (1995). Blood cell dynamics in P-selectin-deficient mice. *Blood*, **86**:1106–1114.

Kunkel, E.J., Jung, U., Bullard, D.C., Normal, K.E., Wolitzky, B.A., Vestweber, D., Beaudet, A.L. and Ley, K. (1996). Absence of trauma-induced leukocyte rolling in mice deficient in both P-selectin and intercellular adhesion molecule-1 (ICAM-1). *J. Exp. Med.*, **183**:57–65.

Kühn, R., Schwenk, F., Aguet, M. and Rajewsky, K. (1995). Inducible gene targeting in mice. *Science*, **269**:1427–1429.

Labow, M.A., Norton, C.R., Rumberger, J.M., Lombard-Gillooly, K.M., Shuster, D.J., Hubbard, J., Bertko, R., Knaack, P.A., Terry, R.W., Harbison, M.L., Kontgen, F., Stewart, C.L., McIntyre, K.W., Will, P.C., Burns, D.K. and Wolitzky, B.A. (1994). Characterization of E-selectin-deficient mice: Demonstration of overlapping function of the endothelial selectins. *Immunity*, **1**:709–720.

Larsen, E., Celi, A., Gilbert, G.E., Furie, B.C., Erban, J.K., Bonfanti, R., Wagner, D.D. and Furie, B. (1989). PADGEM protein: A receptor that mediates the interaction of activated platelets with neutrophils and monocytes. *Cell*, **59**:305–312.

Lasky, L.A. (1992). Selectins: Interpreters of cell-specific carbohydrate information during inflammation. *Science*, **258**:964–969.

Ley, K., Bullard, D., Arbonés, M.L., Bosse, R., Vestweber, D., Tedder, T.F. and Beaudet, A.L. (1995). Sequential contribution of L- and P-selectin to leukocyte rolling *in vivo*. *J. Exp. Med.*, **181**:669–675.

Mayadas, T.N., Johnson, R.C., Rayburn, H., Hynes, R.O. and Wagner, D.D. (1993). Leukocyte rolling and extravasation are severely compromised in P-selectin-deficient mice. *Cell*, **74**:541–554.

McEver, R.P., Moore, K.L., and Cummings, R.D. (1995). Leukocyte trafficking mediated by selectin-carbohydrate interactions. *J. Biol. Chem.*, **270**:11025–11028.

Paigen, B., Plump, A.S. and Rubin, E.M. (1994). The mouse as a model for human cardiovascular disease and hyperlipidemia. *Curr. Opin. Lipidol.*, **5**:258–264.

Sedivy, J.M. and Joyner, A.L. (1992). Gene Targeting (New York: W.H. Freeman & Company).

Sligh, J.E., Ballantyne, C.M., Rich, S.S., Hawkins, H.K., Smith, C.W., Bradley, A. and Beaudet, A.L. (1993). Inflammatory and immune responses are impaired in ICAM-1 deficient mice. *Proc. Natl. Acad. Sci. USA*, **90**:8529–8533.

Smithies, O. and Maeda, N. (1995). Gene targeting approaches to complex genetic disease: Atherosclerosis and essential hypertension. *Proc. Natl. Acad. Sci. USA*, **92**:5266–5272.

Springer, T.A. (1994). Traffic signals for lymphocyte recirculation and leukocyte emigration: The multistep paradigm. *Cell*, **76**:301–314.

Staite, N.D., Justen, J.M., Sly, L.M., Beaudet, A.L. and Bullard, D.C. (1996). Inhibition of delayed-type contact hypersensitivity in mice deficient in both E- and P-selectin in press, *Blood*.

Subramaniam, M., Saffaripour, S., Watson, S.R., Mayadas, T.N., Hynes, R.O. and Wagner, D.D. (1995). Reduced recruitment of inflammatory cells in a contact hypersensitivity response in P-selectin-deficient mice. *J. Exp. Med.*, **181**:2277–2282.

Tedder, T.F., Steeber, D.A., Chen, A. and Engel, P. (1995a). The selectins: vascular adhesion molecules. *FASEB J.*, **9**:866–873.

Tedder, T.F., Steeber, D.A. and Pizcueta, P. (1995b). L-selectin-deficient mice have impaired leukocyte recruitment into inflammatory sites. *J. Exp. Med.*, **181**:2259–2264.

Vora, D.K., Rosenbloom, C.L., Beaudet, A.L. and Cottingham, R.W. (1994). Polymorphisms and linkage analysis for ICAM-1 and the selectin gene cluster. *Genomics*, **2**:473–477.

Watson, M.L., Kingsmore, S.F., Johnston, G.I., Siegelman, M.H., Le Beau, M.M., Lemons, R.S., Bora, N.S., Howard, T.A., Weissman, I.L., McEver, R.P. and Seldin, M.F. (1990). Genomic organization of the selectin family of leukocyte adhesion molecules on human and mouse chromosome 1. *J. Exp. Med.*, **172**:263–272.

Wenzel, K., Felix, S., Kleber, F.X., Brachold, R., Menke, T., Schattke, S., Schulte, K.L., Glaser, C., Rohde, K., Baumann, G. and Speer, A. (1994). E-selectin polymorphism and atherosclerosis: an association study. *Hum. Mol. Genet.*, **3**:1935–1937.

Wooley, P.H. (1988). Collagen-induced arthritis in the mouse. *Methods Enzymol.*, **162**:361–373.

Yang, H., Vora, D.K., Targan, S.R., Toyoda, H., Beaudet, A.L. and Rotter, J.I. (1995). Intercellular adhesion molecule-1 (ICAM-1) gene associations with immunologic subsets of inflammatory bowel disease. *Gastroenterology* (in press).

Zamvil, S.S. and Steinman, L. (1990). The T lymphocyte in experimental allergic encephalomyelitis. *Annu. Rev. Immunol.*, 579–621.

9 The Carbohydrate Components of Selectin Ligands

John B. Lowe

Professor, Department of Pathology, Associate Investigator, the Howard Hughes Medical Institute, University of Michigan Medical School, MSRBI, room 3510, 1150 West Medical Center Drive, Ann Arbor, MI 48109-0650, Tel: 313 647-4779, Fax: 0 313 936-1400, Email: JohnLowe@umich.edu

The surfaces of mammalian cells display large numbers of structurally diverse carbohydrate moieties, termed oligosaccharides or glycoconjugates. These molecules decorate proteins, and are also displayed by membrane-associated lipids, termed glycolipids. Oligosaccharides displayed by proteins are typically attached to specific asparagine residues, termed asparagine-linked glycosylation, or through serine or threonine residues, termed O-linked oligosaccharides. The biosynthetic pathways leading to N- and O-linked oligosaccharide biosynthesis, and glycolipid biosynthesis, have been reviewed (Kornfeld and Kornfeld, 1985; Sadler, 1984). In general, core portions of these oligosaccharide moieties, within a given animal species, are largely identical in structure. By contrast, many of the branching-type structures, and terminal oligosaccharide modifications, are found to be expressed in a cell-type, tissue-specific, or developmentally-regulated manner. The variation in cell surface oligosaccharide structure in association with the state of differentiation of a cell, or in correspondence to its lineage, represents circumstantial evidence that cell surface oligosaccharides maintain important functional roles in the context of an interaction between a cell and its environment. For example, numerous studies have demonstrated dynamic structural changes in cell surface oligosaccharide expression during mammalian embryogenesis (Feizi, 1985). These and other observations imply that mammalian organisms maintain an intricate set of molecules and regulatory mechanisms to regulate such structural changes, and further infer that they maintain important functional attributes. This structural diversity displayed by oligosaccharide moieties, representing potential information diversity, can be potentially "decoded" by members of a growing family of mammalian carbohydrate binding proteins, or lectins (Drickamer, 1988). In many instances, these mammalian lectins have been shown to specifically bind cell surface oligosaccharide structures, and participate as ligands in adhesive processes between cells, for instance, and participate in some instances in mediating interactions between the oligosaccharide structures on specific glycoproteins, and their receptors. These processes include those used by cells to target lysosomal enzymes to the lysosome through the mannose-6-phosphate receptor (Kornfeld, 1987), a process used to clear desialylated serum glycoproteins through an interaction with the asialoglycoprotein receptor expressed by hepatocytes (Ashwell and Harford, 1982), and the process used to modulate clearance rates of circulating pituitary glycoprotein hormones like leutropin (Fiete *et al.*, 1991). As discussed elsewhere in this monograph, experimental work over the past several years has identified three members of a family of mammalian lectins, termed the selectins, that participate in leukocyte adhesion events relevant to leukocyte trafficking and inflammation. This chapter will discuss current information concerning the structure and biosynthesis of the oligosaccharide components of the counter receptors for L-selectin, E-selectin, and P-selectin.

THE OLIGOSACCHARIDE COMPONENTS OF L-SELECTIN LIGANDS DISPLAYED BY LYMPHOID ORGAN HIGH ENDOTHELIAL VENULES

The mammalian lymphoid lineage is characterized by a recirculation process that directs lymphocytes to travel both in the intravascular compartment and in the lymphatic system. Blood-born lymphocytes leave the vasculature within lymph nodes, and other lymphoid organs like Peyer's patches (Rosen, 1993; Springer, 1994; Rosen and Bertozzi, 1994). Extravasated lymphocytes that pass through the parenchyma of the lymphoid organ, and return to the vascular tree through the lymphatic ductal system. This recirculation process enhances the opportunity for lymphocytes to encounter foreign antigens within the lymphatic organs, which represents an essential part of the normal immune response.

Lymphocytes leave the vascular system in the postcapillary venules within lymphoid organs. This process initially involves adhesive events between cell adhesion molecules displayed by lymphocytes, and counter receptors for these adhesion molecules, displayed at the surface of the endothelial cells lining the postcapillary venules (reviewed in Rosen and Bertozzi, 1994; Kansas *et al.*, 1991). These latter endothelial cells maintain a distinct morphology and function, relative to typical vascular endothelial cells. These specialized postcapillary venular endothelial cells are cuboidal in shape, and the venules that are lined with these cells are termed high endothelial venules, or HEVs. Experiments involving immunochemical approaches, and cell homing experiments, demonstrate that the HEV within different lymphatic organs are functionally distinct, and maintain quantitatively and qualitatively distinct sets of cell adhesion counter receptors. For example, lymphocyte homing studies identify homing specificities that are distinct to inflamed synovial tissue, to lymph nodes draining the lung, to the HEV within lymphoid organs in the gut (Peyer's patches), and to the HEV within peripheral lymph nodes (termed PNHEV) (Rosen, 1993; Springer, 1994; Rosen and Bertozzi, 1994). Lymphocyte homing to the peripheral lymph nodes is mediated in large measure by oligosaccharide-dependent adhesive processes that occur between circulating lymphocytes and the PNHEV. This section will deal with this process, and these molecules, in detail.

Functional assays for lymphocyte homing to the PNHEV include the Stamper-Woodruff assay (Stamper and Woodruff, 1977). This procedure involves sampling adhesive events between lymphocytes applied to HEVs exposed on the surface of a frozen section of a peripheral lymph node. Cells remaining attached to the HEV after washing may be quantitated as a measure of adhesion between the applied lymphocytes and the exposed HEV. This assay facilitated the identification of a lymphocyte cell surface molecule that mediates PNHEV-lymphocyte adhesive interactions (Gallatin *et al.*, 1983). This molecule was defined by the monoclonal antibody MEL-14 (Gallatin *et al.*, 1983; Lewinsohn *et al.*, 1987); the MEL-14 monoclonal antibody blocks adhesion of lymphocytes to the PNHEV, diminishes *in vivo* homing of lymphocytes to the peripheral lymph nodes, and the expression level of the MEL-14 epitope on cultured lymphocyte cell lines correlates extremely well with propensity of those lines to home to peripheral lymph nodes, and to bind to peripheral node HEV (Gallatin *et al.*, 1983).

In parallel studies, the nature of the PNHEV determinants that support lymphocyte-HEV adhesive events was explored (Stoolman and Rosen, 1983; Stoolman *et al.*, 1984). This work was stimulated by others' observations that PNHEV are covered by a prominent glycocalyx. As this glycocalyx represents the most likely initial surface to be encountered by a lymphocyte in its interaction with an HEV, this observation suggested the possibility that oligosaccharide components of the glycocalyx are operative in these adhesive events. This notion was supported by the observation that the glutaraldehyde fixation of PNHEV exposed by frozen sections leaves intact lymphocyte-HEV interactions (Stamper and Woodruff, 1977). Since fixation with glutaraldehyde generally denatures peptide antigens, while leaving carbohydrate determinants in their native state, it seemed likely that carbohydrate determinants displayed by the PNHEV glycocalyx maintained an important functional role in lymphocyte-PNHEV interactions (Stoolman and Rosen, 1983).

This hypothesis was explored by testing a spectrum of mono- and polysaccharides for their ability to inhibit lymphocyte-PNHEV interactions (Stoolman and Rosen, 1983). While most monosaccharides were ineffective in this assay, at any concentration, two (D-mannose and L-fucose) could be shown to block lymphocyte-PNHEV interactions, though only at relatively high concentrations (exceeding 75 millimolar). However, the phosphorylated monosaccharides mannose-6-phosphate and fructose-1-phosphate inhibited binding in the low millimolar range. Furthermore, two of a series of polysaccharides examined yielded effective inhibition of lymphocyte-PNHEV interactions in the low nanomolar concentration range. These molecules included fucoidan, a sulfated high molecular weight polymer containing $\alpha(1,2)$- and $\alpha(1,3)$-linked fucose residues (Mian and Percival, 1973), and PPME, a mannan-based molecule rich in mannose-6-phosphate groups (Stoolman *et al.*, 1984; San Blas and Cunningham, 1974). PPME was found to inhibit adhesion when lymphocytes were pre-treated with this substance, whereas pretreatment of PNHEV did not diminish lymphocyte adhesion. This work implied that PPME interacts with a lymphocyte cell surface molecule operative in mediating lymphocyte-PNHEV interactions. These observations were extended in studies demonstrating that fluorescent beads derivatized with PPME bind to lymphoid cell lines that adhere to PNHEV, whereas these beads do not bind to lymphocyte cell lines that do not maintain a PNHEV-binding phenotype (Yednock *et al.*, 1987a). Subsequent work demonstrated that PPME in fact interacts with the lymphocyte polypeptide corresponding to the MEL-14 epitope (Yednock *et al.*, 1987b), lending support to the notion that the MEL-14 epitope is displayed by a lymphocyte cell surface molecule that interacts with a carbohydrate-dependent counter receptor on PNHEV.

Early studies to explore the nature of this carbohydrate-dependent counter receptor provided evidence for an essential role for sialic acid in the function of this counter receptor. These studies included the demonstration that sialidase treatment of paraformaldeyde-fixed

PNHEV blocks lymphocyte adhesion to these HEV (Rosen *et al.*, 1985). Likewise, in an *in vivo* assay of short term lymphocyte homing in the mouse, it was observed that intravenous injection of bacterial sialidases diminishes short term accumulation of lymphocytes in peripheral lymph nodes, and destroys the ability of PNHEV in these animals to support lymphocyte adhesion in standard Stamper-Woodruff assays (Rosen *et al.*, 1989). Considered together, this work implied an essential role for sialic acid and possibly other carbohydrate moieties on the PNHEV as components of the counter receptor for an adhesion molecule on lymphocytes recognized by the MEL-14 monoclonal antibody.

As described in chapter ____ in this monograph, molecular cloning studies demonstrate that the MEL-14 epitope corresponds to a carbohydrate-binding protein termed L-selectin (cluster designation CD62L; Siegelman *et al.*, 1989a; Lasky *et al.*, 1989; Tedder *et al.*, 1989; Camerini *et al.*, 1989; Bowen *et al.*, 1989; Siegelman *et al.*, 1989b). As reviewed elsewhere, the amino terminal domain of L-selectin maintains primary sequence similarity to the other members of the selectin family, and to the larger family of proteins termed C-type lectins originally classified by Drickamer (Drickamer, 1988). The observation that the MEL-14 epitope corresponds to a molecule with sequence similarity to calcium-dependent lectins was satisfyingly consistent with the observation that MEL-14-dependent interactions with PNHEV were themselves calcium-dependent, and were probably mediated at least in part by carbohydrate-dependent mechanisms.

Experimental work completed over the past several years has extended these initial observations substantially, by providing insight into the molecular nature of the endogenous PNHEV ligands for L-selectin. Much of this more recent work has taken advantage of a recombinant L-selectin-IgG molecule, used initially to explore the nature of these ligands on tissue sections, and later in biochemical analyses (Bowen *et al.*, 1990; Watson *et al.*, 1991). In an early set of such studies, Watson *et al.* used the L-selectin-Ig molecule as an immunohistochemical probe to further explore the requirement for sialic acid in interactions between L-selectin and PNHEV ligands (Watson *et al.*, 1991). These studies demonstrated that treatment of PNHEV with sialidases derived from Vibrio cholera or Clostridium perfringens blocks binding of the recombinant L-selectin Ig molecule to PNHEV. Similarly, pre-treatment of PNHEV with Limax flavus agglutinin, a lectin with high specificity for terminal sialic acid residues, blocks lymphocyte-PNHEV interactions. By contrast, a different sialic acid-specific lectin, termed Limulus polyphemus agglutinin, was observed not to block binding of the L-selectin-IgG probe. Considered together, these observations largely confirmed prior work demonstrating a requirement for sialic acid in L-selectin-dependent lymphocyte homing (True *et al.*, 1990), notwithstanding the observation that Limulus polyphemus agglutinin does not block L-selectin IgG-PNHEV interactions. This latter apparently discrepant observation may be accounted for by the possibility that the Limax and Limulus lectins maintain differences in their specificities for the positional linkage of sialic acid modifications (Liener *et al.*, 1986), or by the modifications known to occur on sialic acid residues in mammalian oligosaccharides (i.e. O-acetylation, and N-glycolylation, for example; Schauer, 1985).

Within the past several years, biochemical and molecular cloning approaches have been applied with remarkable success in an effort to define the molecular nature of L-selectin ligands on PNHEV. The L-selectin-Ig chimera was used initially as a biochemical probe to identify glycoproteins synthesized by PNHEV. When used as an immunoprecipitation reagent, the L-selectin-Ig chimera identified 50 kD and 90 kD [^{35}S]-labeled polypeptides in extracts prepared from lymph nodes radiolabeled with [^{35}S]sulfate (Imai *et al.*, 1991; Watson *et al.*, 1990; Watson *et al.*, 1991). These proteins were also immunoprecipitated with a rat monoclonal antibody termed MECA-79, shown previously to inhibit binding of lymphocytes to PNHEV, and thus implicated as capable of recognizing L-selectin counter-receptors (Butcher, 1990; Streeter *et al.*, 1988). Both proteins were identified with these approaches in mesenteric node HEV, but not in Peyer's patches HEV. Evidence supporting the specificity of the L-selectin-Ig chimera for bona fide L-selectin ligands includes the observation that the immunoprecipitation is calcium-dependent, and is inhibited by fucoidan and PPME (as is L-selectin-dependent lymphocyte attachment to PNHEV). Both immunoprecipitated molecules were demonstrated to be sensitive to degradation by trypsin, confirming that they maintain a polypeptide component. Similar experiments demonstrated that these molecules could be radiolabeled with fucose, but not with mannose or glucose. Considered together, these results were interpreted to indicate that the two molecules immunoprecipitated by the L-selectin-Ig chimera, and by the MECA-79 antibody, are sulfated, fucosylated glycoproteins. Absence of labeling with mannose or glucose is consistent with the notion that the glycoproteins contain few, if any, asparagine-linked (N-linked) oligosaccharides, as these would have been evident from the mannose-labeling experiment (Kornfeld and Kornfeld, 1985). Moreover, the electrophoretic mobility of the two proteins was not altered by digestion with N-glycanase, providing additional evidence against their containing an appreciable amount of N-linked oligosaccharides. The publication describing this work mentions unpublished data indicating that reagents capable of releasing serine/threonine-linked oligosaccharides also release radiolabeled fucose and sulfate, again suggesting that the bulk of the oligosaccharides displayed by the two proteins are of the O-linked form. Imai *et al.* (1991), termed these molecules Sgp^{50} and Sgp^{90}, for sulfated glycoprotein of 50 kDa and 90 kDa, respectively. Since the L-selectin-IgG chimera immunoprecipitates both polypeptides even after extraction from denaturing SDS polyacrylamide gels, it seems likely that they are most likely recognized by L-selectin directly through their glycan moieties. More recently, the L-selectin-Ig chimera has been used to identify a third molecule, of Mr 200 kDa, in lymph node extracts (Hemmerich *et al.,* 1994a). This molecule, termed Sgp^{200}, maintains a spectrum of biochemical attributes predicted to be essentially identical to those outlined above for Sgp^{50} and Sgp^{90}.

Both Sgp^{50} and Spg^{90} have been purified, and cDNAs encoding these molecules have been cloned (Lasky *et al.,* 1992; Baumheuter *et al.*, 1993). The cDNA encoding Sgp^{50} predicts a 132 amino acid long polypeptide, termed GlyCAM-1, for *g*lycosylation-dependent *c*ell *a*dhesion *m*olecule. The sequence of GlyCAM-1 predicts a single potential N-linked glycosylation site, but contains numerous serine and threonine residues (Lasky *et al.*, 1992). The large number of serine and threonine residues, together with the large difference in size between the observed (~50,000 Da) and predicted (14,154 Da) molecular masses, imply that O-linked carbohydrate chains account for a large part of the protein's mass, and indicate that the molecule is a mucin-type glycoprotein. There is no obvious transmembrane segment in the sequence of GlyCAM-1; any association that it maintains with the HEV surface may take place via interaction with another membrane-tethered molecule, or through a COOH-terminal 21 residue segment predicted to be helical and amphipathic. The absence of a typical transmembrane domain is consistent with the observation that this molecule is detectable as a soluble entity in the conditioned medium of lymph node organ cultures (Lasky *et al.,* 1992). More recent studies utilizing immunoelectron microscopy demonstrate that little, if any, GlyCAM-1 is displayed at the surface of the endothelial cells lining lymph node HEV, even though the molecule clearly passes through the secretory pathway of these cells, and is released from them as a soluble entity (Kikuta and Rosen, 1994). Consequently, it is likely that GlyCAM-1 does not participate directly in L-selectin-dependent tethering of lymphocytes to the HEV. It remains to be determined if this soluble L-selectin counter-receptor instead operates to modulate the biological properties of lymphocytes, or neutrophils, by interacting with L-selectin expressed on these leukocytes.

Sgp^{90} was affinity purified from mouse peripheral lymph nodes in an effort to isolate cDNAs encoding this molecule (Baumheuter *et al.*, 1993). Protein sequence analysis of the purified molecule identified peptide sequences corresponding to murine CD34, a previously cloned sialomucin molecule. Biochemical and immunohistochemical analyses confirmed that the protein core of Sgp^{90} is in fact CD34. Like GlyCAM-1, the predicted structure of CD34 contains several serine and threonine-rich, mucin-like domains that are heavily O-glycosylated. The biochemical analyses discussed above (Imai *et al.*, 1991; Hemmerich *et al.*, 1994a), involving CD34 synthesized in peripheral lymph node organ culture, indicates that the O-glycans on PLN-derived CD34/Sgp^{90} are sulfated, and likely represent essential components of this molecule that are recognized by L-selectin. Given that CD34 is a transmembrane glycoprotein, it is likely that this L-selectin counter-receptor is importantly involved in tethering lymphocytes to lymph node HEV. CD34 is expressed in other tissues, including vascular endothelia (Fina *et al.*, 1990; Brown *et al.*, 1991; Cheng *et al.*, 1996), and if appropriately glycosylated, could in principle function as an L-selectin counter-receptor to facilitate leukocyte-endothelial cell interactions in acute and chronic inflammation (Lasky, 1992). In this context, it is interesting to note that some glycoforms of CD34 isolated from human tonsils can support L-selectin-dependent leukocyte rolling *in vitro* (Puri *et al.*, 1995). Considered together, these observations suggest that CD34 plays an important role in the display of oligosaccharide ligands for L-selectin by the endothelium.

Other observations indicate that the mucosal vascular addressin molecule known as MAdCAM-1 can also present oligosaccharide ligands to L-selectin (Berg *et al.*, 1993). MAdCAM-1 had been previously identified as the counter-receptor for the lymphocyte homing receptor $\alpha 4\beta 7$, and figures prominently in directing lymphocyte homing to Peyer's patches (Berlin *et al.*, 1993). The biochemical experiments reported by Berg *et al.* demonstrate that MAdCAM-1 isolated from mouse mesenteric lymph nodes displays sialic acid-containing, presumed O-glycans (N-glycanase-resistant), and can support L-selectin-dependent leukocyte rolling *in vitro*. By contrast, MAdCAM-1 isolated from cultured endothelioma cells does not support L-selectin-dependent leukocyte rolling. These observations indicate that organ-specific glycosylation of this molecule dictates its ability to function as an L-selectin counter-receptor, and suggest the interesting possibility that this molecule can subserve two complementary functionalities, namely as a counter-receptor for shear-dependent interactions (mediated by both L-selectin and the $\alpha 4\beta 7$ integrin), and as a counter-receptor for subsequent activation-dependent firm attachment processes.

The early explorations of the nature of the oligosaccharide portions of the L-selectin counter-receptors discussed above indicated that sialic acid is essential to the functionality of L-selectin ligands. Sialidases have been used, for instance, to define the nature of the glycosidic bond(s) that link sialic acid residues within the endogenous L-selectin ligand(s) [i.e. $\alpha(2,3)$, $\alpha(2,6)$, or $\alpha(2,8)$ linkages] (Imai *et al.*, 1992). These studies used GlyCAM-1 as the model for endogenous L-selectin counter-receptors, and demonstrated that a sialidase specific for $\alpha(2,3)$-linked sialic acid residues (Newcastle disease virus neuraminidase) and a broad spectrum sialidase (*Arthrobacter ureafaciens* neuraminidase) are roughly equal in their abilities to inactivate GlyCAM-1-borne ligand activity, and suggested that sialic acid in $\alpha(2,3)$ linkage is essential to L-selectin ligand activity.

Likewise indirect biochemical evidence implied that sulfate is essential for functional L-selectin ligand activity (Imai *et al.,* 1993). These studies demonstrated that an L-selectin-IgG chimera does not bind to GlyCAM-1 when it is purified in a non-sulfated form from PNHEVs treated with chlorate (an inhibitor of sulfation). Subsequent studies demonstrate that L-selectin can bind to both sialyl Lewis x and sialyl Lewis a (Berg *et al.,* 1992; Foxall *et al.*, 1992), and to isomers of these structures wherein the terminal sialic acid residue is replaced by a sulfate residue (Green *et al.*, 1992). Given that these tetrasaccharides can function as ligands for E-selectin, and that the sialyl Lewis x moiety is an essential component of counter-receptors for P-selectin (described in detail below), this work suggested the possibility that the O-glycans implicated in L-selectin ligand activity will bear sulfated, $\alpha(2,3)$sialylated, $\alpha(1,3)$fucosylated structures similar to the sialyl Lewis x determinant. Since the evidence for expression of the sialyl Lewis x or sialyl Lewis a epitopes on human HEV is not especially abundant (Berg *et al.*, 1992; Paavonen and Renkonen, 1992), it seems unlikely that either molecule represents an authentic endogenous ligand for L-selectin.

More recent studies on the structures of the O-linked glycans displayed by GlyCAM-1 have allowed Rosen and colleagues to propose as candidates for L-selectin ligands a pair of sulfated forms of the sialyl Lewis x molecule, as components of a "core 2" type O-linked oligosaccharide displayed by GlyCAM-1, and, by inference, HEV-derived CD34, MAdCAM-1, and Sgp^{200}. In the first of these studies, $^{35}SO_4$-labeled oligosaccharides were prepared from GlyCAM-1 metabolically labeled in lymph node organ culture with $^{35}SO_4$, and were subjected to chemical and enzymatic degradation procedures. High-pH anion exchange chromatographic analysis of the products of these manipulations identified four prominent sulfated moieties (Gal-6-SO_4, GlcNAc-SO_4, [SO_4-6]Galβ(1,4) GlcNac, and Galβ(1,4)[$SO_4$6]GlcNAc) (Hemmerich *et al.*, 1994b). In a companion manuscript, these authors extended these observations using approaches involving with lectin binding studies, together with glycohydrolase digestions, to provide evidence that 6′-sulfated sialyl Lewis x (NeuNAcα(2,3)[SO_4-6]Galβ(1,4)GlcNAc) is represented as a prominent terminal oligosaccharide structure on GlyCAM-1 (Hemmerich *et al.*, 1994c). In related work, Scudder *et al.* utilized a combination of chemical and enzymatic synthesis approaches to construct a pentasaccharide isomer of this molecule in which the N-acetyl-glucosamine moiety is 6-O sulfated (i.e., 6-sulfo sialyl Lewis x; (NeuNAcα(2,3)Galβ(1,4) [SO_4-6][Fucα(1,3)]GlcNAcβ(1,3)Gal) (Scudder *et al.*, 1994). Furthermore, these investigators demonstrated that this molecule can inhibit binding of an L-selectin-IgG chimera to purified peripheral lymph node addressin (PNAd; a heterogeneous mixture of glycoproteins, isolated by affinity chromatography with the monoclonal antibody MECA-79 (Streeter *et al.,* 1988), and containing glycoforms of Sgp^{50}, Spg^{90}, Spg^{200} with L-selectin counter-receptor activity). As the inhibition observed with 6-sulfo sialyl Lewis x (IC_{50} of 0.8 mM) was relatively greater than that observed with a sialyl Lewis x tetrasaccharide

(IC_{50} of 3.2 mM), these authors propose that the 6-sulfo sialyl Lewis x moiety may represent a physiological ligand for L-selectin (Scudder *et al.*, 1994).

More recent experimental results provide additional insight into the possible structures displayed by GlyCAM-1 that are recognized by L-selectin, and about the synthesis of such structures. Using a series of biochemical analyses of β-eliminated oligosaccharide chains displayed by mouse GlyCAM-1, Hemmerich *et al.* (1994c), identified two O-linked glycans encompassing the 6-sulfo sialyl Lewis x and 6′-sulfo sialyl Lewis x moieties show previously by this group to be major capping moieties on GlyCAM-1 (Hemmerich *et al.*, 1994a; 1994b). These molecules (Figure 1) are based on a so-called core 2 substructure. Core 2 structures are characteristic of a number of O-linked oligosaccharides displayed by mucin-type glycoproteins, and are synthesized in part by the core 2 N-acetylglucosaminyltransferase (Bierhuizen *et al.*, 1992; 1995).

The biosynthesis of these candidate L-selectin ligands has been explored to some extent in a recent study involving an analysis of their display by GlyCAM-1 expressed in lymph node organ culture (Crommie and Rosen, 1995). These investigators utilized temperature blocking approaches, and brefeldin A-dependent disruption of the secretory pathway, to deduce the order of addition of the various terminal constituents of these molecules. These studies (i) demonstrate that synthesis of GalNAc-terminating chains (presumably corresponding to

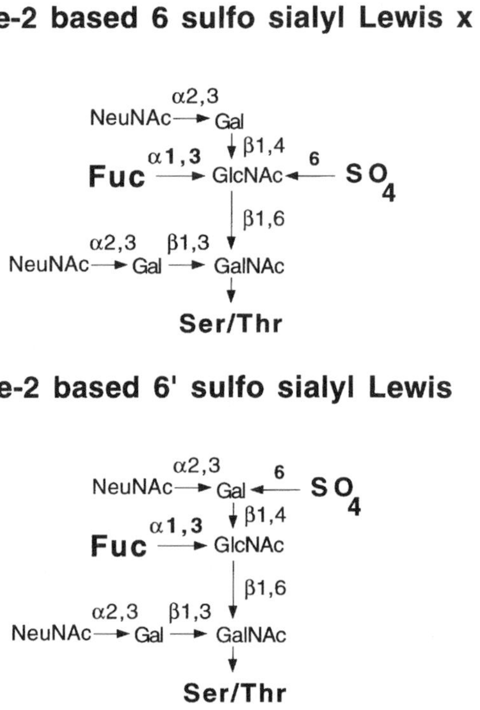

Figure 1 Core 2-based O-linked candidate L-selectin ligands. Proposed structures are derived from biochemical analyses completed on glycans constructed in mouse lymph node organ cultures, and displayed by GlyCAM-1 (Hemmerich *et al.*, 1995c).

GlyCAM-1 modified only by a single GalNAc residue on serines and/or threonines) precedes synthesis of lactosamine-type precursors, (ii) imply that the lactosamine chains are in turn likely sialylated prior to sulfation and fucosylation, and (iii) suggest that sialylation occurs in a pre trans Golgi network compartment. This conclusion is consistent with previous observations indicating that $\alpha(2,3)$sialylation precedes, but cannot follow, $\alpha(1,3)$fucosylation (reviewed in Natsuka and Lowe, 1994). As indicated in Figure 2, the relative order of addition of sulfate and fucose to the $\alpha(2,3)$sialylated precursor is uncertain, and the experiments in Crommie and Rosen were not able to resolve this issue. Specifically, brefeldin A treatment yielded a partial inhibition of fucosylation, but almost complete inhibition of sulfation. This result is consistent with the possibility that sulfation follows fucosylation, and is consistent with others' results demonstrating that several different human $\alpha(1,3)$fucosyltransferases will not utilize NeuNAc$\alpha(2,3)[SO_4$-6]Galβ1, 4GlcNAc to form the 6′-sulfo sialyl Lewis x determinant (Jain *et al.*, 1994). By contrast, however, others have demonstrated that the human $\alpha(1,3)$fucosyltransferase Fuc-TV can utilize the pentasaccharide (NeuNAcα(2,3) Gal$\beta(1,4)[SO_4$-6]GlcNAcβ(1,3)Gal) to form the 6-sulfo sialyl Lewis x determinant (Figure 2) (Scudder *et al.*, 1994), whereas Scudder *et al.* cite their own unpublished data indicating that a partially purified lymph node N-acetylglucosamine-6-O-sulfotransferase does not sulfate oligosaccharides containing $\alpha(1,3)$fucosylated N-acetylglucosamine residues. These results suggest that the formation of 6-sulfo sialyl Lewis x may occur via ordered catalytic events involving $\alpha(1,3)$fucosylation, followed by 6-O-sulfation. However, the physiological relevance of the fucosylation directed by Fuc-TV is not apparent from this work, since the Fuc-TV gene is not known to be expressed in lymph node HEV.

More recent work suggests that Fuc-TVII may be operative in this context, as this gene is expressed in the HEV found in mouse peripheral and mesenteric nodes, and in Peyer's patches (Smith *et al.*, 1996). Furthermore, ablation of this gene in the mouse, through the use of gene "knock out" approaches, yields mice that are deficient in HEV-expressed L-selectin activity, along with a concomitant deficiency in lymphocyte homing to peripheral lymphoid aggregates (Malý *et al.*, 1996). Finally, *in vitro* assays utilizing synthetic acceptor substrates demonstrate that mouse Fuc-TVII can form both sialyl Lewis x and 6-sulfo sialyl Lewis x determinants, but not the 6′-sulfo sialyl Lewis x, from their respective non-fucosylated precursors (Smith *et al.*, 1996; Malý *et al.*, 1996). Considered together, these data (i.) provide additional evidence for the existence of the 6-sulfo sialyl Lewis x determinant on HEV, (ii.) imply that this molecule can be constructed by an ordered synthetic scheme involving sialylation, followed by sulfation, and terminated by $\alpha(1,3)$fucosylation, (iii.) imply that $\alpha(1,3)$fucosylation is essential to L-selectin ligand activity, and (iv.) demonstrate an essential role for Fuc-TVII in this process. Nonetheless, this proposed scheme for 6-sulfo sialyl Lewis x biosynthesis remains to be confirmed with the use of purified sulfotransferase(s), and acceptor substrates that faithfully reflect the oligosaccharide structures endogenous to the HEV, as well as by additional oligosaccharide structural analyses on authentic, HEV-born L-selectin ligands. Likewise, the pathway for construction of 6′-sulfo sialyl Lewis x remains to be established with purified substrates and enzymes, and the relevance of this structure to L-selectin-dependent lymphocyte homing pathways remains an open issue. It should also be noted that a di-sulfated form of the monosulfated sialyl Lewis x moieties may also exist (i.e., 6′,6-disulfo sialyl Lewis x), and may, in principle, function as an important, or essential component of the L-selectin counter-receptor. Indeed, disulfated forms of lactose have been shown to be effective inhibitors of L-selectin-dependent binding

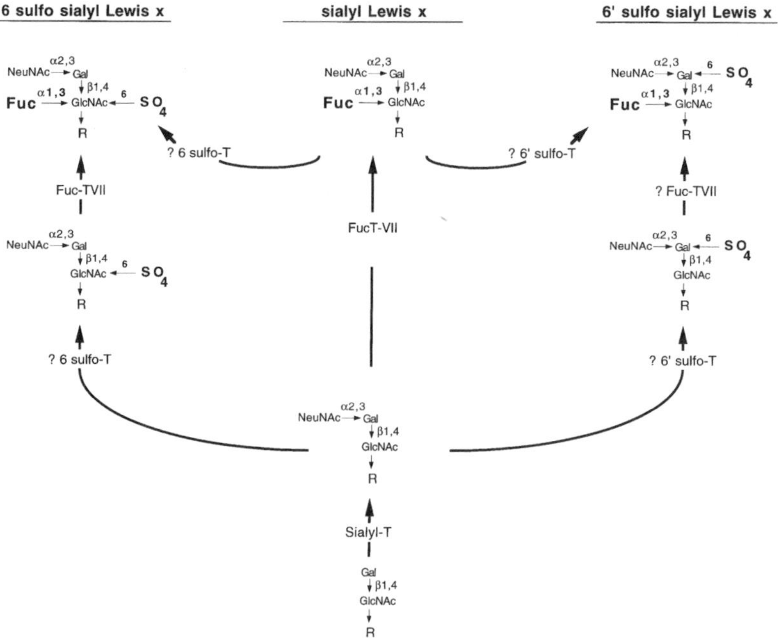

Figure 2 Biosynthesis of candidate L-selectin ligands. The candidate capping structures proposed as possible ligands for L-selectin are displayed at the top (6-sulfo sialyl Lewis x and 6′-sulfo sialyl Lewis x), along with the related structure sialyl Lewis x, an essential component of the human leukocytic ligands for E- and P-selectins. Possible routes for the biosynthesis of these structures are indicated by the bold arrows. These pathways begin with a lactosamine-type structure (bottom) displayed by O-linked glycans of the form illustrated in Figure 1 (i.e., R = core 2, O-linked glycans). These molecules are utilized by one or more α(2,3)sialyltransferases (Sialyl-T) to form an α(2,3)sialylated lactosamine-based molecule proposed as the precursor to the three molecules shown at the top of the figure. This monosialylated precursor may be directly fucosylated by Fuc-TVII, an α(1,3)fucosyltransferase expressed in lymphoid aggregate HEV, to form the sialyl Lewis x moiety (center pathway). Alternatively, this monosialylated precursor may be subjected to one or more distinct sulfation events leading to monosialylated, monosulfated structures. *In vitro* biochemical analyses summarized in the text indicate that Fuc-TVII, and Fuc-TV, can form the 6-sulfo sialyl Lewis x structure from one such precursor (pathway at left). By contrast, compounds representative of the monosialylated, monosulfated potential precursor to the 6′-sulfo sialyl Lewis x structure are not utilized, at least *in vitro*, by any known α(1,3)fucosyltransferase, to form the 6′-sulfo sialyl Lewis x moiety (pathway at the right hand side of the figure). It is possible that the sialyl Lewis x structure is itself a substrate for one or more sulfotransferase activities that may yield the 6-sulfo sialyl Lewis x structure (top left), or the 6′-sulfo sialyl Lewis x structure (top right), or the 6′6-disulfo sialyl Lewis x structure (not shown). It is also possible that a di-sulfated, monosialylated structure may be formed by such enzymes, and that this molecule may then be utilized by Fuc-TVII, or other α(1,3)fucosyltransferases, to form 6′6-disulfo sialyl Lewis x structure (also not shown).

to GlyCAM-1 (Bertozzi *et al.,* 1995). Together, these considerations suggest the possible existence of multiple HEV-expressed sulfotransferase activities, which may operate on α(2,3)sialylated lactosamine-based structures to form sialyl Lewis x precursors, or which utilize the sialyl Lewis x moiety to form the 6′-sulfo sialyl Lewis x and/or 6′,6-disulfo sialyl Lewis x structures, or which may operate on sialyl Lewis x to form the 6-sulfo sialyl Lewis x determinant (Figure 2). Biochemical and molecular cloning studies will be needed to precisely determine which of these possible schemes is physiological.

L-SELECTIN-DEPENDENT ADHESION OF NEUTROPHILS AND MONOCYTES TO ENDOTHELIUM

In addition to its role as an adhesion molecule for lymphocytes, L-selectin is also expressed on neutrophils, eosinophils, and monocytes. Since these latter cells do not traffic to lymph nodes in the same manner that lymphocytes do, it may be inferred that L-selectin on these myeloid-lineage cells may have one or more functions distinct from those involved in lymphocyte homing. One of these functions appears to be to mediate adhesion of these leukocytes to the non-lymphoid vascular endothelium. Early observations relevant to this process indirectly implicated L-selectin as an important component of the process that allows neutrophils to immigrate into inflamed tissues. For instance, it has been demonstrated antibodies against L-selectin can substantially block the immigration of neutrophils into the inflamed peritoneal cavity (Jutila *et al.*, 1989). Similar results with anti-L-selectin antibodies have been observed in experimental models involving neutrophil immigration into inflammatory skin lesions (Lewinsohn *et al.*, 1987). These *in vivo* observations are mirrored by similar results *in vitro*; adhesion of neutrophils to cytokine-activated endothelium *in vitro* is diminished by prior treatment of the neutrophils with anti-L-selectin antibodies (Hallmann *et al.*, 1991; Smith *et al.*, 1991). These observations have been confirmed and extended in other *in vitro* circumstances, by demonstrating that adhesion of monocytes to activated vascular endothelium is dependent upon L-selectin-mediated adhesive processes (Spertini *et al.*, 1992). Similarly, adhesion of monocytes and neutrophils to activated renal glomerular endothelial cells *in vitro* is also an L-selectin-dependent process (Brady *et al.*, 1992).

These early observations demonstrated that L-selectin-dependent adhesion in these circumstances was observable under flow conditions, and were in part responsible for the development of a multi-step, multi-molecule model for leukocyte adhesion and immigration in the vasculature This model has been discussed in detail elsewhere (von Andrian, 1991; Butcher, 1991; Springer, 1994). In particular, these observations were followed up in *in vivo* systems that allow direct visualization of the behavior of leukocytes traveling in the microvasculature (Ley *et al.*, 1991; von Andrian *et al.*, 1991; von Andrian *et al.*, 1992). These intravital microscopy systems document that leukocytes "roll" along the surface of the endothelial cells that line capillaries in inflamed tissues (Ley *et al.*, 1991; Atherton and Born, 1973; Fiebig *et al.*, 1991). This rolling phenomenon is in part a consequence of vasodilatory events that alter the laminar flow characteristics of the fluid in the vessel, in turn causing leukocytes to leave the center of the vessel lumen where they normally travel in the more rapidly flowing part of the stream, to marginal positions near the endothelial monolayer (Nobis *et al.*, 1985). In inflammation, the vessel dilates, and plasma viscosity increases, yielding a slowing of the flow, which is in turn responsible for leukocyte movement to peripheral positions in the stream (Chien, 1982). At marginal positions along the vessel wall, leukocytes are subjected to rotational forces caused by a differential between the rapidly flowing stream near the center of the vessel, and the more slowly flowing or non-flowing stream at the endothelial cell surface. This differential causes the leukocytes to roll along the wall of the capillary, and serves to maximize interactions between endothelial cell surfaces and the leukocyte. Because the velocities of rolling leukocytes are significantly less than those predicted for freely tumbling cells, it can be expected that adhesion is a component of the rolling interactions (Atherton and

Born, 1973). Antibody blocking studies demonstrate that these rolling interactions are at least in part dependent on L-selectin. For example, using intravital microscopy techniques, it is possible to observe diminished leukocyte rolling following intravascular administration of an L-selectin-IgG chimera, or anti-L-selectin antibodies, in both rat and rabbit inflammatory models (Ley *et al.*, 1991; von Andrian *et al.*, 1991; von Andrian, 1992). By contrast, in some of these same models, rolling is not observed to be inhibited when anti-CD18 antibodies are instilled, even though these antibodies are observed to diminish firm attachment of rolling leukocytes (von Andrian *et al.*, 1991; von Andrian *et al.*, 1992). Considered together with other observations, these data imply that leukocyte recruitment occurs through a sequential series of events first requiring leukocyte rolling, followed by firm leukocyte attachment, and followed finally by leukocyte extravasation. Antibody blocking studies imply that rolling requires L-selectin, whereas integrin-dependent adhesive events are necessary for firm attachment. Blocking of the initial, L-selectin-dependent rolling process, is predicted to diminish subsequent events, including extravasation. This prediction has been confirmed, in a mouse model in which recruitment of neutrophils to the inflamed peritoneal cavity is inhibited by intravascular injection of L-selectin-IgG chimera (Watson *et al.*, 1991).

The molecular basis for some of these observations can be accounted for by work demonstrating that human neutrophil L-selectin displays oligosaccharide ligands for endothelial cell expressed E-selectin (Picker *et al.*, 1991a). As discussed elsewhere in this chapter, and in this monograph, E-selectin mediates neutrophil adhesion to vascular endothelium (reviewed in Springer, 1994, and Varki, 1994). These biochemical analyses demonstrate that L-selectin itself represents a relatively abundant component of the glycoprotein complement recognized by E-selectin. These observations suggest that a part of the L-selectin-dependent rolling processes summarized in the previous paragraphs may be accounted for by simple "presentation" by L-selectin of counter receptors for E-selectin.

More recent work has demonstrated that the cellular topography of L-selectin is also essential for L-selectin-dependent leukocyte rolling (von Andrian *et al.*, 1995). This study indicates that L-selectin is preferentially localized to the tips of the microvillous processes that extend above the planar membrane surface of the neutrophil. Given that the microvillous processes are more likely to contact neighboring cells, including endothelial cells, these morphological observations suggested that the particular extracellular topography of L-selectin, at the microvillous tip, was especially relevant for L-selectin-dependent rolling processes. In a series of elegant experiments, these investigators demonstrated that L-selectin is substantially less effective at mediating leukocyte rolling if it is expressed at the planar surface of the leukocyte cell surface, as opposed to the microvillous tip (von Andrian *et al.*, 1995). While these observations suggest an important role for L-selectin in presentation of oligosaccharide counter receptors to E-selectin, and point to the functional importance of L-selectin in leukocyte rolling, by virtue of its extracellular topographical circumstance, they do not address the nature of putative L-selectin counter receptors expressed by the endothelium.

Biochemical approaches have been used in an effort to identify endothelial cell-derived L-selectin counter receptors (Norgard-Sumnicht *et al.*, 1993). These studies identified heparin-containing molecules with a calcium-dependent affinity for L-selectin at intracellular locations within vascular endothelial cells. Whether these molecule represent physiologically-relevant L-selectin counter-receptors remains to be determined, as it is not yet known if they can also be found at the surface of endothelial cells.

Other observations suggest the interesting possibility that leukocytes adherent to the endothelial cell surface via E- and P-dependent adhesion, or other mechanisms, can themselves display oligosaccharide ligands (Bargatze *et al.*, 1994). Using an *in vitro* flow chamber, these investigators demonstrate that adherent leukocytes can support L-selectin-dependent rolling of other leukocytes. The L-selectin "counter receptor" discovered on the adherent leukocytes is sialidase sensitive, and transient in nature. This adhesive mechanism operates in an apparently unidirectional manner, in that it is characterized by interactions between L-selectin displayed by the rolling leukocytes, and the counter receptor activity for L-selectin displayed by the adherent leukocytes, but not the converse. This model may at first appear to be inconsistent with the fact that neutrophils do not ordinarily exhibit homotypic aggregation *in vitro* or *in vivo*, as might be predicted by considering that these cells display both L-selectin, and an L-selectin counter receptor. This apparent conceptual problem with the model proposed by Bargatze *et al.* can be reconciled by considering very recent observations indicating that L-selectin-dependent adhesive processes generally require conditions of flow, and are not detectable under static conditions (Finger *et al.*, 1995). The observations made by Bargatze *et al.* are useful in interpreting, in retrospect, many of the *in vivo* and *in vitro* studies demonstrating a requirement for L-selectin in leukocyte rolling. Specifically, it seems possible that a substantial amount of the L-selectin-dependent leukocyte rolling events observed in these studies can be accounted for by leukocytes rolling on previously pavemented leukocytes. The inducible nature of the L-selectin "ligands" observed in many of these studies may, in fact, be accounted for by the inducible nature of neutrophils adhesion mediated by E- and/or P-selectins, and, subsequently, ICAM-1-dependent adhesive events. Additional studies will be necessary to confirm this possibility, and to identify the nature of the neutrophil ligands for L-selectin.

It should be noted that Bargatze *et al.* also identify an L-selectin-dependent leukocyte adhesion process that is independent of the "leukocyte rolling on leukocyte" process (Bargatze *et al.*, 1994). The nature of the endothelial cell molecule(s) responsible for this portion of L-selectin-dependent leukocyte rolling events remains to be identified. Likewise, it is not yet known if these neutrophil-independent L-selectin ligands correspond to some portion of the "inducible" L-selectin ligands observed previously.

COUNTER-RECEPTORS FOR E-SELECTIN

The biology and biochemistry of the endothelial cell adhesion molecule E-selectin (known formerly as Endothelial Cell Adhesion Molecule 1, or ELAM-1; cluster designation CD62E) has been reviewed (Springer, 1994; Varki, 1994). Like L-selectin, E-selectin maintains an amino terminal domain sharing primary sequence similarity with C-type lectins, including the L-selectin carbohydrate recognition domain. E-selectin is an inducible cell adhesion molecule displayed by endothelial cells, and is expressed by capillary endothelium in a variety of acute and chronic inflammatory conditions. These include immune complex-dependent acute lung injury (Mulligan *et al.*, 1991), rheumatoid arthritis (Koch *et al.*, 1991), sepsis (Redl *et al.*, 1991; Engelberts *et al.*, 1992), skin inflammation (Cotran *et al.*, 1989; Groves *et al.*, 1991; Picker *et al.*, 1991b; Griffiths *et al.*, 1991; Rohd *et al.*, 1992), and in some circumstances involving organ transplantation (Briscoe *et al.*, 1991). E-selectin counter receptors have been identified on neutrophils (Bevilacqua *et al.*, 1985), monocytes (Bevilacqua *et al.*, 1985; Carlos *et al.*, 1990), eosinophils (Weller *et al.*, 1991), memory T-lymphocytes (Picker *et al.*, 1991b; Shimizu *et al.*, 1991; Graber *et al.*, 1990), and natural killer cells (NK cells; Lobb *et al.*, 1991). As these types of cells can be found in inflammatory locations in association with E-selectin expression, it is expected that recruitment of these cells in acute and chronic inflammation involves E-selectin-dependent adhesion of these leukocytes. Analysis of mice deficient in E-selectin have confirmed an important role for E-selectin in some acute inflammatory circumstances (Frennette *et al.*, 1996; Labow *et al.*, 1994). Its role in chronic inflammation remains to be studied with these mice.

Initial clues to the identity of ligands for E-selectin on leukocytes were derived in part from observations concerning possible oligosaccharide ligands for L-selectin. As summarized above, early studies involving HEV-expressed L-selectin ligands implied that these were composed in part of sialylated oligosaccharides. Additional clues to the nature of E-selectin ligands were sought amongst sets of oligosaccharides expressed by cell types that are recognized by E-selectin (i.e., myeloid-lineage cells), but not by other blood cells that do not exhibit E-selectin-dependent adhesion properties (i.e. erythrocytes). Candidate oligosaccharide ligands meeting these criteria were found amongst a group of fucosylated, sialylated oligosaccharides previously identified on myeloid lineage cells by Fukuda and his collaborators (Fukuda *et al.*, 1984; Fukuda *et al.*, 1985a; Spooncer *et al.*, 1984; Fukuda *et al.*, 1985b), and by other investigators (Symington *et al.*, 1985). These oligosaccharides were found as components of glycoprotein and glycolipid molecules containing lactosamine cores [Galβ(1,4)GlcNAcβ(1,3)Gal], as monomers or polymers. Polylactosamine chains are characteristic of a number of glycans on myeloid lineage cells (Fukuda *et al.*, 1984; Fukuda *et al.*, 1985a; Spooncer *et al.*, 1984; Fukuda *et al.*, 1985b; Symington *et al.*, 1985), and are catalyzed by the sequential and alternating actions of two glycosyltransferases, Galβ(1,4) GlcNAc transferase, and GlcNAcβ(1,3)Gal transferase (Figure 3). Polylactosamine chains may be found unmodified on myeloid lineage cells, and many other cell types, and are typically found modified by terminal or subterminal substitution with sialic acid, in either α(2,3)-linkage or α(2,6)-linkage, or by α(1,3)- and/or α(1,2)-linked fucosylation. Since polylactosamine chains modified by α(1,3)-linked residues are typical of myeloid lineage cells, but are not normally found on erythrocytes, it was considered likely that α(1,3)-linked fucose moieties represented an important component of leukocytic E-selectin ligands.

GlcNAc residues within polylactosamine units represent the site for α(1,3)-linked fucosylation of these polymers (Figure 4). This modification is determined by one or more distinct α(1,3)fucosyltransferases (Figure 4, reviewed in Mollicone *et al.*, 1990; Lowe, 1992, and Natsuka and Lowe, 1994). Biochemical and genetic studies demonstrate that α(1,3)fucosylation is generally a terminal event in the biosynthesis of α(1,3)fucosylated molecules. Most, if not all, α(1,3)fucosyltransferases can effectively utilize α(2,3)sialylated polylactosamine substrates, whereas α(2,3)sialylation of α(1,3)fucosylated oligosaccharides is not known to occur (Figure 4).

Lactosamine-based oligosaccharides generated by α(2,3)sialylation, and α(1,3)fucosylation, include the sialyl Lewis x structure (Figure 1, Figure 4), and the neutral monofucosylated moiety Lewis x (also termed SSEA-1, reference 134; or CD15, Figure 4). Isomers of these structures, containing internally or multiply fucosylated moieties include the VIM-2 determinant, wherein an internal GlcNAc residue on a polylactosamine chain is substituted with an α(1,3)-linked fucose residue (Lowe *et al.*, 1991). Alternatively, polyfucosylated moieties have been described, and are presumed to occur through a

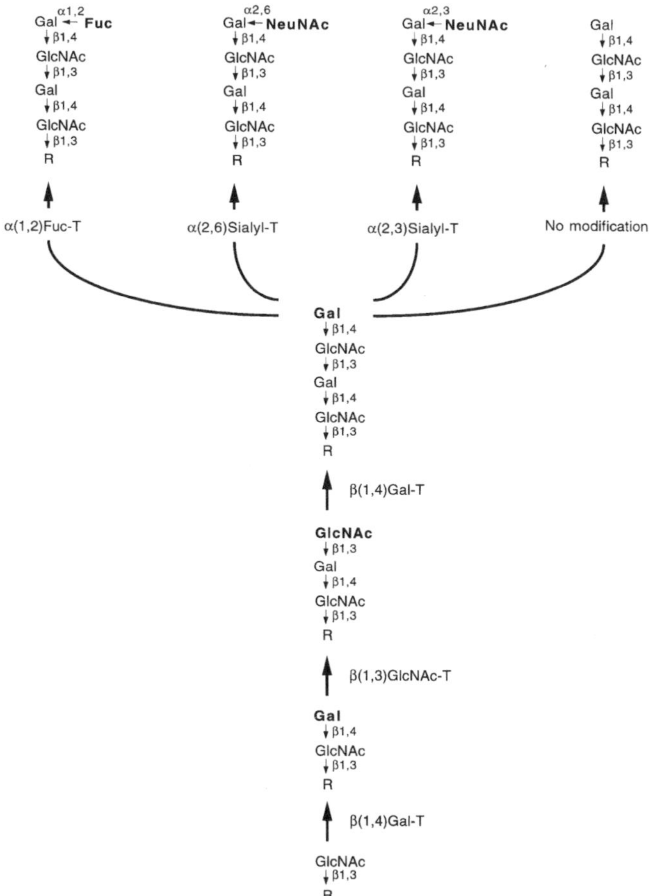

Figure 3 Polylactosamine chain synthesis and termination. Galactose (Gal) and *N*-acetylglucosamine (Glc*N*Ac) residue alternate to form polylactosamine chains, and are linked together through the alternating actions of β(1,4)galactosyltransferase (β1,4Gal-T) and a β(1,3)*N*-acetylglucosaminyltransferase (β(1,3)GlcNAc-T) (Carlos *et al.*, 1990; van den Eijnden *et al.*, 1988). The monosaccharide displayed in bold type corresponds to the sugar added by the preceding enzymatic reaction. Branching of these linear chains can occur via the action of β(1,6)*N*-acetylglucosaminyltransferases that attach N-acetylglucosamine residues to galactosyl units (not shown). Linear or branched polylactosamine chains may be attached to proteins through covalent linkage to serine or threonine residues (O-linked glycans; Sadler, 1984), to asparagine residues (N-linked glycans; Kornfeld and Kornfeld, 1985), or to sphingolipids (glycolipids; Fukuda *et al.*, 1986) (denoted by R in the figure). The terminal lactosamine unit [Galβ(1,4)Glc*N*Acβ(1,3)] may be displayed at the cell surface in an unmodified form (pathway at the top right), or the terminal galactose residue may be subsequently modified by the action of an α(1,2)fucosyltransferase, to yield a terminal α(1,2)-linked fucose residue to form the H blood group determinant (pathway at the top left) (Larsen *et al.*, 1990). H blood group determinants are expressed by erythroid lineage cells (Fukuda *et al.*, 1986), but not normally by myeloid cells (Fukuda *et al.*, 1985b; Fukuda *et al.*, 1986). The terminal galactose may also be modified by sialyltransferases that can attached sialic acid (*N*euNAc; *N*-acetyl neuraminic acid) in either α(2,6)- or in α(2,3)-linkage (top central two pathways) (Weinstein *et al.*, 1982). *N*-acetylglucosamine moieties within the polylactosamine chains may also be modified by the action of α(1,3)fucosyltransferases (see Figure 4).

series of ordered fucosylation events (Fukushi *et al.*, 1984; Howard *et al.*, 1987; Fukuda *et al.*, 1986; Stroud *et al.*, 1996a; Stroud *et al.*, 1996b; Figure 4). These oligosaccharides can in some cases be identified by staining with monoclonal reagents with relative specificity for terminal tri- and tetrasaccharide moieties (Fukushi *et al.*, 1984; Fukushima

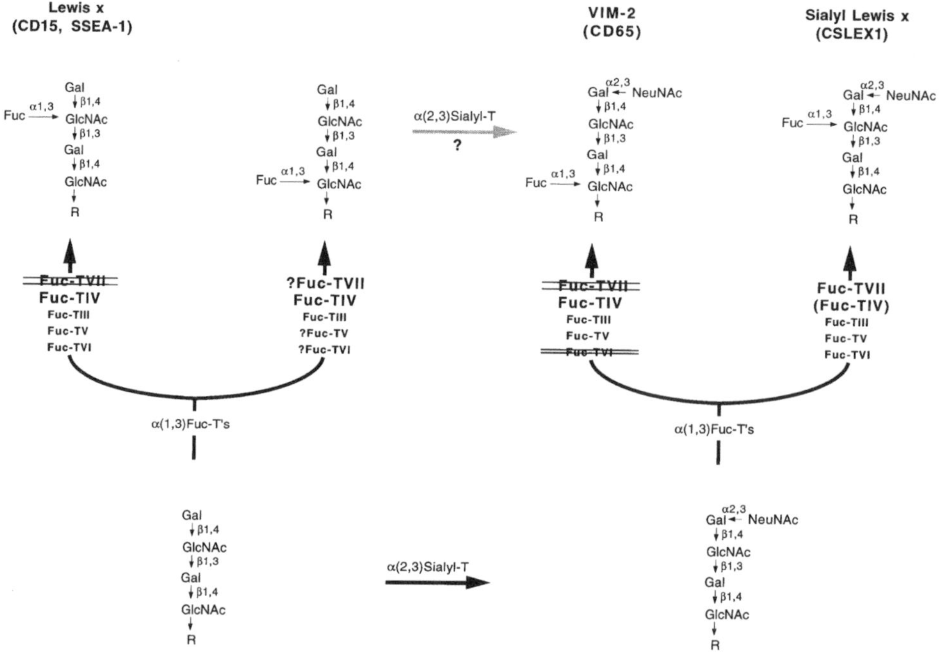

Figure 4 Synthesis of Lewis x, sialyl Lewis x, and VIM-2 determinants. *N*-acetylglucosamine residues within neutral or α(2,3)sialylated polylactosamine chains represent potential precursor substrates for the five known human α(1,3)fucosyltransferases (Fuc-TIII, IV, V, VI, and VII; (Holmes *et al.*, 1985; Howard *et al.*, 1987; Kukowska-Latallo *et al.*, 1990; Goelz *et al.*, 1990; Lowe *et al.*, 1991; Lowe *et al.*, 1990; Natsuka *et al.*, 1994; Sasaki *et al.*, 1994), or the murine homologues of Fuc-TIV and Fuc-TVII (Gersten *et al.*, 1995; Smith *et al.*, 1996).

The Lewis x determinant may be formed by all except Fuc-TVII (struck through with two horizontal lines) from neutral polylactosamine chains. Internally fucosylated neutral polylactosamine chains may also be formed by Fuc-TIII, and to some extent by Fuc-TIV (Sueyoshi *et al.*, 1994). Fuc-TIII can also utilize neutral polylactosamine chains to form a difucosylated molecule (see Fukuda *et al.*, 1986) in which both N-acetylglucosamine residues are modified by α(1,3)fucose residues (Sueyoshi *et al.*, 1994; not shown in the Figure). It is not known if Fuc-TVII, V, or VI can form the internally monofucosylated product, or the difucosylated product (indicated by ?) from neutral chains. The order of fucose addition to form neutral difucosylated chains is also not clearly defined, nor are the relative affinities of the enzymes for non-fucosylated polylactosamine substrate, versus internally or terminally monofucosylated substrates. It should be noted that molecules with more than 2 internal fucosylations have been described (Stroud *et al.*, 1996a; Stroud *et al.*, 1996b). The biosynthesis of these latter molecules is unexplored.

Neutral polylactosamine precursors are also utilized by α(2,3)sialyltransferases (arrow at bottom center), in the formation of α(2,3)sialylated precursor molecules (bottom right). The resulting α(2,3)Sialylated oligosaccharides can also be utilized as acceptor substrates by α(1,3)fucosyltransferases (vertical paths at right). Several α(1,3)fucosyltransferases (Fuc-TIII, Fuc-TV, Fuc-TVI, and Fuc-TVII; Lowe *et al.*, 1991; Kumar *et al.*, 1991; Lowe *et al.*, 1990; Weston *et al.*, 1992a; Weston *et al.*, 1992b; Natsuka *et al.*, 1994; Sasaki *et al.*, 1994) can fucosylate terminal *N*-acetylglucosamine moieties of α(2,3)sialylated precursors to form the monofucosylated sialyl Lewis x molecules detectable by the monoclonal antibody CSLEX1 (path at far right). The ability of Fuc-TIV (also known as ELFT; Lowe *et al.*, 1991) to create CSLEX1-reactive sialyl Lewis x determinants is a function of the glycosylation phenotype of the mammalian host cell line in which this enzyme is expressed (Lowe *et al.*, 1990; Goelz *et al.*, 1990; Lowe *et al.*, 1991; Kumar *et al.*, 1991; Goelz *et al.*, 1994; discussed in Natsuka *et al.*, 1994; indicated by parenthetic display of Fuc-TIV). Fucα(1,3)Glc*N*Ac linkages may also be formed on an internal lactosamine unit (detected by the monoclonal antibody VIM-2) by some of these α(1,3)fucosyltransferases (Fuc-TIII, Lowe *et al.*, 1991; Fuc-TIV or ELFT; Goelz *et al.*, 1990; Lowe *et al.*, 1991; Lowe *et al.*, 1990; Potvin *et al.*, 1990; Fuc-TV; Weston *et al.*, 1992b). VIM-2-reactivity is not found on Fuc-TVI-transfected cells (Weston *et al.*, 1992b). Monofucosylated, α(2,3)sialylated molecules can also be subsequently fucosylated to form difucosylated sialyl Lewis x determinants (Holmes *et al.*, 1986; Howard *et al.*, 1987; not shown in the Figure). Some α(1,3)fucosyltransferases add fucose first to internal lactosamine units (Howard *et al.*, 1987), while others appear to add preferentially to terminal lactosamine units (Howard *et al.*, 1987; Holmes *et al.*, 1986).

et al., 1984). A consideration of previous biochemical and genetic date (reviewed in Natsuka and Lowe, 1994) implies that expression of these α(1,3)fucosylated structures will be controlled by regulating expression of distinct α(1,3)fucosyltransferase genes.

Five distinct human α(1,3)fucosyltransferase genes have now been identified through molecular cloning procedures (Kukowska-Latallo *et al.*, 1990; Kumar *et al.*, 1991; Weston *et al.*, 1992a; Weston *et al.*, 1992b; Natsuka *et al.*, 1994; Sasaki *et al.*, 1994). The ability of these enzymes to construct distinct α(1,3)fucosylated molecules is summarized in Figure 4. The expression patterns of Fuc-TIII, Fuc-TV, and Fuc-TVI have been explored recently (Cameron *et al.*, 1995), but are not necessarily relevant to the synthesis of these molecules in leukocytes. By contrast, Fuc-TIV and Fuc-TVII are expressed in leukocytes (Lowe *et al.*, 1992; Kumar *et al.*, 1991; Goelz *et al.*, 1990; Natsuka *et al.*, 1994; Sasaki *et al.*, 1994; Gersten *et al.*, 1995; Smith *et al.*, 1996). As will be summarized later in this chapter, these two enzymes represent important potential regulatory points in the biosynthesis of leukocytic α(1,3)fucosylated oligosaccharides, and leukocytic ligands for both E-selectin and P-selectin.

Evidence supporting an essential requirement for the sialyl Lewis x determinant as a ligand for E-selectin was published essentially simultaneously by several groups. In one of these studies (Lowe *et al.*, 1990), cloned fucosyltransferase gene segments were used to alter the glycosylation phenotype of cultured Chinese hamster ovary cells (CHO) and COS-1 cell lines, in ways that were informative for determining if the Lewis x, or sialyl Lewis x structures could function as ligands for E-selectin. CHO cells and COS-1 cells do not ordinarily display E-selectin ligands, and are deficient in α(1,3)fucosyltransferase activity. As these cell lines express oligosaccharide structures that can function as precursors to the Lewis x and sialyl Lewis x structures, but do not express the structures themselves, transfection with different α(1,3)fucosyltransferase genes (Fuc-TIV and Fuc-TIII; Figure 4) conferred upon the transfectants the expression of surface-localized Lewis x and/or sialyl Lewis x antigens. These experiments demonstrated that expression of the sialyl Lewis x tetrasaccharide determinant correlates with E-selectin binding, whereas expression of the Lewis x or VIM-2 determinants, in the absence of sialyl Lewis x, is not associated with expression of E-selectin ligand activity (Lowe *et al.*, 1991; Lowe *et al.*, 1990). This work implied that one or more members of the family of α(2,3)sialylated, α(1,3)fucosylated polylactosamines represented by the sialyl Lewis x determinant represent candidate ligands for E-selectin, and inferred that lineage-specific expression of α(1,3)fucosyltransferase genes can control E-selectin ligand expression.

In a contemporaneous publication (Goelz *et al.*, 1990), an expression cloning approach was used to isolate an HL-60 cell-derived cDNA that conferred upon COS cells and CHO cells the ability to bind to E-selectin. Biochemical analyses, and DNA sequence analyses demonstrated that this cDNA encodes a human α(1,3)fucosyltransferase that was termed

Figure 4 *Continued*

None of the known mammalian α(2,3)sialyltransferases have been shown to operate on polylactosamine molecules with a terminal Fucα(1,3)GlcNAc linkage (Holmes *et al.*, 1985; Holmes *et al.*, 1986; Kitagawa and Paulson, 1994), though this remains a possibility (indicated in the Figure by the shaded arrow with question mark). The existence of α(2,3)sialyltransferases that can sialylate monofucosylated precursors to form VIM-2 determinants is also an open question.

ELFT, for *E*LAM-1 *L*igand *F*ucosyl *T*ransferase (Goelz *et al.*, 1990). Sequence comparisons published subsequently demonstrated that ELFT is identical to the human α(1,3)fucosyltransferase Fuc-TIV (Lowe *et al.*, 1992), which, in the previous study (Lowe *et al.*, 1990), and one subsequent publication (Kumar *et al.*, 1991), was shown *not* to confer E-selectin-dependent adhesive properties on transfected CHO cells. However, more recent studies have shown that some strains of Chinese hamster ovary cells, although not others, will express the sialyl Lewis x determinant when transfected with Fuc-TIV/ELFT (Goelz *et al.*, 1990; Goelz *et al.*, 1994; Li *et al.*, 1996b). This work again points to a role for the sialyl Lewis x determinant as a myeloid lineage ligand for E-selectin. Nonetheless, and as discussed in detail below, these studies highlight the uncertain role for ELFT/ Fuc-TIV in the expression of the sialyl Lewis x determinant and E-selectin ligands.

Conclusions concerning a role for sialyl Lewis x as a ligand for E-selectin were derived from experiments published simultaneously, using approaches distinct from the gene transfer experiments summarized above. For instance, liposomes containing purified sialyl Lewis x-containing glycolipids were shown to block E-selectin-dependent cell adhesion, whereas control glycolipids do not (Philips *et al.*, 1990; Polley *et al.*, 1991). This work also reported a correlation between expression of sialyl Lewis x and E-selectin-dependent cell adhesion in mutant Chinese hamster ovary cell lines, as did other studies utilizing a recombinant E-selectin-IgG chimera (Walz *et al.*, 1990). Other early work in this area isolated α(1,3)fucosylated glycolipid structures from myeloid cells, and demonstrated that one of these, the VIM-2 determinant (Figure 4) supports E-selectin-dependent cell adhesion in a solid phase assay system. These studies have been extended to demonstrate that multiply and/or internally fucosylated, α(2,3)sialylated glycolipids (Stroud *et al.*, 1996a; Stroud *et al.*, 1996b), or N-linked glycans (Patel *et al.*, 1994), can serve as effective ligands for E-selectin, when displayed either in solid phase, or in solution, respectively.

Other structural isomers of the sialyl Lewis x determinant have been identified as counter receptors for E-selectin. The sialyl Lewis a structure (Figure 5), for instance, can support E-selectin-dependent cell adhesion (Berg *et al.*, 1991). This molecule was identified as a ligand for E-selectin in a screen of a panel of neoglycoproteins for structures distinct from the sialyl Lewis x determinant, using a monoclonal antibody called HECA-452. HECA-452 is known to recognize oligosaccharide structures on sialyl Lewis x-negative cell types, like skin homing memory T lymphocytes (Picker *et al.*, 1991b; Shimizu *et al.*, 1991), that bind E-selectin. Molecular modeling studies (Berg *et al.*, 1991) demonstrate that the sialyl Lewis x and sialyl Lewis a structures display the fucose and sialic acid moieties in essentially identical positions on the same face of each structure. Nonetheless, skin homing memory T lymphocytes do not display the sialyl Lewis a structure and the nature of the E-selectin ligand on these cells, that reacts with HECA-452, remains unknown.

In parallel, yet independent studies, the sialyl Lewis a determinant, as expressed on a human colorectal carcinoma cell line, COLO201, has been shown to be an effective counter receptor for E-selectin (Takada *et al.*, 1991). Purified sialyl Lewis a-containing molecules have also been demonstrated to support E-selectin-dependent cell adhesion, with an efficiency that is comparable to that manifested by the sialyl Lewis x determinant (Tyrrel *et al.*, 1991). Molecular modeling studies reported here also imply that the sialyl Lewis x and sialyl Lewis a determinants maintain similar molecular shapes (Tyrrel *et al.*, 1991). However, it is clear that the sialyl Lewis a moiety is not typically expressed by any

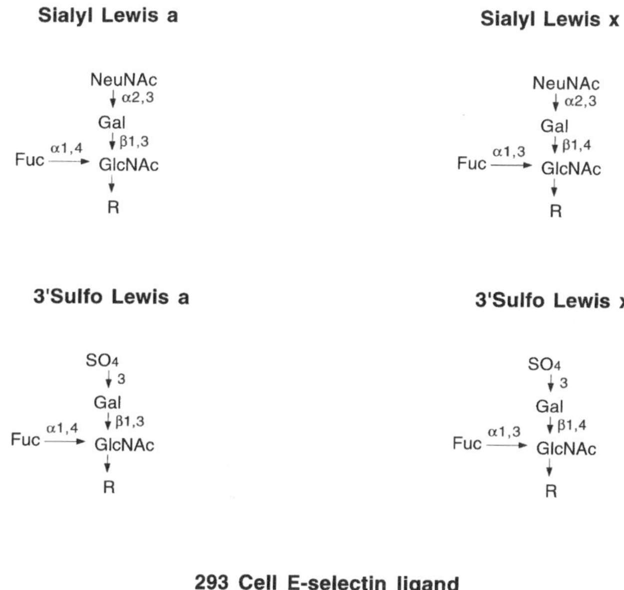

Figure 5 Comparison of sialyl Lewis a, sialyl Lewis x, their 3′ sulfated variants, and a kidney cell E-selectin ligand (Green *et al.*, 1992; Yuen *et al.*, 1992; Yan *et al.*, 1993). The sialyl Lewis a molecule is formed from type I α(2,3)sialylated precursors [i.e., Galβ(1,3)GlcNAc] by α(1,4)fucosylation, catalyzed by Fuc-TIII (Kukowska-Latallo *et al.*, 1990; Lowe *et al.*, 1990; Weston *et al.*, 1992b; Hansson and Zopf, 1985) (not shown in Figure). It is presumed that the 3′ sulfo Lewis a and 3′ sulfo Lewis x are formed by sulfation of neutral precursors, followed by α(1,4)fucosylation, or α(1,3)fucosylation, respectively. It is also presumed that α(1,3)fucosylation of GalNAcβ(1,4)GlcNAc yields the GalNAcβ(1,4)[Fucα(1,3)]GlcNAc structure identified in 293 cells, and proposed as an E-selectin ligand (Yan *et al.*, 1993).

human leukocytes, and consequently does not function as a physiological leukocytic E-selectin ligand.

Work in this latter publication (Tyrrel *et al.*, 1991) also demonstrates that altered sialyl Lewis x structures, in which glucose has been used to substitute for N-acetylglucosamine at the reducing end, or in which this glucose has been reduced with sodium borohydride, maintain high affinities for E-selectin. Similarly, sialyl Lewis x structures containing N-glycolyl or N-acetyl forms of sialic acid, or in which the carbon 8 and/or carbon 9 moieties of the sialic acid have been modified by mild periodate oxidation, continued to maintain E-selectin binding affinities roughly equivalent to that displayed by the authentic sialyl Lewis x tetrasaccharide. By contrast, substitution of the α(2,3)sialic acid linkage with sialic acid in α(2,6)linkage abrogates E-selectin ligand activity.

Structures related to sialyl Lewis x and sialyl Lewis a, but in which the sialic acid moiety has been replaced by a sulfate group, also display substantial E-selectin binding capabilities (Yuen *et al.*, 1992). Molecular modeling studies again suggest that the sulfate moiety occupies a position in space similar to the carboxylate group of the sialic acid in sialyl Lewis a and sialyl Lewis x tetrasaccharides. Most of the modeling studies, using E-selectin, together with mutational analyses of this molecule, imply that three positively charged residues in the E-selectin carbohydrate recognition domain may interact with the negatively charged segment of sialic acid in sialyl Lewis x, and specifically point to interactions between the carboxylate moiety of the sialic Lewis x determinant and Lys^{113} within E-selectin (Erbe *et al.*, 1992). Additional modeling studies (Cooke *et al.*, 1994); Kogan *et al.*, 1995) derive from more recent work based on the three dimensional structure of the E-selectin carbohydrate recognition domain (solved by x-ray crystallography; Graves *et al.*, 1994), and on the solution structure of the sialyl Lewis x tetrasaccharide (solved by NMR spectroscopy; Cooke *et al.*, 1994). These additional modeling studies, supported by mutagenesis approaches (Kogan *et al.*, 1995), suggest that the sialyl Lewis x tetrasaccharide interacts with its binding site on E-selectin through Ca^{2+}-dependent coordination with the fucose residue of the sialyl Lewis x determinant, in a manner analogous to that used by the rat mannose binding protein to bind to mannose (Weis *et al.*, 1992). Moreover, the results reported in these more recent studies yield conclusions about the nature of the specific charged residues relevant to E-selectin-sialyl Lewis x interactions that are at odds with those reported earlier (Erbe *et al.*, 1992), by suggesting, for example, that the charge imparted by Lys^{113} is not necessary for ligand recognition. It seems likely that a final understanding of the molecular details of E-selectin-sialyl Lewis x interactions will require solution of the structure of ligand-replete E-selectin.

The apparent Kd for binding of E-selectin to sialyl Lewis x is approximately 0.1 to 1.0 mM (Jacob *et al.*, 1995; Cooke *et al.*, 1994). These observations suggest that leukocyte-E-selectin interactions are inherently weak, and/or that the leukocyte cell surface molecules that display sialyl Lewis x impart some additional specificity and adherence strength to the interaction. Biochemical data support the concept that effective sialyl Lewis x "presentation" by a leukocyte is in part a function of specific leukocyte cell surface molecules. For example, neutrophil L-selectin displays the sialyl Lewis x determinant, and maintains a relatively high affinity for E-selectin (Picker *et al.*, 1991a). Furthermore, there is evidence that L-selectin is expressed by the neutrophil microvilli, and that this location affords an extraordinary means of "presenting" oligosaccharide ligands to E-selectin in a manner that will facilitate neutrophil rolling (von Andrian *et al.*, 1995). These observations suggest the possibility that the ability of a specific leukocyte cell surface glycoprotein to effectively present a ligand to E-selectin may be a function of the abundance of that protein, its topographical location, and other properties inherent to that protein, including the number, locations, and detailed structures of the glycans displayed by that protein.

Two other specific glycoproteins have been identified as counter-receptors for E-selectin. These include PSGL-1, a polypeptide identified as a leukocyte counter-receptor for P-selectin discussed later in this chapter (Moore *et al.*, 1994; Sako *et al.*, 1993; Lenter *et al.*, 1994; Asa *et al.*, 1995), and ESL-1 (E-selectin ligand 1; Levinovitz *et al.*, 1993; Lenter *et al.*, 1994; Steegmaler *et al.*, 1995), a widely-expressed murine polypeptide with sequence similarity to a Golgi-localized polypeptide capable of binding fibroblast growth factor (Mourelatos *et al.*, 1995). PSGL-1 apparently shares the microvillous position occupied by L-selectin on leukocytes (Moore *et al.*, 1995), and recently also ESL-1 was shown to be enriched on microvilli (Steegmaier, Borges, Berger, Schwarz and Vestweber, submitted).

Nonetheless, it is clear that neutrophil PSGL-1 displays sialylated, fucosylated, O-linked poly-N-acetyllactosamine chains (Moore *et al.*, 1994), that effective PSGL-1-mediated E-selectin-dependent adhesion requires α(1,3)fucosylation of the component oligosaccharides (Sako *et al.*, 1993; Li *et al.*, 1996a; Vachino *et al.*, 1995), and that PSGL-1 maintains an extended conformation that may facilitate display of these molecules to E-selectin (Li *et al.*, 1996b). Likewise, effective ESL-1 function requires α(1,3)fucosylation of its component N-linked oligosaccharides (Steegmaler *et al.*, 1995).

COUNTER-RECEPTORS FOR P-SELECTIN

P-selectin is a 140 kDa glycoprotein localized to platelet α-granules (Stenberg *et al.*, 1985; Berman *et al.*, 1986), megakaryocytes (Beckstead *et al.*, 1986; McEver *et al.*, 1987), and the Weibel-Palade bodies of vascular endothelial cells (Bonfanti *et al.*, 1989; McEver *et al.*, 1989). This glycoprotein was originally designated granule membrane protein (GMP-140; Stenberg *et al.*, 1985), or platelet activation-dependent granule to external membrane protein (PADGEM, Berman *et al.*, 1986), and has been given the cluster designation CD62P. P-selectin is rapidly released (i.e., within about 10 minutes) from its intracellular locations following thrombin stimulation (platelets and endothelial cells, Berman *et al.*, 1986), or after stimulation with histamine, phorbol ester, and the calcium ionophore A23187 (endothelial cells, Hattori *et al.*, 1989). P-selectin synthesis may be induced by tumor necrosis factor α (Weller *et al.*, 1992) and other inflammatory mediators (reviewed in McEver, 1994). There is also evidence that P-selectin may be constitutively expressed by unstimulated vascular endothelium *in vivo*, at levels capable of mediating rolling-type leukocyte adhesion (Nolte *et al.*, 1994). P-selectin has been implicated in the adhesion of a variety of leukocytes to vascular endothelium, and in a variety of circumstances associated with acute and chronic inflammation (reviewed in McEver, 1994). Analysis of mice deficient in P-selectin have demonstrated that this molecule plays an important role in leukocyte emigration in a number of these circumstances (Mayadas *et al.*, 1993; Frennette *et al.*, 1996; Johnson *et al.*, 1995; Bullard *et al.*, 1995; Kunkel *et al.*, 1996).

Human and murine P-selectin cDNAs have been cloned (Johnston *et al.*, 1989; Weller *et al.*, 1992). Sequence analysis predicts an amino terminal domain with sequence similarity to the L- and E-selectin carbohydrate recognition domains, as well as an overall sequence organization similar to these other two selectins. A function for P-selectin in leukocyte adhesion was implied by observations indicating that activated platelets bind to neutrophils, monocytes, and the HL-60 and U937 myeloid cell lines, whereas non-activated platelets do not (Junji *et al.*, 1986; Silverstein and Nachman, 1987). The P-selectin-dependency of this adhesion was demonstrated in an assay whereby thrombin-activated platelets were allowed to form microscopically visible rosettes around myeloid cell types (Larsen *et al.*, 1989). Rosette formation was shown to be EDTA-sensitive, and to be blocked specifically with an anti-P-selectin antisera, or with purified P-selectin. Finally, vesicles reconstituted with purified P-selectin exhibited specific and saturable binding to U937 cells. Similar results have been reported by others (Hamburger and McEver, 1990), and extended in experiments showing that COS cells expressing a transfected P-selectin cDNA bind specifically to HL-60 cells (Geng *et al.*, 1990). Importantly, formalin fixation of HL-60 cells and neutrophils did not diminish binding efficiency (Geng *et al.*, 1990). These observations were reminiscent of experiments reporting retention of L-selectin-dependent binding of lymphocytes to paraformaldehyde-fixed HEV (Stamper and Woodruff, 1977), and suggested that carbohydrate may be an important component of the P-selectin counter-receptor(s).

Evidence for a role for endothelial cell P-selectin in leukocyte adhesion was provided in experiments where neutrophils were shown to adhere to endothelial cell monolayers following histamine- of PMA-induced expression of P-selectin (Geng *et al.*, 1990). Subsequent work has demonstrated that P-selectin-dependent leukocyte adhesion is calcium-dependent (Geng *et al.*, 1990; Geng *et al.*, 1991; Gamble *et al.*, 1990; Skinner *et al.*, 1991), and involves two high affinity calcium binding sites on P-selectin (Geng *et al.*, 1990; Geng *et al.*, 1991). Upon occupancy, these sites direct a conformational change in P-selectin that facilitates leukocyte adhesion.

The structural similarity of P-selectin to L-selectin suggested a role for oligosaccharides a components of leukocytic ligands for P-selectin, and prompted initial work that tested the role of the Lewis x determinant in P-selectin-dependent adhesion (Larsen *et al.*, 1990). These studies demonstrated inhibition of P-selectin-dependent cell adhesion with a Lewis x-active oligosaccharide known as lacto-N-fucopentaose III (LNFIII), and with monoclonal anti-Lewis x antibodies, but not with control reagents. LNFIII inhibited P-selectin-dependent leukocye adhesion half maximally at 50 μg/ml; the relatively high concentration required for inhibition prompted the authors to conclude that the Lewis x/ CD15 determinant may be only a part of the endogenous ligand for P-selectin. Subsequent attempts to duplicate these observations were initiated using an anti-CD15 antibody (82H5), and a multivalent Lewis x-albumin conjugate. Neither reagent was found to block P-selectin-neutrophil interactions (Moore *et al.*, 1991).

Other early work implied a requirement for sialic acid in P-selectin-dependent cell adhesion. For instance, P-selectin-dependent adhesion of HL-60 cells, and of neutrophils, is reduced by prior treatment of these cells with any one of three different broad-spectrum neuraminidases (*Clostridium perfringens*, *Vibrio cholera*, or *Arthrobacter ureafaciens*) (Corral *et al.*, 1990). To further determine if putative P-selectin ligands on myeloid cells include sialic acid in α(2,3) or in α(2,6) linkage (Fukuda *et al.*, 1985a; Spooncer *et al.*, 1984; Fukuda *et al.*, 1985b), cells were treated with a neuraminidase that cleaves *N*euNAcα(2,3) or *N*euNAcα(2,8) linkages, but not *N*euNAcα(2,6) linkages (Newcastle disease virus; NDV; Paulson *et al.*, 1982), and were then subjected to assay for P-selectin-dependent adhesion. Since NDV neuraminidase pre-treatment did not decrease binding, under conditions where essentially all cell surface *N*euNAcα(2,3) linkages were removed, the authors of this study concluded that *N*euNAcα(2,3) moieties are not required for P-selectin-dependent leukocyte adhesion, and suggested that NeuNAcα(2,6) linkages may represent important components of P-selectin ligands.

However, different conclusions were reached by other investigators, who demonstrated that (i) purified, soluble P-selectin binds specifically, and saturably, to myeloid cells, with half maximal binding at concentrations between 0.9 nM and 2.2 nM P-selectin, (ii), binding is diminished by protease treatment, and (iii), binding is diminished more by pretreatment with a broad spectrum sialidase (*Vibrio cholera*; 28% reduction in binding) than by pretreatment with NDV neuraminidase (52% reduction in binding) (Moore *et al.*, 1991). These observations consequently implicate *N*euNAcα(2,3) linkages as important components of P-selectin ligand activity.

Subsequent studies demonstrate that the sialyl Lewis x moiety represents an important component of P-selectin counter-receptor activity, and in part supports the prior evidence for involvement of both sialic acid and the Lewis x structure. These subsequent studies include a demonstration that (i) anti-sialyl Lewis x antibodies can inhibit P-selectin-dependent platelet rosetting, (ii) a sialyl Lewis x-positive CHO cell mutant (LEC11 cells) binds activated platelets, whereas a Lewis x-positive, sialyl Lewis x-negative CHO mutant LEC12 does not, and (iii) sialyl Lewis x-positive glycolipids, and a sialyl Lewis x-positive hexasaccharide, can each efficiently diminish P-selectin-dependent adhesion (Polley *et al.*, 1991). Similar conclusions were reached in other early studies, in which a correlation was made between P-selectin-dependent cell adhesion and cell surface expression of the sialyl Lewis x moiety in a panel of cultured cell lines (Zhou *et al.*, 1991). This study was the first to point to a requirement for specific cell surface polypeptides in the display of P-selectin ligands. Specifically, quantitative determinations of binding site number and affinity on a pair of the cell lines tested (HL-60 cells and Fuc-TIII-transfected CHO cells) indicated that the HL-60 cells maintained a low number of high affinity P-selectin binding sites, whereas the transfected CHO cells expressed a large number of low affinity binding sites. The simplest explanation of these results is that one or more specific cell surface glycoproteins displayed by HL-60 cells, but absent from CHO cells, modulates, in a positive manner, the efficiency with which the sialyl Lewis x moiety is recognized by P-selectin.

This notion is supported in part by experiments demonstrating the protease-sensitive nature of leukocyte P-selectin ligands (Larsen *et al.*, 1992). These observations were extended in efforts to identify the protein(s) involved (Moore *et al.*, 1992). Affinity chromatography procedures and blotting studies identified a 240 kDa polypeptide (120 kDa under reducing conditions) in HL-60 cells interacts with P-selectin in a calcium-dependent manner (Moore *et al.*, 1992). Subsequent molecular cloning efforts disclose that this molecule, termed P-selectin glycoprotein ligand (PSGL-1; discussed above), encodes a transmembrane polypeptide rich in proline, serine, and threonine residues typical of mucin-type glycoproteins. The PSGL-1 cDNA was isolated with an approach in which a mammalian cDNA expression library was screened for clones that impart P-selectin-dependent binding upon COS cells whose glycosylation phenotype had been modified by a co-transfected glycosyltransferase (the α(1,3/4)fucosyltransferase Fuc-TIII) predicted to be necessary, but not sufficient, for P-selectin-dependent cell adhesion (Sako *et al.*, 1993). These studies demonstrate that post-translational modification of PSGL-1 by α(1,3)fucosylation is required for effective recognition by P-selectin, and imply that sialylated, α(1,3)fucosylated O-linked glycans displayed by PSGL-1 are essential to this protein's P-selectin recognition characteristics.

Subsequent studies demonstrate that PSGL-1 is widely expressed by blood cells (Vachino *et al.*, 1995), but is not functional as a P-selectin counter-receptor unless its O-linked glycans are appropriately modified by the action of the core 2 GlcNAc transferase, by α(2,3)sialylation, and by α(1,3)fucosylation (Moore *et al.*, 1994; Li *et al.*, 1996b). However, the precise structures maintained by the oligosaccharides on functional leukocyte PSGL-1 remain to be determined. By contrast, recent studies demonstrate that sulfation of one or more of the three tyrosine residues near the NH2-terminus of PSGL-1 is essential for recognition by P-selectin (Wilkins *et al.*, 1995; Sako *et al.*, 1995; Pouyani and Seed, 1995). It is tempting to use these observations to help account for previous results demonstrating that sulfatides (3-sulfated galactosylceramides) can support (Aruffo *et al.*, 1991) or inhibit (Todderud *et al.*, 1992) P-selectin-dependent cell adhesion. Nonetheless, it is well-known that a variety of cells that express sulfatides, including erythrocytes and platelets, do not bind to P-selectin. Furthermore, the sulfatide binding properties and cell adhesion properties of P-selectin have been dissociated using mutant P-selectin molecules (Erbe *et al.*, 1993). Consequently, it is not yet clear that interactions between P-selectin and sulfatides have physiological relevance, or are related to protein tyrosine sulfation of PSGL-1.

Recent work indicates that a leukocyte molecule termed heat stable antigen (cluster designation CD24) can function as a counter-receptor for PSGL-1 (Sammar *et al.*, 1994; Aigner *et al.*, 1995). It is not yet known if this molecule maintains a microvillous topography, or if it contains sulfated tyrosine residues essential to P-selectin ligand activity.

FUCOSYLTRANSFERASES RELEVANT TO LEUKOCYTE E- AND P-SELECTIN LIGAND SYNTHESIS

It should be clear from the summary above that E- and P-selectin recognize specific cell surface glycoproteins on leukocytes. It should also be clear, however, that these molecules must maintain proper post-translational modifications in order to be recognized by these two selectins. In particular, it is implicitly and explicitly evident that α(1,3)fucosylation of the component oligosaccharide displayed by these glycoproteins is essential to their function as E- and/or P-selectins (Sako *et al.*, 1993; Pouyani and Seed, 1995, Moore *et al.*, 1994; Li *et al.*, 1996b; Steegmaier *et al.*, 1995). It is therefore of some substantial interest to define which α(1,3)fucosyltransferase(s) participate in this essential fucosylation event in leukocytes, and/or their progenitors. In particular, given the important role for E- and/or P-selectins in pathological inflammation (Seekamp *et al.*, 1994), and the demonstrated anti-inflammatory effectiveness of agents that block selectin-dependent leukocyte adhesion (Mulligan *et al.*, 1993; Buerke *et al.*, 1994), it becomes relevant to consider approaches that will inhibit leukocyte-specific α(1,3)fucosyltransferases, and thus lead to an anti-inflammatory outcome by virtue of decreasing expression of functional leukocyte selectin ligands.

As noted above, there are five known human α(1,3)fucosyltransferase loci (Fuc-TIII, IV, V, VI, and VII). It is possible to exclude an essential role for Fuc-TIII and Fuc-TVI in leukocyte selectin ligand fucosylation since humans homozygous for null alleles at these loci (the Lewis blood group locus, and the "plasma-type" fucosyltransferase locus, respectively) are without obvious deficiency in leukocytic selectin ligands (Mollicone *et al.*, 1994a; Mollicone *et al.*, 1994b). Likewise, these genes are not normally expressed in leukocytes (Cameron *et al.*, 1995; Postigo *et al.*, 1994; Watkins, W.M., personal communication). Similarly, Fuc-TV is not generally expressed in normal leukocytes (Yago *et al.*, 1993; Cameron *et al.*, 1995; Postigo *et al.*, 1994; Koszdin and Bowen, 1992; Clarke *et al.*, 1996). By contrast, leukocytes express both Fuc-TIV (Goelz *et al.*, 1990; Lowe *et al.*, 1991; Kumar *et al.*, 1991; Yago *et al.*, 1993; Gersten *et al.*, 1995) and Fuc-TVII (Natsuka *et al.*, 1994; Sasaki *et al.*, 1994; Smith *et al.*, 1996), suggesting these enzymes as candidates for participation in leukocyte selectin ligand biosynthesis.

Fuc-TVII is clearly able to form the sialyl Lewis x determinant *in vitro*, and in cultured cell lines transfected with expression vectors encoding this enzyme (Natsuka *et al.*, 1994; Sasaki *et al.*, 1994; Smith *et al.*, 1996), and is capable of directing expression of ligands for E- and/or P-selectin on transfected cells (Sasaki *et al.*, 1994; Pouyani and Seed, 1995; Sako *et al.*, 1995; Li *et al.*, 1996b; Knibbs *et al.*, 1996). There is evidence that this enzyme controls inducible expression of E-selectin ligands on human T-lymphocytes (Knibbs *et al.*, 1996). An essential role for this enzyme in the expression of functional leukocyte E- and P-selectin activity is evident from an analysis of mice made deficient in this enzyme through "gene knock out" technology (Maly *et al.*, 1996). While these Fuc-TVII (–/–) mice develop and breed normally, their blood neutrophils and monocytes do not bind to chimeric mouse E-selectin-human IgM or mouse P-selectin-human IgM histochemical probes. The leukocytes do not maintain normal "rolling" properties when analyzed by intravital microscopy approaches, and are profoundly deficient in their ability to mobilize to extravascular sites in association with inflammatory challenges. It can be concluded that these deficits are a consequence of deficient fucosylation of the glycans displayed by leukocyte E- and/or P-selectin counter-receptor (ESL-1, PSGL-1, L-selectin, and CD24), because the leukocytes in these mice display essentially normal levels of these four glycoproteins. Considered together, these observations demonstrate an essential role for Fuc-TVII in the construction of functional E- and P-selectin ligands in murine leukocytes, and a requirement for α(1,3)linked fucose as a component of functional E- and P-selectin counter-receptors.

These mice also maintain a circulating leukocytosis characterized by a 7-fold increase in the number of circulating neutrophils, and substantial increases in the numbers of other blood leukocytes. When considered together with the observation that a circulating leukocytosis is also evident in P-selectin-deficient mice (Mayadas *et al.*, 1993, and in mice made deficient in both E-selectin and P-selectin (Frenette *et al.*, 1996), these results support the hypothesis that the E- and P-selectins, and their leukocytic counter-receptors, function to maintain homeostasis in the numbers of myeloid lineage leukocytes circulating in the blood (Johnson *et al.*, 1995).

A role for Fuc-TIV in selectin ligand synthesis is less clear. As discussed above, this enzyme, also known as ELAM-1 Ligand Fucosyl Transferase (ELFT; Goelz *et al.*, 1990), is not generally able to direct cell surface expression of the sialyl Lewis x determinant and E-selectin ligands when expressed in cultured cell lines (for examples, see Lowe *et al.*, 1990; Lowe *et al.*, 1991; Kumar *et al.*, 1991). Nonetheless, it is clear that under some circumstances, Fuc-TIV can direct expression of sialyl Lewis x and E-selectin ligand activity (Goelz *et al.*, 1994; Li *et al.*, 1996b). The variable ability of this enzyme to exhibit this phenotype is believed to be a function of the glycosylation phenotype of the host cell line in which it is expressed (Goelz *et al.,* 1994). However, the biochemical attributes of the glycans, or other molecules, necessary to support this property are not known, and it is not known if leukocytes maintain the glycosylation machinery necessary to support Fuc-TIV-dependent selectin ligand expression. Preliminary analysis of mice made deficient in Fuc-TIV suggest that this enzyme may play a contributory, but perhaps non-essential, role in E- and P-selectin ligand expression (Thall and Lowe, unpublished observations).

While the observations made with these fucosyltransferase-deficient mice may provide insight into the relevance of Fuc-TIV and Fuc-TVII in selectin ligand synthesis, a full understanding of the consequences of these genetic manipulations will require a knowledge of the structures of the oligosaccharides displayed by E- and P-selectin counter-receptors on the leukocytes of wild type mice, and in the mutant strains. In this context it is noteworthy that murine leukocytes do not express CSLEX-1-active sialyl Lewis x determinants (Ito *et al.*, 1994; Malý *et al.*, 1996), nor an antigen corresponding to an oligosaccharide ligand for E-selectin identified recently in a cultured murine myeloid cell line (Osanai *et al.*, 1996)(Lowe, unpublished observations). Thus, while it seems clear that the murine E- and P-selectin counter-receptors will bear $\alpha(1,3)$fucosylated oligosaccharides, the precise structures of these molecules remains to be defined.

Acknowledgements

Dr. Lowe is an Associate Investigator of the Howard Hughes Medical Institute. This work was supportedby NIH grant R01GM47455.

References

Aigner, S., Ruppert, M., Hubbe, M., Sammar, M., Sthoeger, Z., Butcher, E.C., Vestweber, D. and Altevogt, P. (1995). Heat stable antigen (mouse CD24) supports myeloid cell binding to endothelial and platelet P-selectin. *Int. Immunol.*, **7**:1557–1565.

Arbones, M.L., Ord, D.C., Ley, K., Ratech, H., Maynard-Curry, C., Otten, G., Capon, D.J. and Tedder, T.F. (1994). Lymphocyte homing and leukocyte rolling and migration are impaired in L-selectin-deficient mice. *Immunity*, **1**:247–260.

Aruffo, A., Kolanus, W., Walz, G., Fredman, P. and Seed, B. (1991). CD62/P-selectin recognition of myeloid and tumor cell sulfatides. *Cell*, **67**:35–44.

Asa, D., Raycroft, L., Ma, L., Aeed, P.A., Kaytes, P.S., Elhammer, A.P. and Geng, J.G. (1995). The P-selectin glycoprotein ligand functions as a common human leukocyte ligand for P- and E-selectins. *J. Biol. Chem.*, **270**:11662–11670.

Ashwell, G. and Hartford, J. (1982). Carbohydrate-specific receptors of the liver. *Annu. Rev. Biochem.*, **51**:531–554.

Atherton, A. and Born, G.V. (1973a). Proceedings: effects of neuraminidase and N-acetyl neuraminic acid on the adhesion of circulating granulocytes and platelets in venules. *J. Physiol. (London)*, **234**:66P–67P.

Atherton, A. and Born, G.V. (1973b). Relationship between the velocity of rolling granulocytes and that of the blood flow in venules. *J. Physiol.*, **233**:157–165.

Bargatze, R.F., Kurk, S., Butcher, E.C. and Jutila, M.A. (1994). Neutrophils roll on adherent neutrophils bound to cytokine-induced endothelial cells via L-selectin on the rolling cells. *J. Exp. Med.*, **180**:1785–1792.

Baumhueter S., Singer, M.D., Henzel, W., Hemmerich, S., Renz, M., Rosen, S.D. and Lasky, L.A. (1993). Binding of L-selectin to the vascular sialomucin CD34. *Science*, **262**:436–438.

Beckstead, J.H., Stenberg, P.E., McEver, R.P., Shuman, M.A. and Bainton, D.F. (1986). Immunohistochemical localization of membrane and alpha-granule proteins in human megakaryocytes: application to plastic-embedded bone marrow biopsy specimens. *Blood*, **67**:285–293.

Berg, E.L., Robinson, M.K., Mansson, O., Butcher, E.C. and Magnani, J.L. (1991). A carbohydrate domain common to both sialyl Le^a and sialyl Le^x is recognizd by the endothelial cell leukocyte adhesion molecule ELAM-1. *J. Biol. Chem.*, **266**:14869–14872.

Berg, E.L., Magnani, J., Warnock, R.A., Robinson, M.K. and Butcher, E.C. (1992). Comparison of L-selectin and E-selectin ligand specificites: the L-selectin can bind to the E-selectin ligands sialyl Le^x and sialyl Le^a. *Biochem. Biophys. Res. Commun.*, **184**:1048–1055.

Berg, E.L., McEvoy, L.M., Berlin, C., Bargatze, R.F. and Butcher, E.C. (1993). L-selectin-mediated lymphocyte rolling on MAdCAM-1. *Nature*, **366**:695–698.

Berlin, C., Berg, E.L., Briskin, M.J., Andrew, D.P., Kilshaw, P.J., Holzmann, B., Weissman, I.L., Hamann, A. and Butcher, E.C. (1993). Alpha 4 beta 7 integrin mediates lymphocyte binding to the mucosal vascular addressin MAdCAM-1. *Cell*, **74**:185–195.

Berman, C.L., Yeo, E.L., Wencel Drake, J.D., Furie, B.C., Ginsberg, M.H. and Furie, B. (1986). A platelet alpha granule membrane protein that is associated with the plasma membrane after activation. Characterization and subcellular localization of platelet activation-dependent granule-external membrane protein. *J. Clin. Invest.*, **78**:130–137.

Bertozzi, C.R., Fukuda, S. and Rosen, S.D. (1995). Sulfated disaccharide inhibitors of L-selectin: deriving structural leads from a physiological selectin ligand. *Biochemistry*, **34**:14271–14278.

Bevilacqua, M.P., Pober, J.S., Wheeler, M.E., Cotran, R.S. and Gimbrone, M.A., Jr. (1985a). Interleukin 1 acts on cultured human vascular endothelium to increase the adhesion of polymorphonuclear leukocytes, monocytes, and related leukocyte cell lines. *J. Clin. Invest.*, **76**:2003–2011.

Bevilacqua, M.P., Pober, J.S., Wheeler, M.E., Cotran, R.S. and Gimbrone, M.A., Jr. (1985b). Interleukin-1 activation of vascular endothelium. Effects on procoagulant activity and leukocyte adhesion. *Am. J. Pathol.*, **121**:394–403.

Bierhuizen, M.F. and Fukuda, M. (1992). Expression cloning of a cDNA encoding UDP-GlcNAc:Gal beta 1-3-GalNAc-R (GlcNAc to GalNAc) beta 1-6GlcNAc transferase by gene transfer into CHO cells expressing polyoma large tumor antigen. *Proc. Natl. Acad. Sci. USA*, **89**:9326–9330.

Bierhuizen, M.F., Maemura, K., Kudo, S. and Fukuda, M. (1995). Genomic organization of core 2 and I branching beta-1,6-N acetylglucosaminyltransferases. Implication for evolution of the beta-1,6-N-acetylglucosaminyltransferase gene family. *Glycobiology*, **5**:417–425.

Bonfanti, R., Furie, B.C., Furie, B. and Wagner, D.D. (1989). PADGEM is a component of Weibel-Palade bodies in endothelial cells. *Blood*, **73**:1109–1112.

Bosse, R. and Vestweber, D. (1994). Only simultaneous blocking of the L- and P-selectin completely inhibits neutrophil migration into mouse peritoneum. *Eur. J. Immunol.*, **24**:3019–3024.

Bowen, B.R, Nguyen, T. and Lasky, L.A. (1989). Characterization of a human homologue of the murine peripheral lymph node homing receptor. *J. Cell Biol.*, **109**:421–427.

Bowen, B.R., Fennie, C. and Lasky, L.A. (1990). The Mel 14 antibody binds to the lectin domain of the murine peripheral lymph node homing reeptor. *J. Cell Biol.*, **110**:147–153.

Brady, H.R., Spertini, O., Jiminez, W., Brenner, B.M., Marsden, P.A. and Tedder, T.F. (1992). Neutrophils, monocytes, and lymphocytes bind to cytokine–activated kidney glomerular endothelial cells through L-selectin (LAM-1) *in vitro*. *J. Immunol.*, **49**:2437–2444.

Briscoe, D.M., Schoen, F.J., Rice, G.E., Bevilacqua, M.P., Ganz, P. and Pober, J.S. (1991). Induced expression of endothelial-leukocyte adhesion molecules in human cardiac allografts. *Transplantation*, **51**:537–539.

Brown, J., Greaves, M.F. and Molgaard, H.V. (1991). The gene encoding the stem cell antigen, CD34, is conserved in mouse and expressed in haemopoietic progenitor cell lines, brain, and embryonic fibroblasts. *Int. Immunol.*, **3**:175–184.

Buerke, M., Weyrich, A.S., Zheng, Z., Gaeta, F.C., Forrest, M.J. and Lefer, A.M. (1994). Sialyl Lewis^{x}-containing oligosaccharide attenuates myocardial reperfusion injury in cats. *J. Clin. Invest.*, **93**:1140–1148.

Bullard, D.C., Qin, L., Lorenzo, I., Quinlin, W.M., Doyle, N.A., Bosse, R., Vestweber, D., Doerschuk, C.M. and Beaudet, A.L. (1995). P-selectin/ICAM-1 double mutant mice; acute emigration of neutrophils into the peritoneum is completely absent but is normal into pulmonary alveoli. *J. Clin. Invest.*, **95**:1782–1788.

Butcher, E.C. (1990). Cellular and molecular mechanisms that direct leukocyte traffic. *Am. J. Pathol.*, **136**:3–11.

Butcher, E.C. (1991). Leukocyte-endothelial cell recognition: three (or more) steps to specificity and diversity. *Cell*, **67**:1033–1036.

Camerini, D., James, S.P., Stamenkovic, I. and Seed, B. (1989). Leu-8/TQ1 is the human equivalent of the Mel-14 lymph node homing receptor. *Nature*, **342**:78–82.

Cameron, H.S., Szczepaniak, D. and Weston, B.W. (1995). Expression of human chromosome 19p alpha(1,3)-fucosyltransferase genes in normal tissues. Alternative splicing, polyadenylation, and isoforms. *J. Biol. Chem.*, **270**:20112–20122.

Carlos, T.M., Dobrina, A., Ross, R. and Harlan, J.M. (1990). Multiple receptors on human monocytes are involved in adhesion to cultured human endothelial cells. *J. Leukocyte Biol.*, **48**:451–456.

Cheng, J., Baumhueter, S., Cacalano, G., Carver-Moore, K., Thibodeaux, H., Thomas, R., Broxmeyer, H.E., Cooper, S., Hague, N., Moore, M. and Lasky, L.A. (1996). Hematopoietic defects in mice lacking the sialomucin CD34. *Blood*, **87**:479–490.

Chien, S. (1982). Rheology in the microcirculation in normal and low flow states. *Adv. Shock. Res.*, **8**:71–80.

Clarke, J.L. and Watkins, W.M. (1996). α1,3-Fucosyltransferase expression in developing human myeloid cells. Antigenic, enzymatic, and in RNA analyses. *J. Biol. Chem.*, **271**:10317–10328.

Cooke, R.M., Hale, R.S., Lister, S.G., Shah, G. and Weir, M.P. (1994). The conformation of the sialyl Lewis X ligand changes upon binding to E-selectin. *Biochemistry*, **33**:10591–10596.

Corral, L., Singer, M.S., Macher, B.A. and Rosen, S.D. (1990). Requirement for sialic acid on neutrophils in a GMP-140 (PADGEM) mediated adhesive interaction with activated platelets. *Biochem. Biophys. Res., Comm.*, **172**:1349–1356.

Cotran, R.S., Gimbrone, M.A., Jr., Bevilacqua, M.P., Mendrick, D.L. and Pober, J.S. (1989). Induction and detection of a human endothelial activation antigen *in vivo*. *J. Exp. Med.*, **164**:661–666.

Crommie, D. and Rosen, S.D. (1995). Biosynthesis of GlyCAM-1, a mucin-like ligand for L-selectin. *J. Biol. Chem.*, **270**:22614–22624.

Drickamer, K. (1988). Two distinct classes of carbohydrate-recognition domains in animal lectins. *J. Biol. Chem.*, **263**:9557–9560.

Engelberts, I., Samyo, S.K., Leeuwenberg, J.F.M., van der Linden, C.J. and Buurman, W.A. (1992). A role for ELAM-1 in the pathogenesis of MOF during septic shock. *J. Surg. Res.*, **53**:136–144.

Erbe, D.V., Wolitzky, B.A., Presta, L.G., Norton, C.R., Ramos, R.J. and Burns, D.K. (1992). Identification of an E-selectin region critical for carbohydrate recognition and cell adhesion. *J. Cell Biol.*, **119**:215–217.

Erbe, D.V., Watson, S.R., Presta, L.G., Wolitzky, B.A., Foxall, C., Brandley, B.K. and Lasky, L.A. (1993). P- and E-selectin use common sites for carbohydrate ligand recognition and cell adhesion. *J. Cell Biol.*, **120**:1227–1235.

Feizi, T. (1985). Demonstration by monoclonal antibodies that carbohydrate structures of glycoproteins and glycolipids are onco-developmental antigens. *Nature*, **314**:53–57.

Fiebig, E., Ley, K. and Arfors, K.E. (1991). Rapid leukocyte accumulation by "spontaneous" rolling and adhesion in the exteriorized rabbit mesentery. *Int. J. Microcirc. Clin. Exp.*, **10**:127–144.

Fiete, D., Srivastava, V., Hindsgaul, O. and Baenziger, J.U. (1991). A hepatic reticuloendothelial cell receptor specific for SO4-4GalNAc beta 1,4GlcNAc beta 1,2Man alpha that mediates rapid clearance of lutropin. *Cell*, **67**:1103–1110.

Fina, L., Molgaard, H.V., Robertson, D., Bradley, N.J., Monaghan, P., Delia, D., Sutherland, D.R., Baker, M.A. and Greaves, M.F. (1990). Expression of the CD34 gene in vascular endothelial cells. *Blood*, **75**:2417–2426.

Finger, E.B., Puri, K.D., Alon, R., Lawrence, M.B., von Andrian, U.H. and Springer, T.A. (1995). Adhesion through L-selectin requires a threshold hydrodynamic shear. *Nature*, **379**:266–269.

Foxall, C., Watson, S.R., Dowbenko, D., Fennie, C., Lasky, L.A., Kiso, M., Hasegawa, A., Asa, D. and Brandley, B.K. (1992). The three members of the selectin receptor family recognize a common carbohydrate epitope, the sialyl Lewis X oligosaccharide. *J. Cell Biol.*, **4**:895–902.

Frenette, P.S., Mayadas, T.N., Rayburn, H., Hynes, R.O. and Wagner, D.D. (1996). Susceptibility to infection and altered hematopoiesis in mice deficient in both P- and E-Selectins. *Cell*, **84**:563–574.

Fukuda, M., Spooncer, E., Oates, J.E., Dell, A. and Klock, J.C. (1984). Structure of sialylated fucosyl lactosaminoglycan isolated from human granulocytes. *J. Biol. Chem.*, **259**:10925–10935.

Fukuda, M., Bothner, B., Ramsamooj, P., Dell, A., Tiller, P.R., Varki, A. and Klock, J.C. (1985a). Structures of sialylated fucosyl polylactosaminoglycans isolated from chronic myelogenous leukemia cells. *J. Biol. Chem.*, **260**:12957–12967.

Fukuda, M.N., Dell, A., Oates, J.E., Wu, P., Klock, J.C. and Fukuda, M. (1985b). Structures of glycosphingolipids isolated from human granulocytes. The presence of a series of linear poly-N-acetyllactosaminylceramide and its significance in glycolipids of whole blood cells. *J. Biol. Chem.*, **260**:1067–1082.

Fukuda, M.N., Dell, A., Tiller, P.R., Varki, A., Klock, J.C. and Fukuda, M. (1986). Structure of a novel sialylated fucosyl lacto-N-norhexaosylceramide isolated from chronic myelogenous leukemia cells. *J. Biol. Chem.*, **261**:2376–2383.

Fukushi, Y., Hakomori, S., Nudelman, E. and Cochran, N. (1984). Novel fucolipids accumulating in human adenocarcinoma II. Selective isolation of hybridoma antibodies that differentially recognize mono-, di-, and trifucosylated type 2 chain. *J. Biol. Chem.*, **259**:4681–4685.

Fukushima, K., Hirota, M., Terasaki, P.I., Wakisaka, A., Togashi, H., Chia, D., Suyama, N., Fukushi, Y., Nudelman, E. and Hakomori, S. (1984). Characterization of sialosylated Lewis x as a new tumor-associated antigen. *Cancer Res.*, **44**:5279–5285.

Gallatin, W.M., Weissman, I.L. and Butcher, E.C. (1983). A cell-surface molecule involved in organ-specific homing of lymphocytes. *Nature*, **304**:30–34.

Gamble, J.R., Skinner, M.P. Berndt, M.C. and Vadas, M.A. (1990). Prevention of activated neutrophil adhesion to endothelium by soluble adhesion protein GMP-140. *Science*, **249**:414–417.

Geng, J.G., Bevilacqua, M.P., Moore, K.L, McIntyre, T.M., Prescott, S.M., Kim, J.M., Bliss, G.A., Zimmerman, G.A. and McEver, R.P. (1990). Rapid neutrophil adhesion to activated endothelium mediated by GMP-140. *Nature*, **343**:757–760.

Geng, J.G., Moore, K.L., Johnson, A.E. and McEver, R.P. (1991). Prevention of activated neutrophil adhesion to endothelium by soluble adhesion protein GMP-140. *Science*, **249**:414–417.

Gersten, K.M., Natsuka, S., Trinchera, M., Petryniak, B, Kelly, R.J., Hiraiwa, N., Jenkins, N.A., Gilbert, D.J., Copeland, N.G. and Lowe, J.B. (1995). Molecular cloning, expression, chromosomal assignment, and tissue-specific expression of a murine α-(1,3)-fucosyltransferase locus corresponding to the human ELAM-1 ligand fucosyl transferase. *J. Biol. Chem.*, **270**:25047–2056.

Goelz, S.E., Hession, C., Goff, D., Griffiths, B., Tizard, R. and Newman, B. (1990). ELGF: a gene that directs the expression of an ELAM-1 ligand. *Cell*, **63**:1349–1356.

Goelz, S., Kumar, R., Potvin, B., Sundaram, S., Brickelmaier, M. and Stanley, P. (1994). Differential expression of an E-selectin ligand (SLe^x) by two Chinese hamster ovary cell lines transfected with the same α(1,3)-fucosyltransferase gene (ELFT). *J. Biol. Chem.*, **269**:1033–1040.

Graber, N., Gopal, T.V., Wilson, D., Beall, L.D., Polte, T. and Newman, W. (1990). T cells bind to cytokine-activated endothelial cells via a novel, inducible sialoglycoprotein and endothelial leukocyte adhesion molecule-1. *J. Immunol.*, **145**:819–830.

Green, P.J., Tamatani, T., Watanabe, T., Miyasaka, M., Hasegawa, A., Kiso, M., Yuen, C-T., Stoll, M.S. and Feizi, T. (1992). High affinity binding of leukocyte adhesion molecule L-selectin to 3′-sulphated-Le^a and — Le^x oligosaccharides and the predominance of sulphate in this interaction deminstrated by binding studies with a series of lipid-linked oligosaccharides. *Biochem. Biophys. Res. Commun.*, **188**:244–251.

Griffiths, C.E., Barker, J.N., Kunkel, S. and Nickoloff, B.J. (1991). Modulation of leukocyte adhesion molecules, a T-cell chemotaxin (IL-8) and a regulatory cytokine (TNF-alpha) in allergic contact dermatitis (rhus dermatitis). *Br. J. Dermatol.*, **124**:519–526.

Groves, R.W., Allen, M.H., Barker, J.N., Haskard, D.O. and MacDonald, D.M. (1991). Endothelial leukocyte adhesion molecule-1 (E-selectin) expression in cutaneous inflammation. *Br. J. Dermatol.*, **124**:117–123.

Hallmann, R., Jutila, M.A., Smith, C.W., Anderson, D.C., Kishimoto, T.K. and Butcher, E.C. (1991). The peripheral lymph node homing receptor, L-selectin, is involved in CD18-independent adhesion of human neutrophils to endothelium. *Biochem. Biophys. Res. Commun.*, **174**:236–243.

Hamburger, S.A. and McEver, R.P. (1990). GMP-140 mediates adhesion of stimulated platelets to neutrophils. *Blood*, **75**:550–554.

Hansson, G.C. and Zopf, D. (1985). Biosynthesis of the cancer-associated sialyl-Lea antigen. *J. Biol. Chem.*, **260**:9388–9382.

Hattori, R., Hamilton, K.K., Fugate, R.D., McEver, R.D. and Sims, P.J. (1989). Stimulated secretion of endothelial von Willebrand factor is accompanied by rapid redistribution to the cell surface of the intracellular granule membrane protein GMP-140. *J. Biol. Chem.*, **264**:7768–7771.

Hemmerich, S. and Rosen, S.D. (1994c). 6′-Sulfated sialyl Lewis x is a major capping group of GlyCAM-1. *Biochemistry*, **33**:4830–4835.

Hemmerich, S., Bertozzi, C.R., Leffler, H. and Rosen, S.D. (1994b). Identification of the sulfated monosaccharides of GlyCAM-1, an endothelial-derived ligand for L-selectin. *Biochemistry*, **33**:4820–4829.

Hemmerich, S., Butcher, E.C. and Rosen, S.D. (1994a). Sulfation-dependent recognition of HEV-ligands by L-selectin and MECA-79, and adhesion blocking mAb. *J. Exp. Med.*, **180**:2219–2226.

Hemmerich, S., Leffler, H. and Rosen, S.D. (1995). Structure of the O-Glycans in GlyCAM-1, an endothelial-derived ligand for L-selectin. *J. Biol. Chem.*, **270**:12035–12047.

Holmes, E.H., Ostrander, G.K. and Hakomori, S. (1985). Enzymatic basis for the accumulation of glycolipids with X and dimeric X determinants in human lung cancer cells (NCI-H69). *J. Biol. Chem.*, **260**:7619–7627.

Holmes, E.H., Ostrander, G.K. and Hakomori, S. (1986). Biosynthesis of the sialyl-Lex determinants carried by type 2 chain glycosphingolipids (IV3NeuAcIII3FucnLc4, V13neuAcV3FucnLc6, and V13NeuAcIII3V3Fuc2nLc6) in human lung carcinoma PC9 cells. *J. Biol. Chem.*, **261**:3737–3743.

Howard, D.R., Fukuda, M., Fukuda, M.N. and Stanley, P. (1987). The GDP-fucose: N-acetylglucosaminide 3-α-L-fucosyltransferases of LEC11 and LEC12 Chinese hamster ovary mutants exhibit novel specificities for glycolipid substrates. *J. Biol. Chem.*, **262**:16830–16837.

Imai, Y., Singer, M.S., Fennie, C., Lasky, L.A. and Rosen, S.D. (1991). Identification of a carbohydrate-based endothelial ligand for a lymphocyte homing receptor. *J. Cell Biol.*, **113**:1213–1221.

Imai, Y., Lasky, L.A. and Rosen, S.D. (1992). Further characterization of the interaction between L-selectin and its endothelial ligands. *Glycobiology*, **2**:373–381.

Imai, Y., Lasky, L.A. and Rosen, S.D. (1993). Sulphation requirement for GlyCAM-1, an endothelial ligand for L-selectin. *Nature*, **361**:555–557.

Ito, K., Handa, K. and Hakomori S. (1994). Species-specific expression of sialosyl-Le(x) on polymorphonuclear leukocytes (PMN), in relation to selectin-dependent PMN responses. *Glycoconj. J.*, **11**:232–237.

Jacob, G.S., Kirmaier, C., Abbas, S.Z., Howard, S.C., Steininger, C.N., Welply, J.K. and Scudder, P. (1995). Binding of sialyl Lewis x to E-selectin as measured by fluorescence polarization. *Biochemistry*, **34**:1210–1217.

Jain, R.K., Vig, R., Rampal, R., Chandrasekaran, E.V. and Matta, K.L. (1994). Total synthesis of 3′-O-sialyl, 6′O-sulfo Lewis x, NeuAcα2,3(6-O-SO3Na) Galβ1,4(Fucα1,3)-GlcNAc-β-OMe: A major capping group of GlyCAM-1. *J. Am. Chem. Soc.*, **116**:12123–12124.

Johnson, R.C., Mayadas, T.N., Frenette, P.S., Mebius, R.E., Subramaniam, M., Lacasce, A., Hynes, R.O. and Wagner, D.D. (1995). Blood cell dynamics in P-selectin-deficient mice. *Blood*, **86**:1106–1114.

Johnston, G.I., Cook, R.G. and McEver, R.P. (1989). Cloning of GMP-140, a granule memrane protein of platelets and endothelium: sequence similarity to proteins involved in cell adhesion and inflammation. *Cell*, **56**:1033–1044.

Jungi, T.W., Spycher, M.O., Nydegger, U.E. and Barandun, S. (1986). Platelet-leukocyte interaction: selective binding of thrombin-stimulated platelets to human monocytes, polymorphonuclear leukocytes, and related cell lines. *Blood*, **67**:629–636.

Jutila, M.A., Rott, L., Berg, E.L. and Butcher, E.C. (1989). Function and regulation of the neutrophil MEL-14 antigen *in vivo*: comparison with LFA-1 and Mac-1. *J. Immunol.*, **143**:3318–3324.

Kansas, G.S., Spertini, O. and Tedder, T.F. (1991). Leukocyte adhesion molecule-1 (LAM-1): structure, function, genetics, and evolution. In *Cellular and molecular mechanisms of inflammation*, edited by C.G. Cochrane and M.A. Gimbone, Jr., p. 31. San Diego: Academic Press.

Kikuta, A. and Rosen, S.D. (1994). Localization of ligands for L-selectin in mouse peripheral lymph node high endothelial cells by colloidal gold conjugates. *Blood*, **84**:3766–3775.

Kitagawa, H. and Paulson, J.C. (1994). Cloning of a novel alpha 2,3-sialyltransferase that sialylates glycoprotein and glycolipid carbohydrate groups. *J. Biol. Chem.*, **269**:1394–401.

Knibbs, R.N., Craig, R.A., Natsuka, S., Chang, A., Cameron, M., Lowe, J.B. and Stoolman, L.M. (1996). The fucosyltransferase Fuc-TVII regulates E-selectin ligand synthesis in human T-cells. *J. Cell. Biol.*, **133**:911–920.

Koch, A.E., Burrows, J.C., Haines, G.K., Carlos, T.M., Harlan, J.M. and Leibovich, S.J. (1991). Immunolocalization of endothelial and leukocyte adhesion moelcules in human rheumatoid and osteoarthritic synovial tissues. *Lab. Invest.*, **64**:313–320.

Kogan, T.P., Revelle, B.M., Tapp, S., Scott, D. and Beck, P.J. (1995). A single amino acid residue can determine the ligand speificity of E-selectin. *J. Biol. Chem.*, **270**:14047–14055.

Kornfeld, R. and Kornfeld, S. (1985). Assembly of asparagine-linked oligosaccharides. *Annu. Rev. Biochem.*, **54**:631–634.

Kornfeld, S. (1987). Trafficking of lysosomal enzymes. *FASEB J.*, **1**:462–468.

Koszdin, K.L. and Bowen, B.R. (1992). The cloning and expression of a human alpha-1,3 fucosyltransferase capable of forming the E-selectin ligand. *Biochem. Biophys. Res. Commun.*, **187**:152–157.

Kukowska-Latallo, J.F., Larsen, R.D., Nair, R.P. and Lowe, J.B. (1990). A cloned human cDNA determines expression of a mouse stage-specific embryonic antigen and the Lewis blood group α (1,3/1,4) fucosyltransferase. *Genes Dev.*, **4**:1288–1303.

Kumar, R., Potvin, B., Muller, W.A. and Stanley, P. (1991). Cloning of a human α(1,3)-fucosyltransferase gene that encodes ELFT but does not confer ELAM-1 recognition of Chinese hamster ovary cell transfectants. *J. Biol. Chem.*, **266**:21777–21783.

Kunkel, E.J., Jung, U., Bullard, D.C., Norman, K.E., Wolitzky, B.A., Vestweber, D., Beaudet, A.L. and Ley, K. (1996). Absence of trauma-induced leukocyte rolling in mice deficient in both P-selectin and intercellular adhesion molecule 1. *J. Exp. Med.*, **183**:57–65.

Labow, M.A., Norton, C.R., Rumberger, J.M., Lombard-Gillooly, K.M., Shuster, D.J., Hubbard, J., Bertko, R., Knaack, P.A., Terry, R.W., Harbison, M.L., Kontgen, F., Stewart, C.L., McIntyre, K.W., Will, P.C., Burns, D.K. and Wolitzky, B.A. (1994). Characterization of E-selectin-deficient mice: demonstration of overlapping function of the endothelial selectins. *Immunity*, **1**:709–720.

Larsen, E., Celi, A., Gilbert, G.E., Furie, B.C., Erban, J.K., Bonfanti, R., Wagner, D.D. and Furie, B. (1989). PADGEM protein: a receptor that mediates the interaction of activated platelets with neutrophils and monocytes. *Cell*, **59**:305–312.

Larsen, E., Palabrica, T., Sajer, S., GIlbert, G.E., Wagner, D.D., Furie, B.C. and Furie, B. (1990). PADGEM-dependent adhesion of platelets to monocytes and neutrophils is mediated by a lineage-specific carbohydrate, LNF III (CD15). *Cell*, **63**:467–474.

Larsen, G.R., Sako, D., Ahern, T.J., Shaffer, M., Erban, J., Sajer, S.A., Gibson, R.M., Wagner, D.D., Furie, B.C. and Furie, B. (1992). P-selectin and E-selectin. Distinct but overlapping leukocyte ligand specificities. *J. Biol. Chem.*, **267**:11104–11110.

Larsen, R.D., Ernst, L.K, Nair, R.P. and Lowe, J.B. (1990). Molecular cloning, sequence, and expression of human GDP-L-Fucose:β-D-galacoside 2-α-L-fucosyltrnasferase cDNA that can form the H blood group antigen. *Proc. Natl. Acad. Sci. USA*, **87**:6674–6678.

Lasky, L.A. (1992a). Selectins: interpreters of cell-specific carbohydrate information during inflammation. *Science*, **258**:964–969.

Lasky, L.A., Singer, M.S., Yednock, T.A., Dowbenko, D., Fennie, C. and Rodriguez, H. (1989). Cloning of a lymphocyte homing receptor reveals a lectin domain. *Cell*, **56**:1045–1055.

Lasky, L.A., Singer, M.S., Dowbenko, D., Imai, Y., Henzel, W.J., Grimley, C., Fennie, C., Gillett, N., Watson, S.R. and Rosen, S.D. (1992b). An endothelial ligand for L-selectin in a novel mucin-like molecule. *Cell*, **69**:927–938.

Lenter, M., Levinovitz, A., Isenmann, S. and Vestweber, D. (1994). Monospecific and common glycoprotein ligands for E- and P-selectin on myeloid cells. *J. Cell. Biol.*, **125**:471–481

Levinovitz, A., Muhlhoff, J., Isenmann, S. and Vestweber, D. (1993). Identification of a glycoprotein ligand for E-selectin on mouse myeloid cells. *J. Cell Biol.*, **121**:449–459.

Lewinsohn, D.M., Bargatze, R.F. and Butcher, E.C. (1987). Leukocyte-endothelial cell recognition: evidence of a common molecular mechanism shared by neutrophils, lymphocytes, and other leukocytes. *J. Immunol.*, **138**:4313–4321.

Ley, K., Gaehtgens, P., Fennie, C., Singer, M.S., Lasky, L.A. and Rosen, S.D. (1991). Lectin-like cell adhesion molecule 1 mediates leukocyte rolling in mesenteric venules *in vivo*. *Blood*, **77**:2553–2555.

Li, F., Erickson, H.P., James, J.A., Moore, K.L., Cummings, R.D. and McEver, R.P. (1996a). Visualization of P-selectin glycoprotein ligand-1 as a highly extended molecule and mapping of protein epitopes for monoclonal antibodies. *J. Biol. Chem.*, **271**:6342–6348.

Li, F., Wilkins, P.P., Crawley, S., Weinstein, J., Cummings, R.D. and McEver, R.P. (1996b). Post-translational modifications of recombinant P-selectin glycoprotein ligand-1 required for binding to P- and E-selectin. *J. Biol. Chem.*, **271**:3255–3264.

Liener, I.E., Sharon, N. and Goldstein, I.J. (1986). *The lectins, properties, functions, and applications in biology and medicine*. Orlando, FL: Academic Press.

Lobb, R.R., Chi-Rosso, G., Leone, D.R., Rosa, M.D., Bixler, S., Newman, B.M., Luhowskyj, S., Benjamin, C.D., Dougas, I.G. and Goelz, S.E. (1991). Expression and functional characterization of a soluble form of endothelial-leukocyte adhesion molecule 1. *J. Immunol.*, **147**:124–129.

Lowe, J.B. (1992). Molecular cloning, expression, and uses of mammalian glycosyltransferses. *Seminars in Cell Biology*, **2**:289–307.

Lowe, J.B., Stoolman, L.M., Nair, R.P., Larsen, R.D., Berhend, T.L. and Marks, R.M. (1990). ELAM-1-dependent cell adhesion to vascular endothelium determined by a tranfected human fucosyltransferase cDNA. *Cell*, **63**:475–484.

Lowe, J.B., Kukowska-Latallo, J.F., Nair, R.P., Larsen, R.D., Marks, R.M., Macher, B.A., Kelly, R.J. and Ernst, L.K. (1991). Molecular cloning of a human fucosyltransferase gene that determines expression of the Lewis x and VIM-2 epitopes but not ELAM-1-dependent cell adhesion. *J. Biol. Chem*, **266**:17467–17477.

Malý, P., Thall, A.D., Marks, R.M., Petryniak, B., Rogers C.E., Smith, P.L., Gersten, K., M., Cheng, G., Saunders, T. L., Camper, S. A., Isogai, Y., Camphausen, R. T., Sullivan, F. X., Hindsgaul, O. von Andrian, U. H. and Lowe, J. B. (1996). The Fuc-TVII a(1,3)fucosyltransferase locus controls leukocyte trafficking through an essential role in L-, E-, and P-selectin ligand expression. *Cell*, **86(3)**:643–653.

Mayadas, T.N., Johnson, R.C., Rayburn, H., Hynes, R.O. and Wagner, D.D. (1993). Leukocyte rolling and extravasation are severely compromised in P selectin-deficient mice. *Cell*, **74**:541–554.

McEver, R.P. (1994). Selectins. *Curr. Opin. Immunol.*, **6**:75–84.

McEver, R.P. (1989). GMP-140, a platelet α-granule membrane protein, is also synthesized by vascular endothelial cells and is localized in Weibel-Palade bodies. *J. Clin. Invest.*, **84**:92–99.

McEver, R.P., Marshall-Carlson, L. and Beckstead, J.H. (1987). The platelet α-granule membrane protein GMP140 is also synthesized by human vascular endothelial cells and is present in blood vessels of diverse tissues. *Blood*, **70**:355a.

Mian, A.J. and Percival, E. (1973). Carbohydrates of the brown seaweed *Himanthalia lorea* and *Bifurcaria bifucata*. Part II. Structural studies of the "fucans". *Carbohyd. Res.*, **26**:147–153.

Mollicone, R., Gibaud, A., Francois, A., Ratcliffe, M. and Oriol, R., (1990). Acceptor specificity and tissue distribution of three human alpha-3-fucosyltransferases. *Eur. J. Biochem.*, **191**:169–176.

Mollicone, R., Reguigne, I., Fletcher, A., Aziz, A., Rustam, M., Weston, B.W., Kelly, R.J., Lowe, J.B. and Oriol, R. (1994a). Molecular basis for plasma α(1,3)fucosyltransferase deficiency in Indonesian pedigrees. *J. Biol. Chem.*, **269**:12662–12671.

Mollicone, R., Reguigne, I., Kelly, R.J., Fletcher, A., Watt, J,. Chatfield, S., Aziz, A., Cameron, H.S., Weston, B.W., Lowe, J.B. and Oriol, R. (1994b). Molecular basis for Lewis α(1,3/1,4)fucosyltransferase gene deficiency. *J. Biol. Chem.*, **269**:20987–20994.

Moore, K.L., Varki, A. and McEver, R.P. (1991). GMP-140 binds to a glycoprotein receptor on human neutrophils: evidence for a lectin-like interaction. *J. Cell Biol.*, **112**:491–499.

Moore, K.L., Stults, N.L., Diaz, S., Smith, D.F., Cummings, R.D., Varki, A. and McEver, R.P. (1992). Identification of a specific glycoprotein ligand for P-selectin (CD62) on myeloid cells. *J. Cell Biol.*, **118**:445–456.

Moore, K.L., Eaton, S.F., Lyons, D.E., Lichenstein, H.S., Cummings, R.D. and McEver, R.P. (1994). The P-selectin glycoprotein ligand from human neutrophils displays sialylated, fucosylated, O-linked poly-N-acetyllactosamine. *J. Biol. Chem.*, **269**:23318–23327.

Moore, K.L., Patel, K.D., Bruehl, R.E., Li, F., Johnson, D.A., Lichenstein, H.S., Cummings, R.D., Bainton, D.F. and McEver, R.P. (1995). P-selectin glycoprotein ligand-1 mediates rolling of human neutrophils on P-selectin. *J. Cell. Biol.*, **128**:661–671.

Mourelatos, Z., Gonatas, J.O., Nycum, L.M., Gonatas, N.K. and Biegel, J.A. (1995). Assignment of the GLG1 gene for MGF-160, a fibroblast growth factor and E-selectin binding membrane sialoglycoprotein of the Golgi apparatus, to chromosome 16q22-q23 by fluorescence in situ hybridization. *Genomics*, **28**:354–355.

Mulligan, M.S., Varani, J., Dame, M.K., Lane, C.L., Smith, C.W., Anderson, D.C. and Ward, P.A. (1991). Role of endothelial-leukocyte adhesion molecule 1 (ELAM-1) in neutrophil-mediated lung injury in rats. *J. Clin. Invest.*, **88**:1396–1406.

Mulligan, M.S., Paulson, J.C., DeFrees, S., Zheng, Z-L., Lowe, J.B. and Ward, P.A. (1993). Protective effects of oligosaccharides in P-selectin-dependent lung injury. *Nature*, **364**:149–151.

Natsuka, S. and Lowe, J.B. (1994). Glycosyltransferases in oligosaccharide biosynthesis. *Current Opinion in Structural Biology*, **4**:683–691.

Natsuka, S., Gersten, K.M., Zenita, K., Kannagi, R. and Lowe, J.B. (1994). Molecular cloning of a cDNA encoding a novel human leukocyte α-1,3-fucosyltransferase capable of synthesizing the sialyl Lewis x determinant. *J. Biol. Chem.*, **269**:16789–16794.

Nobis, U., Pries, A.R., Cokelet, R. and Gaehtgens, P. (1985). Radial distribution of white cells during blood flow in small tubes. *Microvasc. Res.*, **29**:295–304.

Nolte, D., Schmid, P., Jager, U., Botzlar, A., Roesken, F., Hecht. R., Uhl, E., Messmer, K. and Vestweber, D. (1994). Leukocyte rolling in venules of striated muscle and skin is mediated by P-selectin, not by L-selectin. *Am. J. Physiol.*, **267**:H1637–42

Norgard-Sumnicht, K.E., Varki, N.M. and Varki, A. (1993). Calcium-dependent heparin-like ligands for L-selectin in nonlymphoid endothelial cells. *Science*, **261**:480–483.

Osanai, T., Feizi, T., Chai, W., Lawson, A.M., Gustavsson, M.L., Sudo, K., Araki, M., Araki, K. and Yuen, C-T. (1996). Two families of murine carbohydrate ligands for E-selectin. *Biochem. Biophys. Res. Commun.*, **218**:610–615.

Paavonen, T. and Renkonen, R. (1992). Selective expression of sialyl-Lewis X and Lewis A epitopes, putative ligands for L-selectin, on peripheral lymph-node high endothelial venules. *Am. J. Pathol.*, **141**:1259–1264.

Patel, T.P., Goelz, S.E., Lobb, R.R. and Parekh, R.B. (1994). Isolation and characterization of natural protein-associated carbohydrate ligands for E-selectin. *Biochemistry*, **33**:14815–14824.

Paulson, J.C., Weinstein, J., Dorland, L., van Halbeek, H. and Vliegenthardt, J.F. (1982). Newcastle disease virus contains a linkage-specific glycoprotein sialidase. Application to the localization of sialic acid residues in N-linked oligosaccharides of alpha 1-acid glycoprotein. *J. Biol. Chem.*, **257**:12734–12738.

Phillips, M.L., Nudelman, E., Gaeta, F.C., Perez, M., Singhal, A.K., Hakomori, S. and Paulson, J.C. (1990). ELAM-1 mediates cell adhesion by recognition of a carbohydrate ligand, sialyl-Lex. *Science*, **250**:1130–1132.

Picker, L.J., Kishimoto, T.K., Smith, C.W., Warnock, R.A. and Butcher, E.C. (1991a). ELAM-1 is an adhesion molecule for skin-homing T cells. *Nature*, **349**:796–799.

Picker, L.J., Warnock, R.A., Burns, A.R., Doerschuk, C.M., Berg, E.L. and Butcher, E.C. (1991b). The neutrophil LECAM-1 presents carbohydrate ligands to the vascular selectins ELAM-1 and GMP-140. *Cell*, **66**:921–933.

Polley, M.J., Phillips, M.L., Wayner, E., Nudelman, E., Singhal, A.K., Hakomori, S. and Paulson, J.C. (1991). CD62 and endothelial cell-leukocyte adhesion molecule 1 (ELAM-1) recognize the same carbohydrate ligand, sialyl-Lewis x. *Proc. Natl. Acad. Sci. USA*, **88**:6224–6228.

Potvin, B., Kumar, R., Howard, D.R. and Stanley, P. (1990). Transfection of a human α-(1,3)fucosyltransferse gene into Chinese hamster ovary cells. Complications arise from activation of endogenous α-(1,3)fucosyltransferases. *J. Biol. Chem.*, **265**:1615–1622.

Pouyani, T. and Seed, B. (1995). PSGL-1 recognition of P-selectin is controlled by a tyrosine sulfation consensus at the PSGL-1 amino terminus. *Cell*, **83**:333–343.

Puri, K.D., Finger, E.B., Gaudernack, G. and Springer, T.A. (1995). Sialomucin CD34 is the major L-selectin ligand in human tonsil high endothelial venules. *J. Cell Biol.*, **131**:261–270.

Redl, H., Dinges, H.P. Buurman, W.A., van der Linden, C.J., Pober, J.S., Cotran, R.S. and Schlag, G. (1991). Expression of endothelial leukocyte adhesion molecule-1 in septic but not traumatic/hypovolemic shock in the baboon. *Am. J. Pathol.*, **139**:461–466.

Rohd, D., Schluter-Wigger, W., Mielke, V., von den Driesch, P., von Gaudecker, B. and Sterry, W. (1992). Infiltration of both T cells and neutrophils in the skin is accompanied by the expression of endothelial leukocyte adhesion molecule-1 (ELAM-1: an immunohistochemical and ultrastructural study. *J. Invest. Dermatol.*, **98**:794–799.

Rosen, S.D. (1993). Cell surface lectins in the immune system. *Seminars in Immunology*, **5**:237–247.

Rosen, S.D., Bertozzi, C.R. (1994). The selectins and their ligands. *Curr. Opin. Cell Biol.*, **6**:663–673.

Rosen, S.D., Singer, M.S., Yednock, T.A. and Stoolman, L.M. (1985). Involvement of sialic acid on endothelial cells in organ-specific lymphocyte recirculation. *Science*, **228**:1005–1007.

Rosen, S.D., Ch, S-I., True, D.D., Singer, M.S. and Yednock, T.A. (1989). Intravenously injected sialidase inactivates attachment sites for lymphocytes on high endothelial venules. *J. Immunol.*, **142**:1895–1902.

Postigo, A.A., Marazuela, M., Sanchez-Madrid, F. and de Landazuri, M.O. (1994). B lymphocyte binding to E- and P-selectins is mediated through the de novo expression of carbohydrates on *in vitro* and *in vivo* activated human B cells. *J. Clin. Invest.*, **94**:1585–1596.

Sadler, J.E. (1984). Biosynthesis of glycoproteins: formation of O-linked oligosaccharides. In *Biology of Carbohydrates*, edited by V. Ginsburg and P.W. Robbins, pp. 200–287. New York, New York: John Wiley and Sons.

Sako, D., Chang, X.J., Barone, K.M., Vachino, G., White, H.M., Shaw, G., Veldman, G.M., Bean, K.M., Ahern, T.J., Furie, B., Cumming, D.A. and Larsen, G.R. (1993). Expression cloning of a functional glycoprotein ligand for P-selectin. *Cell*, **75**:1179–1186.

Sako, D., Comess, K.M., Barone, K.M., Camphausen, R.T., Cumming, D.A. and Shaw, G.D. (1995). A sulfated peptide segment at the amino terminus of PSGL-1 is critical for P-selectin binding. *Cell*, **83**:323–331.

Sammar, M., Aigner, S., Hubbe, M., Schirrmacher, V., Schachner, M., Vestweber, D. and Altevogt, P. (1994). Heat-stable antigen (CD24) as ligand for mouse P-selectin. *Int. Immunol.*, **6**:1027–1036.

San Blas, G. and Cunningham, W.L. (1974). Structure of cell wall and exocellular mannans from the yeast *Hansenula holstii*. I. Mannans produced in phosphate-containing medium. *Biochem. Biophys. Acta.*, **354**:233–246.

Sasaki, K., Kurata, K., Funayama, K., Nagata, M., Watanabe, E., Ohta, S., Hanai. N. and Nishi, T. (1994). Expression cloning of a novel α1,3-fucosyltransferase that is involved in biosynthesis of the sialyl Lewis x carbohydrate determinants in leukocytes. *J. Biol .Chem.*, **269**:14730–14737.

Schauer, R. (1985). Sialic acids and their roles as biological masks. *Trends Biochem. Sci.*, **10**:357–361.

Scudder, P.R., Shailubhai, K. Duffin, K.L., Streeter, P.R. and Jacob, G.S. (1994). Enzymatic synthesis of a 6′-sulphated sialyl-Lewis x which is an inhibitor of L-selectin binding to peripheral addressin. *Glycobiology*, **4**:929–933.

Seekamp, A., Till, G.O., Mulligan, M.S., Paulson, J.C., Anderson, D.C., Miyasaka, M. and Ward, P.A. (1994). Role of selectins in local and remote tissue injury following ischemia and reperfusion. *Am. J. Pathol.*, **144**:592–598.

Shimizu, Y., Shaw, S., Graber, N., Gopal, T.V., Horgan, K.J., Van Seventer, G.A. and Newman, W. (1991). Activation-independent binding of human memory T cells to adhesion molecule ELAM-1. *Nature*, **349**:799–802.

Siegelman, M.H., van de Rijn, M. and Weissman, I.L. (1989a). Mouse lymph node homing receptor cDNA clone encodes a glycoprotein revealing tandem interaction domains. *Science*, **243**:1165–1172.

Siegelman, M.H. and Weissman, I.L. (1989b). Human homologue of mouse lymph node homing receptor: evolutionary consideration at tandem cell interaction domain. *Proc. Natl. Acad. Sci. USA*, **86**:5562–5566.

Silverstein, R.L. and Nachan, R.L. (1987). Thrombospondin binds to monocytes-macrophages and mediates platelet-monocyte adhesion. *J. Clin. Invest.*, **79**:867–874.

Skinner, M.P., Lucas, C.M., Burns, G.F., Chesterman, C.N. and Berndt, M.C. (1991). GMP-140 binding to neutrophils is inhibited by sulphated glycans. *J. Biol. Chem.*, **266**:5371–5374.

Smith, C.W., Kishimoto, T.K., Abbassi, O., Hughes, B., Rohlein, R., McIntire, L.V., Butcher, E., Anderson, D.C. and Abbass, O. (1991). Chemotactic factors regulate lectin adhesion molecules 1 (LECAM-1)-dependent neutrophil adhesion to cytokine-stimulated endothelal cells *in vitro*. *J. Clin. Invest.*, **87**:609–618.

Smith, P.L., Gersten, K.M., Petryniak, B., Kelly, R.J., Rogers, C., Natsuka, Y., Alford, J.A., III, Scheidegger, E.P., Natsuka, S. and Lowe, J.B. (1996). Expression of the α(1,3)fucosyltransferase Fuc-TVII in lymphoid aggregate high endothelial venules correlates with expression of L-selectin ligands. *J. Biol. Chem.*, **279**:8250–8259.

Spertini, O., Luscinskas, F.W., Gimbrone, M.A., Jr. and Tedder, T.F. (1992). Monocyte attachment to activated vascular endothelium *in vitro* is mediated by leukocyte adhesion molecule-1 (L-selectin) under non-static conditions. *J. Exp. Med.*, **175**:1789–1792.

Spooncer, E., Fukuda, M., Klock, J.C., Oates, J.E. and Dell, A. (1984). Isolation and characterization of polyfucosylated lactosaminoglycan from human granulocytes. *J. Biol. Chem.*, **259**:4792–4801.

Springer, T.A. (1994). Traffic signals for lymphocyte recirculation and leukocyte emigration: The multistep paradigm. *Cell*, **76**:301–314.

Stamper, H.B., Jr. and Woodruff, J.J. (1977). An *in vitro* model of lymphocyte homing. I. Characterization of the interaction between thoracic duct lymphocytes and specialized high-endothelial venules of lymph nodes. *J. Immunol.*, **119**:772–780.

Steegmaier, M., Levinovitz, A, Isenmann, S., Borges, E., Lenter, M., Kocher, H.P., Kleuser, B. and Vestweber, D. (1995). The E-selectin-ligand ESL-1 is a variant of a receptor for fibroblast growth factor. *Nature*, **373**:615–620.

Stenberg, P.E., McEver, R.P., Shuman, M.A., Jacques, Y.V. and Bainton, D.F. (1985). A platelet alpha-granule membrane protein (GMP-140) is expressed on the plasma membrane after activation. *J. Cell Biol.*, **101**:880–886.

Stoolman, L.M. and Rosen, S.D. (1983). Possible role for cell-surface carbohydrate-binding molecules in lymphocyte recirculation. *J. Cell. Biol.*, **96**:722–729.

Stoolman, L.M., Tenforde, T.S. and Rosen, S.D. (1984). Phosphomannosyl receptors may participate in the adhesive interaction between lymphocytes and high endothelial venules. *J. Cell Biol.*, **99**:1535–1540.

Streeter, P.R., Rouse, B.T.N. and Butcher, E.C. (1988). Immunohistologic and functional characterization of a vascular addressin involved in lymphocyte homing into peripheral lymph nodes. *J. Cell Biol.*, **107**:1853–1862.

Stroud, M.R., Handa, K., Salyan, M.E.K., Ito, K., Levery, S.B., Hakomori, S-I., Reinhold, B.B. and Reinhold, V.N. (1996a). Monosialogangliosides of human myelogenous leukemia HL60 cells and normal human leukocytes. 1. Separation of E-selectin binding from nonbinding gangliosides, and absence of sialosyl-le^x having tetraosyl to octasyl core. *Biochemistry*, **35**:758–769.

Stroud, M.R., Handa, K., Salyan, M.E.K., Ito, K., Levery, S.B., Hakomori, S-I., Reinhold, B.B. and Reinhold, V.N. (1996b). Monosialogangliosides of human myelogenous leukemia HL60 cells and normal human leukocytes. 2. Characterization of E-selectin binding fractions, and structural requirements for physiological binding to E-selectin. *Biochemistry*, **35**:770–778.

Sueyoshi, S., Tsuboi, S. Sawada-Hirai, R., Dang, U.N., Lowe, J.B. and Fukuda, M. (1994). Expression of distinct fucosylated oligosaccharides and carbohydrate-mediated adhesion efficiency directed by two different α-1,3-fucosyltransferases. Comparison of E- and L-selectin-mediated adhesion. *J. Biol. Chem.*, **269**:32342–32350.

Symington, F.W., Hedges, D.L. and Hakomori, S-I. (1985). Glycolipid antigens of human polymorphonuclear neutrophils and the inducible HL-60 myeloid leukemia line. *J. Immunol.*, **134**:2498–2506.

Takada, A., Ohmori, K., Takahashi, N., Tsuyuoka, K., Yago, K., Zenita, K., Hasegawa, A. and Kannagi, R., (1991). Adhesion of human cancer cells to vascular endothelium mediated by a carbohdyrate antigen, sialyl Lewis A. *Biochem. Biophys. Res. Commun.*, **179**:713–719.

Tedder, T.F, Isaacs, C.M., Ernst, T.J., Demetri, G.D., Adler, D.A. and Disteche, C.M. (1989). Isolation and chromosomal localization of cDNAs encoding a novel human lymphocyte cell surface molecule, LAM-1. *J. Exp. Med.*, **170**:123–133.

Todderud, G., Alford, J., Millsap, K.A., Aruffo, A. and Tramposch, K.M. (1992). PMN binding to P-selectin is inhibited by sulfatide. *J. Leukoc. Biol.*, **52**:85–88.

True, D.D., Singer, M.S., Lasky, L.A. and Rosen, S.D. (1990). Requirement for sialic acid on the endothelial ligand of a lymphocyte homing receptor. *J. Cell Biol.*, **111**:2757–2764.

Tyrrell, D., James, P., Rao, N., Foxall, C., Abbas, S., Dasgupa, F., Nashed, M., Hasegawa, A., Kiso, M. and Asa, D. (1991). Structural requirements for the carbohydrate ligand of E-selectin. *Proc. Natl. Acad. Sci. USA*, **88**:10372–10376.

Vachino, G., Chang, X.J., Veldman, G.M., Kumar, R., Sako, D., Fouser, L.A., Berndt, M.C. and Cumming, D.A. (1995). P-selectin glycoprotein ligand-1 is the major counter-receptor for P-selectin on stimulated T cells and is widely distributed in non-functional form on many lymphocytic cells. *J. Biol. Chem.*, **270**:21966–21974.

van den Eijnden, D.H., Koenderman, A.H. and Schiphorst, W.E. (1988). Biosynthesis of blood group i-active polylactosaminoglycans. Partial purification and properties of an UDP-GlcNAc:N-acetyllactosaminide beta 1---3-N-acetylglucosaminyltransferase from Novikoff tumor cell ascites fluid. *J. Biol. Chem.*, **263**:12461–12471.

Varki, A. (1994). Selectin ligands. *Proc. Natl. Acad. Sci. USA*, **91**:7390–7397.

von Andrian, U.H., Chambers, J.D., McEvoy, L., Bargatze, R.F., Arfors, K.E. and Butcher, E.C. (1991). A two step model of leukocyte-endothelial cell interaction in inflammation: distinct roles for LECAM-1 and the leukocyte $\alpha 2$ integrins *in vivo*. *Proc. Natl. Acad. Sci. USA*, **88**:7538–7542.

von Andrian, U.H., Hansell, P., Chambers, J.D., Berger, E.M., Torres-Filho, I., Butcher, E.C. and Arfors, K.E. (1992). L-selectin function is required for beta 2-integrin-mediated neutrophil adhesion at physiological shear rates *in vivo*. *Am. J. Physiol.*, **263**:H1034–H1044.

von Andrian, U.H., Hasslen, S.R., Nelson, R.D., Erlandsen, S.L. and Butcher, E.C. (1995). A central role for microvillous receptor presentation in leukocyte adhesion under flow. *Cell*, **82**:989–999.

Walz, G., Aruffo, A., Kolanus, W., Bevilacqua, M. and Seed, B. (1990). Recognition by ELAM-1 of the sialyl-Le^x determinant on myeloid and tumor cells. *Science*, **250**:1132–1135.

Watson, S.R., Imai, Y., Fennie, C., Geoffrey, J.S., Rosen, S.D. and Lasky, L.A. (1990). A homing receptor-IgG chimera as a probe for adhesive ligands of lymph node high endothelial venules. *J. Cell Biol.*, **110**:2221–2229.

Watson, S.R., Fennie, C. and Lasky, L.A. (1991a). Neutrophil influx into an inflammatory site inhibited by a soluble homing receptor-IgG chimera. *Nature*, **349**:164–167.

Watson, S.R., Imai, Y., Fennie, C., Geoffrey, J., Singer, M., Rosen, S.D. and Lasky, L.A. (1991b). The complement binding-like domains of the murine homing receptor facilitate lectin activity. *J. Cell Biol.*, **115**:235–243.

Weinstein, J., de Souza-e-Silva, U. and Paulson, J.C. (1982). Purification of a Gal β1->4GlcNAc α2->6sialyltransferase and a Gal β1->3(4)GlcNAc α2->3sialyltransferase to homogeneity from rat liver. *J. Biol. Chem.*, **247**:13835–13844.

Weis, W.I., Drickamer, K. and Hendrickson, W.A. (1992). Structure of a C-type mannose-binding protein complexed with an oligosaccharide. *Nature*, **360**:127–134.

Weller, A., Isenmann, S. and Vestweber, D. (1992). Cloning of the mouse endothelial selectins. Expression of both E- and P-selectin is inducible by tumor necrosis factor alpha. *J. Biol. Chem.*, **267**:15176–83.

Weller, P.F., Rand, T.H., Goelz, S.E., Chi-Rosso, G. and Lobb, R.R. (1991). Human eosinophil adherence to vascular endothelium mediated by binding to vascular cell adhesion molecule 1 and endothelial leukocyte adhesion molecule 1. *Proc. Natl. Acad. Sci. USA*, **88**:7430–7433.

Weston, B.W., Nair, R.P., Larsen, R.D. and Lowe, J.B. (1992a). Isolation of a novel human α(1,3)fucosyltransferase gene and molecular comparison to the human Lewis blood group α(1,3/1,4)fucosyltransferase gene. Syntenic, homologous, nonallelic genes encoding enzymes with distinct acceptor substrate specificities. *J. Biol. Chem.*, **267**:4152–4160.

Weston, B.W., Smith, P.L., Kelly, R.J. and Lowe, J.B. (1992b). Molecular cloning of a fourth member of a human α(1,3)fucosyltransferase gene family: multiple homologous sequences that determine expression of the Lewis x, sialyl Lewis x, VIM-2, and difucosyl sialyl Lewis x epitopes. *J. Biol. Chem.*, **267**:24575–24584.

Wilkins, P.P., Moore, K.L., McEver, R.P. and Cummings, R.D. (1995). Tyrosine sulfation of P-selectin glycoprotein ligand-1 is required for high affinity binding to P-selectin. *J. Biol. Chem.*, **270**:22677–22680.

Yago, K., Zenita, K., Ginya, H., Sawada, M., Ohmori, K., Okuma, M., Kannagi, R. and Lowe, J.B. (1993). Expression of α(1,3)fucosyltransferases which synthesize sialyl Lewis x and sialyl Lewis a, the carbohydrate ligands for E- and P-selectins in human malignant cell lines. *Cancer Res.*, **53**:5559–5565.

Yan, S.B., Chao, Y.B. and van Halbeek, H. (1993). Novel Asn-linked oligosaccharides terminating in GalNAc beta (1->4) [Fuc alpha (1->3)]GlcNAc beta (1->) are present in recombinant human protein C expressed in human kidney 293 cells. *Glycobiology*, **3**:597–608.

Yednock, T.A., Butcher, E.C., Stoolman, L.M. and Rosen, S.D. (1987b). Receptors involved in lymphocyte homing: relationship between a carbohydrate-binding receptor and the MEL-14 antigen. *J. Cell Biol.*, **104**:725–731.

Yednock, T.A., Stoolman, L.M. and Rosen, S.D. (1987a). Phosphomannosyl-derivatized beads detect a receptor involved in lymphocyte homing. *J. Cell Biol.*, **104**:713–724.

Yuen, C-T., Lawson, A.M., Chai, W., Larkin, M., Stoll, M.S., Stuart, A.C., Sullivan, F.X., Ahern, T.J. and Feizi, T. (1992). Novel sulfated ligands for the cell adhesion molecule E-selectin revealed by the neoglycolipid technology among O-linked oligosaccharides on an ovarian cystadenoma glycoprotein. *Biochemistry*, **31**:9126–9131.

Zhou, Q., Moore, K.L., Smith, D.F., Varki, A., McEver, R.P. and Cummings, R.D. (1991). The selectin GMP-140 binds to sialylated, fucosylated lactosaminoglycans on both myeloid and nonmyeloid cells. *J. Cell Biol.*, **115**:557–564.

10 Glycoprotein Ligands for L-selectin

Susan R. Watson

Nexstar Pharmaceuticals, Boulder, CO80301, USA

The selectins are a relatively recently identified and much reviewed family of adhesion molecules (Lasky 1992; Stoolman 1992; Bevilacqua and Nelson 1993; Kishimoto 1993; Rosen and Bertozzi 1994). Initially, L-selectin, also known as gp90^{MEL}, LAM 1 (Tedder *et al.*, 1989), CD62L and TQ1 (Camerini *et al.*, 1989), was described as the peripheral lymph node homing receptor and was thought to be involved only with the homing of naive lymphocytes to peripheral lymph nodes (Gallatin *et al.*, 1983). Subsequent studies showed that the receptor was constitutively expressed on virtually all leukocytes with the exception of memory T cells and was involved in the extravasation of neutrophils, monocytes and possibly eosinophils into inflammatory sites (Lewinsohn *et al.*, 1987; Jutila *et al.*, 1989; Knol *et al.*, 1993). The discovery that the selectins had a C-type lectin domain at the amino terminus of the molecule caused a resurgence in the field of carbohydrate chemistry and confirmed the idea that carbohydrate moieties play a critical role in cell-cell recognition and adhesion events (Stoolman, 1992). This review will focus on the glycoprotein ligands identified for L-selectin and the role that these molecules play in biology.

BACKGROUND

The Role of L-selectin in Lymphocyte Trafficking

The lymphoid tissues are divided into the primary lymphoid organs, bone marrow and thymus, the secondary lymphoid tissues such as the peripheral lymph nodes (PLN), mesenteric lymph nodes (MLN), Peyer's patches (PP), tonsils, etc., and the tertiary lymphoid tissues or the sites where the immune cells mediate their effects. In the bone marrow and thymus B and T cells, respectively, reach maturity. In the case of T cells in the thymus the cells are "educated" to discriminate between self and non-self (Adkins *et al.*, 1987; Kappler *et al.*, 1988; von Boehmer, 1988; Kincade *et al.*, 1989; Uckun, 1990). The mature cells then seed the secondary tissues. Once in the secondary tissues the lymphocytes are not sessile but constantly recirculate. In the case of naive lymphocytes entering PLN, MLN or PP from the blood stream they do so at specialised endothelial cell sites called high endothelial venules or HEV (Thome, 1898; Gowans, 1957; Gowans, 1959; Marchesi and Gowans, 1964). These endothelial cells, unlike other endothelium in the body, are raised and puffy in appearance hence the name "high". This "high" modification of endothelium is not a requirement for lymphocyte trafficking as the LN of sheep, SCID mice or mice that have been subjected to irradiation to deplete circulating lymphocytes (900Rads) do not possess HEV. However, lymphocytes still traffic to the LN in these animals and the endothelium of these LN bear L-selectin ligands (Miller, 1969; Mebius *et al.*, 1991; Mackay *et al.*, 1992; Mebius *et al.*, 1993). Early work studying lymphocyte trafficking showed that the process was selective in that only 25% of the lymphocytes entering a PLN by way of the circulation extravasated into that node (Ford, 1975). Furthermore unless the node was inflamed neutrophils and monocytes rarely if ever interacted with the HEV of LN (Kamperdijk *et al.*, 1978; Lewinsohn *et al.*, 1987; Jutila *et al.*, 1989).

The trafficking of lymphocytes is essential for efficient antigenic recognition. Thus in a LN, antigen, associated with professional antigen presenting cells such as dendritic cells, drains into the tissue by way of the afferent lymphatics and naive cells are constantly trafficking into the lymphoid tissue from the circulation. This serves to increase the chance that a lymphocyte of appropriate specificity may be present to recognise an antigen and elicit an immune response. If the cells do not get triggered by antigen they drain from the LN into the efferent lymphatics and return to the circulation via the thoracic duct to start the odyssey once more. Cells that respond to antigen in a LN also exit in the efferent lymphatics but generally exhibit altered trafficking patterns and are capable of going to sites such as skin and synovial tissues to perform, for better or worse, their effector functions.

The molecular mechanisms governing lymphocyte trafficking began to be elucidated with the development of an *in vitro* binding assay, the Stamper-Woodruff assay, that allowed the visualisation of the binding of lymphocytes to HEV (Stamper and Woodruff, 1976). In this assay, frozen sections of LN are overlayed with a suspension of lymphocytes, often labeled with a fluorochrome. The sections are then incubated at 7°C under gentle rotation. At the end of the period allowed for binding, microscopic observation shows that the lymphocytes have bound to the section but only on the HEV. Using this assay Butcher and his colleagues were able to characterise cell-lines that bound to the HEV of PLN and MLN but not to PP HEV (Butcher *et al.*, 1980). One of these cell-lines, 38C13, was used as an immunogen to raise an antibody, MEL 14, that blocked the *in vitro* binding of lymphocytes to PLN HEV but not to PP HEV. This antibody also blocked greater than 95% of lymphocyte trafficking to PLN *in vivo* (Gallatin *et al.*, 1983). The antibody's effect on trafficking of cells to PP has been a debated issue. The initial *in vivo* experiments suggested that it had no effect on trafficking to this lymphoid organ however subsequent experiments from several groups demonstrated that it was capable of partial blocking to PP (Hamann *et al.*, 1991; Mebius and Watson, 1993). Recently elegant studies have dissected the system involved in lymphocyte homing to PP and shown that it is a multifactorial process with both L-selectin and $\alpha_4\beta_7$ being involved in the initial interaction of lymphocytes with PP HEV (Bargatze *et al.*, 1995).

Using the Stamper-Woodruff assay, Rosen and his colleagues also began to probe the molecular mechanism of lymphocyte interaction with HEV. They observed that this interaction was dependent on the presence of divalent cations, calcium in particular, and that sugars such as fucoidan and polyphosphomannan ester (PPME) were able to inhibit this binding. Furthermore treatment of the PLN sections, before use in the Stamper-Woodruff assay, with sialidase reduced the binding of lymphocytes to the HEV to background levels. These results taken together suggested that the binding involved carbohydrates, was calcium dependent and thus possibly involved a lectin-like molecule (Stoolman and Rosen, 1983; Stoolman *et al.*, 1984; Rosen *et al.*, 1985; Yednock *et al.*, 1987; Yednock and Rosen, 1989). This hypothesis was substantiated when the molecule recognised by the MEL 14 antibody was cloned by several groups. The molecular cloning revealed that L-selectin is a classical type 1 membrane protein that has at its amino-terminal a calcium-dependent lectin domain followed by an epidermal growth factor-like (EGF) domain and then two short consensus repeats (SCRs) similiar to those found in members of complement regulatory proteins (Bowen *et al.*, 1989; Camerini *et al.*, 1989; Lasky *et al.*, 1989; Siegelman *et al.*, 1989; Siegelman and Weissman, 1989; Tedder *et al.*, 1989). Thus the presence of the calcium dependent lectin domain at the amino terminus of the molecule explained the *in vitro* findings that the interactions of lymphocytes with HEV were inhibitable both by the addition of carbohydrates and the removal of calcium. Later studies mapping the binding site of the antibody, MEL 14, on the L-selectin molecule demonstrated that it bound to the lectin domain again implicating the importance of protein-carbohydrate interactions in the binding of L-selectin positive lymphocytes to HEV (Bowen *et al.*, 1990).

The Role of L-selectin in Acute and Chronic Inflammation

Acute inflammation

As previously stated, L-selectin was initially thought to be a receptor involved only in the trafficking of naive lymphocytes to PLN. However, early studies by Marchesi and Gowans (Marchesi and Gowans, 1964) had shown that neutrophils could bind HEV and it was later demonstrated that circulating neutrophils bound to HEV more readily following antigenic challenge of the LN and that this binding was L-selectin dependent (Lewinsohn *et al.*, 1987). In contrast, monocytes bind poorly, if at all, to HEV of unstimulated PLN but bind actively and in an L-selectin dependent manner to the HEV of PLN taken from animals 3 days after antigen priming (Jutila *et al.*, 1989). *In vivo* it was shown that animals pretreated with MEL 14 before intraperitoneal injection of thioglycollate, which induces an early influx of neutrophils followed by a delayed appearance of monocytes, showed dramatic reductions in the number of neutrophils and monocytes found in the peritoneal influx (Jutila *et al.*, 1989). Many lines of evidence now suggest that L-selectin, along with the other two members of the selectin family, E- and P-selectin, plays a role in the extravasation of neutrophils and monocytes to sites of acute inflammation (Mayadas *et al.*, 1993; Arbones *et al.*, 1994; Labow *et al.*, 1994).

The paradigm of selectin-mediated adhesive events was formulated for the neutrophil. Extravasation of the neutrophil from the vasculature is initiated in the venules. Here the cell starts to "roll" along the endothelial cell surface at speeds now much lower than that of the blood flow and this initial low affinity interaction has been shown to be mediated by all three members of the selectin family (Ley *et al.*, 1991; von Andrian *et al.*, 1991; Abbassi *et al.*, 1993; Berg *et al.*, 1993; Dore *et al.*, 1993; Ley *et al.*, 1993; Mayadas *et al.*,

1993; Von Andrian *et al.*, 1993; Springer, 1994). Subsequently the neutrophil becomes activated by members of the C-X-C chemokine family such as IL-8 and, finally, the cell binds tightly to the endothelium by means of the integrin family of adhesion molecules and extravasates into the tissue (Butcher, 1990; von Andrian *et al.*, 1991; Springer, 1994). Selectins have also been shown to mediate the rolling of lymphocytes (Berg *et al.*, 1993; Jutila *et al.*, 1994; Bargatze *et al.*, 1995) suggesting that this three step model maybe a general paradigm for how cells extravasate from the blood into the tissues.

These observations lead to idea that ligands for L-selectin are expressed not only on the surface of HEV but on the venules of the peripheral vasculature. Furthermore *in vivo* rolling experiments show that L-selectin mediated rolling, unlike P-selectin mediated rolling, does not occur immediately following an inflammatory insult (Ley *et al.*, 1995). This suggests that the L-selectin ligands require either modification before functional activity or possibly de novo synthesis. Tantalizing data on the role of L-selectin ligands in the periphery has emerged from a study looking at the circulating levels of soluble selectins in patients at risk of ARDS. The investigators found that the level of soluble L-selectin was significantly lower in those patients who progressed to ARDS and that the decrease in circulating levels was correlated with the number of organs that failed in the patients (Donnelly *et al.*, 1994). These data could be interpreted as suggesting that in this patient population there is the induction of a ligand for L-selectin on the "damaged" endothelium that binds the soluble L-selectin. If this ligand is not GlyCAM 1 or CD34, two of the molecularly identified L-selectin ligands that will be discussed later, it should be of interest to characterise it and understand the mechanism(s) by which it is regulated.

Chronic inflammation

Data is beginning to accumulate from both *in vitro* and *in vivo* sources concerning the role of L-selectin in chronic inflammatory responses. Using either MECA 79, an antibody that recognises L-selectin ligands on the surface of HEV, or an L-selectin-IgG chimera, a recombinant protein that also recognises L-selectin ligands, induction of L-selectin ligands has been demonstrated at a variety of sites such as inflamed skin, rheumatoid synovium and human inflammatory bowel disease (Mackay *et al.*, 1992; Michie *et al.*, 1993). *In vivo* L-selectin has been shown to be involved in the accumulation of cells in a mouse DTH response (Dawson *et al.*, 1992) and protect against the development of insulitis in NOD mice (Yang *et al.*, 1993). It should be noted, however, that neither of these animal models truly reflect the human condition. Finally, the role of L-selectin in memory T cell localization is debatable as it is not clear whether memory cells remain L-selectin negative or whether at least a small percentage of memory cells can re-express L-selectin with time after antigenic stimulation (Bradley *et al.*, 1992; Mackay, 1992; Picker *et al.*, 1993; Picker *et al.*, 1993).

METHODS FOR IDENTIFYING SELECTIN LIGANDS

Antibodies

Several very different approaches have been used to successfully identify ligands for L-selectin. Phil Streeter working in Eugene Butcher's lab took the then classical approach of making monoclonal antibodies to stromal elements from peripheral and mesenteric lymph nodes and screening by immunohistochemistry for staining of high endothelial venules in lymphoid tissues. This lead to the development of the MECA series of antibodies. Of particular interest are MECA79 and MECA367 (Streeter *et al.*, 1988; Streeter *et al.*, 1988). MECA79 recognises HEV in peripheral lymph nodes and weakly recognises HEV in Peyer's patches. Immunoprecipitation of surface-labeled material from PLN has shown that MECA79 recognises at least 7 major bands ranging in molecular weight from 50 kD to 200 kD and these materials have been shown to be ligands for L-selectin (Berg *et al.*, 1991; Berg *et al.*, 1993). Furthermore, immunohistochemical analysis has shown this antigenic determinant to be present on venules at sites of chronic inflammation suggesting that L-selectin participates in the recruitment of mononuclear cells to sites of chronic inflammation (Mackay *et al.*, 1992; Michie *et al.*, 1993). In contrast MECA367 recognises the HEV of Peyer's patches strongly and immunoprecipitates a single 58–66 K band (Berg *et al.*, 1993). This band has been cloned and designated MadCAM 1 (Briskin *et al.*, 1993) and is a molecule that can function as a ligand for both L-selectin and $\alpha_4\beta_7$ bearing lymphocytes (Berg *et al.*, 1993; Berlin *et al.*, 1993; Hamann *et al.*, 1994; Bargatze *et al.*, 1995).

Immunoadhesins or Immunoglobulin Chimaeras

The second approach, which has also become established, involved the construction of chimaeric molecules where the extra cellular domains of L-selectin were linked to the CH2 and CH3 domains of human IgG_1 (Capon *et al.*, 1989; Watson *et al.*, 1990; Watson *et al.*, 1991). These engineered proteins or "immunoadhesins" have several advantages. As the construct is made in such a way as to include the "hinge" region of the immunoglobulin parent the resulting protein is a dimer and, generally, has a higher affinity for its ligand than a monomeric form (Erbe and Lasky, 1993). The presence of the Ig tail allows for a single-step purification of the chimaeric molecule on either Protein A or Protein G Sepharose and it also allows for detection of binding of the chimaeric molecule to its ligand(s) as there are a wide variety of secondary reagents available that recognise either human or mouse Fc. Thus the molecules can be used to immunohistochemically recognise ligands, function in flow-cytometric assays and block cell-cell interactions (Watson *et al.*, 1990; Erbe *et al.*, 1993).

The L-selectin-IgG stains HEV (Figure 1), inhibits lymphocyte binding to HEV *in vitro* (Watson *et al.*, 1990) and lymphocyte and leukocyte trafficking *in vivo*, (Watson *et al.*, 1991) and has allowed the molecular identification of two L-selectin ligands (Lasky *et al.*, 1992; Baumhueter *et al.*, 1993). The identification of these two ligands, GlyCAM 1 and CD34, was facilitated by earlier observations made by Paul Andrews in Bill Ford's laboratory. He showed that if pieces of rat PLN were cultured in radiolabelled inorganic sulphate the radio-label was incorporated into high molecular weight, secreted macromolecules (Andrews *et al.*, 1980; Andrews *et al.*, 1982; Andrews *et al.*, 1983). Thus Yasuyuki Imai in Steve Rosen's laboratory used the same organ culture system and initially identified two sulfated glycoproteins of 50 kD and 90 kD molecular weight (Imai *et al.*, 1991). The 50 kD molecule has been molecularly cloned and designated GlyCAM 1 (Lasky *et al.*, 1992) and the 90 kD species was shown to be CD34 (Baumhueter *et al.*, 1993). These endothelial ligands, like MadCAM 1, contain highly O-glycosylated mucin-like domains that appear to serve as scaffolds for the presentation of carbohydrate ligands

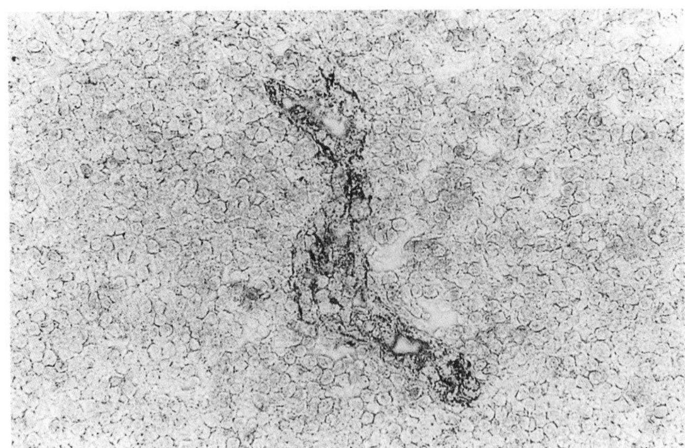

Figure 1 *L-selectin-IgG stains high endothelial venules.* Six micron sections of mouse PLN were stained with gold-conjugated L-selectin and the staining enhanced with silver (Mebius and Watson, 1993). As can be seen the staining is restricted to the area of the HEV.

to the lectin domain of L-selectin (Lasky *et al.*, 1992; Baumhueter *et al.*, 1993; Briskin *et al.*, 1993; Shimizu and Shaw, 1993).

The murine L-selectin-IgG chimera also recognises a fourth sulphated ligand that is approximately 200 kD that has yet to be molecularly identified (Hemmerich *et al.*, 1994; Hoke *et al.*, 1995). It does appear to be similiar to the molecule described by Andrews as it is a high molecular-weight secreted molecule. There also appears to be a form that is associated with cell lysates (Hoke *et al.*, 1995). Another group using a rat L-selectin-IgG chimaera have also identified a ligand of approximately 190 kD (Tamatani *et al.*, 1993) and a 250 kD ligand has been described in human tonsil tissue. In the latter experiments the probe was an antibody that recognised a complex form of sialyl Lewis X (Sawada *et al.*, 1993). There are also reports of inducible ligands for L-selectin on activated vascular endothelium but little is known about the biochemical properties of these molecules (Spertini *et al.*, 1991; 1992).

MUCINS AS LIGANDS FOR SELECTINS

The first selectin ligand to be molecularly identified was GlyCAM 1. This molecule turned out to be a sialo-mucin (Shimizu and Shaw, 1993). Although the mature molecule is approximately 50 kD, the polypeptide backbone is only 15–17 kD and is 60% threonine/serine rich (Lasky *et al.*, 1992). Thus there are many sites available for O-linked glycosylation suggesting that the polypeptide core serves to present carbohydrates at high density to L-selectin. GlyCAM 1 was purified from the supernatants of PLN organ cultures and the cloning showed that this molecule lacked a classical transmembrane domain and instead possessed an amphipathic helix at the carboxyl terminus (Lasky *et al.*, 1992). *In vivo* studies have shown that GlyCAM 1 is functionally present in mouse serum, can be precipitated by L-selectin-IgG and block the binding of murine lymphocytes to HEV (Brustein *et al.*, 1992). Recent calculations by Mark Singer in Steve Rosen's lab have shown that there is between 0.5–1.0 μg/ml GlyCAM 1 in mouse serum (Abstract # 4771. 9th International Congress Immunology, 23–29 July, 1995).

The polypeptide backbone of GlyCAM 1 is also expressed in the lactating murine mammary gland and is secreted into the milk. In this setting the molecule is not recognised by L-selectin-IgG suggesting that the molecule can be glycosylated differently in different sites (Dowbenko *et al.*, 1993). CD34 was identified as the 90 kD molecule precipitated by the L-selectin-IgG chimera and it is also a member of the sialomucin family of molecules. This molecule is a classical type I membrane protein and is expressed on all endothelial cells as well as being a marker for stem cells (Baumhueter *et al.*, 1993; Baumhueter *et al.*, 1994).

MadCAM 1, the molecule recognised by MECA 367, has recently been cloned and shown to have both a mucin-like domain and immunoglobulin domains similiar to those in molecules belonging to the super-immunoglobulin family (Briskin *et al.*, 1993). Furthermore, MadCAM functions as a ligand for both L-selectin and $\alpha_4\beta_7$ bearing lymphocytes and available data suggests that MadCAM can support the initial rolling phase of lymphocytes by way of its mucin-like domain and moreover mediate tight attachment through integrins (Berg *et al.*, 1993; Berlin *et al.*, 1993; Hamann *et al.*, 1994; Bargatze *et al.*, 1995). It is apparent, however, that not all MadCAM is modified to support L-selectin interactions and the regulation of this modification is not clear (Berg *et al.*, 1993). Peyer's patches labelled *in vivo* with ^{35}sulphate incorporate large amounts of the label but the L-selectin-IgG chimera does not precipitate a molecule of a molecular weight appropriate for MadCAM (Brustein *et al.*, 1992).

In closing this section it should be noted that the third sulphated ligand for L-selectin made by PLN, sgp200, has not been identified molecularly. It appears to exist as both a soluble molecule and associated with cell lysates (Hoke *et al.*, 1995) and has been shown to require sialic acid, fucose and sulphate for its binding to L-selectin (Hemmerich *et al.*, 1994). It is hypothesised that this molecule will be found to be a mucin.

CARBOHYDRATES ON L-SELECTIN LIGANDS

Recently, reviews of carbohydrates involved in selectin interactions has been published (Feizi, 1993; Varki, 1993; Rosen and Bertozzi, 1994). This review will concentrate on what is known about the biology of selectin ligands. However, it should be pointed out that in a variety of *in vitro* assays all three selectins have been shown to bind to simple sugar structures such as sialyl Lewis X and sialyl Lewis A (Aruffo *et al.*, 1991; Asa *et al.*, 1992; Berg *et al.*, 1992; Foxall *et al.*, 1992; Needham and Schnaar, 1993; Nelson *et al.*, 1993). However, it is only gradually becoming apparent what gives specificity in this system of cell/cell binding. Thus it now seems clear that the hallmark of an L-selectin ligand (Imai *et al.*, 1993) and, possibly a P-selectin ligand (Aruffo *et al.*, 1991; Needham and Schnaar, 1993) is the presence of sulphate.

The characterisation of the carbohydrates present on one of the naturally occurring ligands for L-selectin, GlyCAM 1, showed that 6′-sulphated sialyl Lewis X was a major capping group (Hemmerich *et al.*, 1994; Hemmerich and Rosen, 1994). These data clarified experiments in which an antibody directed against a complex form of sialyl Lewis X stained human HEV (Paavonen and Renkonen, 1992). The observation that both an L- and

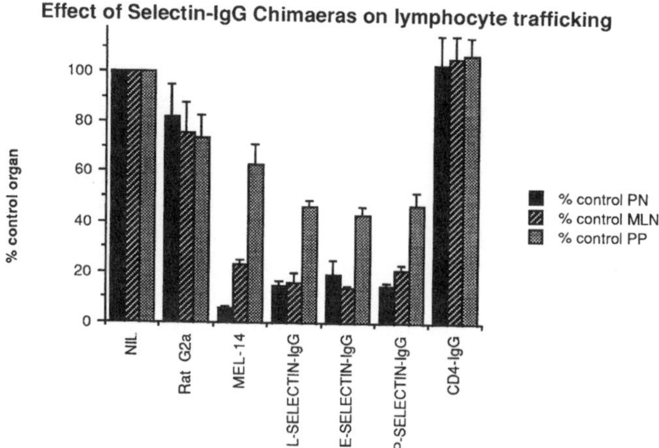

Figure 2 *All three selectin chimaeras block lymphocyte trafficking in vivo.* Lymphocytes from the PLN and MLN of 5–6 week old Balb/C female mice were labeled with ^{51}chromium. Two million cells mixed with 100 μg of antibody (MEL 14 or rat IgG_{2a}), selectin chimaeras or CD4-IgG were injected intravenously into syngeneic recipients. Two hours later the animals were sacrificed and PLN, MLN, PP, spleen, liver, lungs, thymus and bone marrow removed. The counts in the organs were expressed as a percentage of the counts in the untreated control animals.

an E-selectin-IgG immunoadhesin stained murine HEV in a calcium dependent fashion is also clarified by this characterisation of GlyCAM 1's carbohydrates (Mebius and Watson, 1993). In these experiments both L-selectin-IgG and E-selectin-IgG precipitated sulphated molecules from lymph nodes and inhibited lymphocyte trafficking *in vivo*. Since only L-selectin would be expected to bind high endothelial venules in this way, we proposed that the glycoprotein(s) recognized by L-selectin are decorated with both sulfated carbohydrates and sialyl Lewis X or sialyl Lewis X-like structures. In these studies we also showed that a P-selectin-IgG chimera precipitated sulphated molecules from PLN (Mebius and Watson-unpublished data). This reaction was calcium dependent and inhibitable by the addition of fucoidan. Furthermore the P-selectin-IgG construct was able to block *in vivo* lymphocyte trafficking (Figure 2; Watson and Mebius-unpublished data). These data suggest that P-selectin recognises the sulphated portions of the L-selectin ligands.

MODULATION OF L-SELECTIN LIGANDS

Ablation of Afferent Lymphatics

As L-selectin is so pivotal to the trafficking of naive lymphocytes to PLN (Bradley *et al.*, 1994) and as the ligands for L-selectin are expressed on HEV it was possible that they were subject to modulation. Studies had demonstrated that ligation of the afferent lymphatic vessels of a popliteal lymph node, a PLN, had effects on both the morphology and function of the HEV (Mebius *et al.*, 1991). Thus following deafferentiation, HEV become flat in appearance and can no longer bind lymphocytes in the Stamper-Woodruff assay *in vitro* and *in vivo* ^{51}chromium-labeled lymphocytes did not localize to operated lymph nodes. High endothelial venules from operated animals stain only faintly and ablumenally with MECA 79, the monoclonal antibody that recognizes PLN ligands for L-selectin (Streeter *et al.*, 1988; Streeter *et al.*, 1988). This result was interesting in the context of the 50, 90 and 200 kD sulfated ligands for L-selectin, since all of these glycoproteins were shown to contain an epitope(s) recognized by MECA 79 (Imai *et al.*, 1993; Hemmerich *et al.*, 1994). Thus, it appeared plausible that deafferentiation of the afferent lymphatics down-regulated one or more ligands for L-selectin. Studies showed that following deafferentiation the operated LN incorporated less sulphate per unit weight of tissue, the L-selectin-IgG chimera failed to stain the HEV and none of the three sulphated glycoproteins could be precipitated by L-selectin-IgG (Mebius *et al.*, 1993). A polyclonal anti-GlyCAM-1 antiserum failed to stain the operated LN whilst CD34 staining remained robust and message levels for GlyCAM 1 were virtually undetectable 4–7 days after deafferentiation whilst those for CD34 were unaffected (Baumhueter *et al.*, 1994). Thus ablation of afferent lymphatics leads to decreased sulphate incorporation into the operated LN, lack of staining of the endothelium with an L-selectin-IgG and absence of message for GlyCAM 1.

Induction of an Immune Response

In another series of experiments we studied the effects of an immune response on the three sulphated ligands synthesised by PLN. The initial experiments showed that the sgp200, like GlyCAM 1, is secreted and found in normal mouse serum but is also cell-associated (Hoke *et al.*, 1995). However, quantitation of the radio-immunoprecipitates suggests that either the concentration of sgp200 in the blood is significantly less than that of GlyCAM 1 (sgp50) or its circulating half-life is much shorter (Watson-unpublished data). Levels of GlyCAM 1 and secreted sgp200 are down-regulated 3–4 days after the induction of an immune response while the levels of cell associated L-selectin ligands, CD34 and sgp200, remain largely unaltered throughout the immune response. The decrease in GlyCAM 1 and secreted sgp200 is only seen in LN exposed to antigen and the nature of the antigen appears to be immaterial. Thus animals can be primed with particulate antigens such as sheep red blood cells, soluble proteins such as keyhole limpet haemocyanin (KLH) or contact sensitising agents such as oxazolone and all stimuli lead to a decrease in the levels of GlyCAM 1 and the secreted form of sgp200. An increased number of naive lymphocytes traffic to the primed PLN in an L-selectin-dependent manner following the induction of an immune response and elevated levels of other cell types, such as monocytes and neutrophils, are also found in the primed LN. Thus it is possible that the secreted forms of L-selectin ligands, sgp200 and GlyCAM 1, act as modulators of cell adhesion and that cell-associated CD34 and sgp200 are the ligands that mediate the initial loose binding of lymphocytes to HEV (Hoke *et al.*, 1995).

GlyCAM 1 has been identified as a secreted molecule that is synthesised by PLN (or the lactating mammary gland (Dowbenko *et al.*, 1993)) and is present in mouse serum (Brustein *et al.*, 1992). During an immune response we have shown that synthesis of GlyCAM 1 by the antigenically stimulated PLN is dramatically down-regulated. Analysis of GlyCAM 1 in the serum of animals undergoing an immune response shows that if the response is localised to a single PLN, as can be achieved by priming subcutaneously with KLH, the circulating levels of GlyCAM 1 remain unaltered. However, if the animal is primed with oxazolone such that 3–4 pairs of PLN are involved then a significant decrease in the amount of GlyCAM 1 precipitated from the serum can be observed. This decrease occurs 24–36 h after that in the PLN suggesting that the circulating half-life of GlyCAM 1 is 24 h or greater (Watson- unpublished data).

SPECULATION

With several L-selectin ligands having been described it is of interest to speculate how they might all function. Analysis of trafficking in GlyCAM 1 and CD34 knock-out mice (made by L. Lasky and M. Moore at Genentech) has shown that neither phenotype leads to a defect in lymphocyte trafficking to the PLN or MLN (Lasky, Moore and Watson-unpublished data). This suggests that either neither ligand is solely required for lymphocytes to access lymphoid tissue or that the KO animals have developed compensatory mechanisms. It is possible that more than one L-selectin ligand is needed to facilitate the entry of cells into tissue and this hypothesis can be evaluated by either making a double KO mouse or by biologically "knocking-out" GlyCAM 1 in CD34 KO mice by the induction of an immune response. Another hypothesis, favoured by this author, is that the cognate ligand for L-selectin-dependent PLN trafficking is sgp200. This ligand has not yet been molecularly identified and so reagents to test this hypothesis are limited. I also speculate that sgp200 will be found to be the L-selectin ligand that is induced on peripheral venules and at sites of chronic inflammation. Thus understanding the molecular nature of this ligand and how it is regulated may provide new insights into regulating both acute and chronic inflammation.

The secreted forms of sgp200 and GlyCAM 1 possibly act as decoys so that only 1 in 4 lymphocytes and no neutrophils or monocytes enter a resting PLN. Whilst sgp200 can be found in serum quantitation of the radio-immunoprecipitates suggests that the concentration of sgp200 in the blood is significantly less than that of GlyCAM 1 (Watson-unpublished data). Thus the serum form of sgp200 is possibly degraded much faster in serum than is GlyCAM 1. Another possible role for GlyCAM 1 is that it is involved in regulating the transit time of lymphocytes as they traffic through a PLN. It is also possible that GlyCAM 1 is meant to act in the serum to prevent chance binding of leukocytes to venule endothelium in the periphery. The data showing GlyCAM 1, CD34 and L-selectin-IgG staining at sites of chronic inflammation such as the pancreata of NOD mice would suggest at first glance that this molecule, possibly in conjunction with CD34, is serving as a mechanism for the entry of leukocytes into the inflammatory site (Baumhueter *et al.*, 1994). However, it is also possible that the converse is true and that GlyCAM 1 is induced as an attempt to down-regulated an inflammatory response.

The role of CD34 as an L-selectin ligand is made more interesting as it is found on all endothelial cells as well as stem cells. This leads to the idea that there are of isoforms of CD34 and that only sulphated forms of the molecule can function as L-selectin ligands in the PLN and MLN (Baumhueter *et al.*, 1993; Baumhueter *et al.*, 1994). Data from the deafferentiated mouse experiment have shown that whilst the L-selectin chimaera does not stain such tissues both staining for CD34 and its message level remain robust suggesting that the turnover rate for sulphated isoform of CD34 is high relative to the turnover rate for other isoforms of CD34. It has also been hypothesised that it is by induction of the sulphation of CD34 on the peripheral venules that leads to its function as an L-selectin ligand at sites of acute and chronic inflammation.

In all fairness to the reader several caveats should be made about these speculations. Many are based on observations made using inorganic sulphate incorporation into organ cultures. Thus only molecules that incorporate sulphate readily will be represented. The truth of this can be seen when other labels such as tritiated threonine and serine are used in organ culture. I have suggested that there will turn-out to be only one cognate ligand for L-selectin, sgp200. The other hypothesis is that several of these ligands will serve as a collective ligand(s) for L-selectin. Thus deletion of GlyCAM 1 or CD34 either by knock-out technology or the induction of an immune response will not inhibit L-selectin dependent trafficking as CD34, sgp200 and other as yet unidentified L-selectin ligands can still function in this role.

Acknowledgment

I should like to acknowledge Clayton E. Smith for his ever-present support and encouragement in this the hardest year of my life.

MODULATION OF L-SELECTIN LIGANDS — SUMMARY

1. GLYCAM-1
   - A. SYNTHESISED BY PLN AND MLN HEV.
   - B. FOUND IN SERUM.
   - C. DOWN-REGULATED BY DEAFFERENTIATION OF AFFERENT LYMPHATICS AND IMMUNE RESPONSE.
   - D. LEVELS NORMAL IN X-IRRADIATED MICE (900RADS), SCIDS, nu/nu and GNOTOBIOTICS.
   - E. PRESENT AT SITES OF INFLAMMATION eg NOD PANCREAS.
   - F. EXPRESSED BY HEV AT DAY 1 OR 2 AFTER BIRTH.
   - G. GENETIC KO HAS NO DEFECT IN LYMPHOCYTE TRAFFICKING.
   - F. POLYPEPTIDE ALSO EXPRESSED IN LACTATING MAMMARY GLAND AND FOUND IN MILK BUT NOT FUNCTIONAL L-SELECTIN LIGAND.

2. CD34
   - A. EXPRESSED ON ALL ENDOTHELIAL CELLS AND STEM CELLS.
   - B. UNAFFECTED BY DEAFFERENTIATION OR IMMUNE RESPONSE.
   - C. LEVELS NORMAL IN X-IRRADIATED MICE (900RADS), SCIDS, nu/nu and GNOTOBIOTICS.
   - D. GENETIC KO HAS NO DEFECT IN LYMPHOCYTE TRAFFICKING.

3. SGP200
   - A. APPEARS TO HAVE BOTH A SECRETED AND A MEMBRANE ASSOCIATED FORM AND ONLY SYNTHESISED BY PLN AND MLN.
   - B. SECRETED FORM DOWN-REGULATED BY DEAFFERENTIATION OF AFFERENT LYMPHATICS AND IMMUNE RESPONSE.
   - C. MEMBRANE ASSOCIATED FORM UNAFFECTED BY IMMUNE RESPONSE.

References

Abbassi, O., Kishimoto, T.K., McIntire, L.V., Anderson, D.C. and Smith, C.W. (1993). E-selectin supports neutrophil rolling *in vitro* under conditions of flow. *J. Clin. Invest.*, **92**:2719.

Adkins, B., Mueller, C., Okada, C., Reichert, R., Weissman, I. and Spangrude, G. (1987). Early events in T-cell maturation. *Ann. Rev. Immunol.*, **5**:325.

Andrews, P., Ford, W. and Stoddart, R. (1980). Metabolic studies of high-walled endothelium of post-capillary venules in rat lymph nodes. *Blood Cell and Vessel Walls: Functional Interactions, Ciba Found*, **71**:211.

Andrews, P., Milsom, D. and Ford, W. (1982). Migration of lymphocytes across specialised endothelium. V. Production of a sulphated macromolecule by high endothelial cells of rat lymph nodes. *J. Cell Sci.*, **57**:277.

Andrews, P., Milsom, D. and Stoddart, R. (1983). Glycoconjugates from high endothelial cells. I. Partial characterisation of a sulphated glycoconjugate from the high endothelial cells of rat lymph nodes. *J. Cell Sci.*, **59**:231.

Arbones, M.L., Ord, D.C., Ley, K., Ratech, H., Maynard-Curry, C., Otten, G., Capon, D.J. and Tedder, T.F. (1994). Lymphocyte homing and leukocyte rolling and migration are impaired in L-selectin deficient mice. *Immunity*, **1**:247.

Aruffo, A., Kolanus, W., Walz, G., Freedman, P. and Seed, B. (1991). CD62/P-selectin recognition of myeloid and tumor sulfatides. *Cell*, **67**:35.

Asa, D., Gant, T., Oda, Y. and Brandley, B.K. (1992). Evidence for two classes of carbohydrate binding sites on selectins. *Glycobiology*, **2**:395.

Bargatze, R., Jutila, M. and Butcher, E. (1995). Distinct roles of L-selectin and integrins $\alpha 4\beta 7$ and LFA-1 in lymphocyte homing to Peyer's Patch HEV *in situ*: The multistep hypothesis confirmed and refined. *Immunity*, **3**:99.

Baumhueter, S., Dybdal, N., Kyle, C. and Lasky, L. (1994). Global vascular expression of murine CD34, a sialomucin-like endothelial ligand for L-selectin. *Blood*, **84**:2554.

Baumhueter, S., Singer, M.S., Henzel, W., Hemmerich, S., Renz, M., Rosen, S.D. and Lasky, L.A. (1993). Binding of L-selectin to the vascular sialomucin, CD34. *Science*, **262**:436.

Berg, E.J., Magnani, J., Warnock, R.A., Robinson, M.K. and Butcher, E.C. (1992). Comparison of L-selectin and E-selectin ligand specificities: L-selectin can bind the E-selectin ligands sialyl Le^x and sialyl Le^a. *Biochem. Biophys. Res. Comm.*, **184**:1048.

Berg, E.L., McEvoy, L.M., Berlin, C., Bargatze, R.F. and Butcher, E.C. (1993). L-selectin-mediated lymphocyte rolling on MAdCAM-1. *Nature*, **366**:695.

Berg, E.L., Robinson, M.K., Warnock, R.A. and Butcher, E.C. (1991). The human peripheral lymph node vascular addressin is a ligand for LECAM-1, the peripheral homing receptor. *J. Cell Biol.*, **114**:343.

Berlin, C., Berg, E.L., Briskin, M.J., Andrew, D.P., Kilshaw, P.J., Holzmann, B., Weissman, I.L., Hamann, A. and Butcher, E.C. (1993). $\alpha 4\beta 7$ integrin mediates lymphocyte binding to the mucosal vascular addressin MadCAM-1. *Cell*, **74**:185.

Bevilacqua, M.P. and Nelson, R.M. (1993). Selectins. *J. Clin. Invest.*, **91**:379.

Bowen, B., Fennie, C. and Lasky, L.A. (1990). The MEL-14 antibody binds to the lectin domain of the murine peripheral lymph node homing receptor. *J. Cell Biol.*, **110**:147.

Bowen, B.R., Nguyen, T. and Lasky, L.A. (1989). Characterization of a human homologue of the murine peripheral lymph node homing receptor. *J. Cell Biol.*, **109**:421.

Bradley, L.M., Atkins, G. and Swain, S. (1992). Long-term CD4 memory T cells from the spleen lack MEL-14, the lymph node homing receptor. *J. Immunol.*, **148**:324.

Bradley, L.M., Watson, S.R. and Swain, S. (1994). Entry of naive CD4 T cells into peripheral lymph nodes requires L-selectin. *J. Exp. Med.*, **180**:2401.

Briskin, M.J., McEvoy, L.M. and Butcher, E.C. (1993). MadCAM-1 has homology to immunoglobulin and mucin-like adhesion receptors and to IgA1. *Nature*, **363**:461.

Brustein, M., Kraal, G., Mebius, R.E. and Watson, S.R. (1992). Identification of a soluble form of a ligand for the lymphocyte homing receptor. *J. Exp. Med.*, **176**:1415.

Butcher, E.C. (1990). Cellular and molecular mechanisms that direct leukocyte traffic. *Am. J. Pathol.*, **136**:3.

Butcher, E.C., Scollay, R.G. and Weissman, I.L. (1980). Organ Specificity of lymphocyte migration: mediation by highly selective lymphocyte interaction with organ-specific determinants on high endothelial venules. *Eur. J. Immunol.*, **10**:556.

Camerini, D., James, S.P., Stamenkovic, I. and Seed, B. (1989). Leu 8/TQ1 is the human equivalent of the MEL-14 lymph node homing receptor. *Nature*, **342**:78.

Capon, D.J., Chamow, S.M., Mordenti, J., Marsters, S.A., Gregory, T., Mitsuya, H., Byrn, R.A., Lucas, C., Wurm, F.M., Groopman, J.E., Broder, S. and Smoth, D.H. (1989). Designing CD4 immunoadhesins for AIDS therapy. *Nature*, **337**:525.

Dawson, J., Sedgwick, A., Edwards, J. and Lees, P. (1992). The monoclonal antibody MEL-14 can block lymphocyte migration into a site of chronic inflammation. *Eur. J. Immunol.*, **22**:1647.

Donnelly, S., Haslett, C., Dransfield, I., Robertson, C., Carter, D., Ross, J., Grant, I. and Tedder, T. (1994). Role of selectins in the development of adult respiratory distress syndrome. *Lancet*, **344**:215.

Dore, M., Korthuis, R.J., Granger, D.N., Entman, M.L. and Smith, C.W. (1993). P-selectin mediates spontaneous leukocyte rolling *in vivo*. *Blood*, **82**:1308.

Dowbenko, D., Kikuta, A., Fennie, C., Gillett, N. and Lasky, L. (1993). Glycosylation-dependent cell adhesion molecule 1 (GlyCAM 1) mucin is expressed by lactating mammary gland epithelial cells and is present in milk. *J. Clin. Invest.*, **92**:952.

Erbe, D.V. and Lasky, L.A. (1993). Application of immunoglobulin chimeras to receptor-ligand studies. In *Methods in Molecular Genetics Vol. 2*, Academic Press.

Erbe, D.V., Watson, S.R., Presta, L.G., Wolitzky, B.A., Foxall, C., Brandley, B.K. and Lasky, L.A. (1993). P- and E-selectin use common sites for carbohydrates ligand recognition and cell adhesion. *J. Cell Biol.*, **120**:1227.

Feizi, T. (1993). Oligosaccharides that mediate mammalian cell-cell adhesion. *Curr. Opin. Struct. Biol.*, **3**:701.

Ford, W.L. (1975). Lymphocyte migration and immune responses. *Prog. Allergy*, **19**:1.

Foxall, C., Watson, S.R., Dowbenko, D., Fennie, C., Lasky, L.A., Kiso, M., Hasegawa, A., Asa, D. and Brandley, B.K. (1992). The three members of the selectin receptor family recognize a common carbohydrate epitope, the sialyl Lewis^x oligosaccharide. *J. Cell Biol.*, **117**:895.

Gallatin, W.M., Weissman, I.L. and Butcher, E.C. (1983). A cell surface molecule involved in organ-specific homing of lymphocyte. *Nature*, **303**:30.

Gowans, J. (1957). The effect of continuous reinfusion of lymph and lymphocytes from the thoracic duct of unanaesthetized rats. *Br. J. Pathol.*, **38**:67.

Gowans, J. (1959). Lymphocyte recirculation. *Br. Med. Bull.*, **15**:50.

Hamann, A., Andrew, D.P., Jablonski-Westrich, D., Holzmann, B. and Butcher, E.C. (1994). Role of α4-integrins in lymphocyte homing to mucosal tissues *in vivo*. *J. Immunol.*, **152**:3282.

Hamann, A., Jablonski-Westrich, D. and Thiele, H.G. (1991). Homing receptors re-examined: mouse LECAM-1 (MEL-14 antigen) is involved in lymphocyte migration into gut-associated lymphoid tissue. *Eur. J. Immunol.*, **21**:2925.

Hemmerich, S., Bertozzi, C., Leffler, H. and Rosen, S.D. (1994). Identification of the sulfated monosaccharides of GlyCAM-1, an endothelial ligand for L-selectin. *Biochemistry*, **33**:4820.

Hemmerich, S., Butcher, E. and Rosen, S. (1994). Sulfation-dependent recognition of high endothelial venules (HEV)- ligands by L-selectin and MECA 79, an adhesion blocking monoclonal antibody. *J. Exp. Med.*, **180**:2219.

Hemmerich, S. and Rosen, S.D. (1994). 6′-sulfated sialyl Lewis X is a major capping group of GlyCAM-1. *Biochemistry*, **33**:4830.

Hoke, D., Mebius, R., Dybdal, N., Dowbenko, D., Gribling, P., Kyle, C., Baumhueter, C. and Watson, S. (1995). Selective modulation of the expression of L-selectin ligands by an immune response. *Current Biol.*, **5**:670.

Imai, Y., Lasky, L.A. and Rosen, S.D. (1993). Sulphation requirement for GlyCAM-1, an endothelial ligand for l-selectin. *Nature*, **361**:555.

Imai, Y., Singer, M.S., Fennie, C., Lasky, L.A. and Rosen, S.D. (1991). Identification of a carbohydrate-based endothelial ligand for a lymphocyte homing receptor. *J. Cell Biol.*, **113**:1213.

Jutila, M., Rott, L., Berg, E. and Butcher, E. (1989). Function and regulation of the neutrophil MEL 14 antigen *in vivo*: comparison with LFA-1 and MAC-1. *J. Immunol.*, **143**:3318.

Jutila, M.A., Bargatze, R.F., Kurk, S., Warnock, A., Ehsani, N., Watson, S.R. and Walcheck, B. (1994). Cell surface P- and E-selectin support shear-dependent rolling of γ/δ T cells. *J. Immunol.*, **153**:3917.

Jutila, M.A., Berg, E.L., Kishimoto, T.K., Picker, L.J., Bargatze, R.F., Bishop, D.K., Orosz, C.G., Wu, N.W. and Butcher, E.C. (1989). Inflammation-induced endothelial cell adhesion to lymphocytes, neutrophils and monocytes. *Transplantation*, **48**:727.

Jutila, M.A., Rott, L., Berg, E.L. and Butcher, E.C. (1989). Function and regulation of the neutrophil MEL-14 antigen *in vivo*: Comparison with LFA-1 and MAC-1. *J. Immunol.*, **143**:3318.

Kamperdijk, E.W.A., Raaymakers, E.M., De Leeuw, J.H.S. and Hoefsmit, E.C.M. (1978). Lymph node macrophages and reticulum cells in the immune response. *Cell Tissue Res.*, **192**:1.

Kappler, J., Staerz, U., White, J. and Marrack, P. (1988). Self-tolerance eliminates T cells specific for Mis-modified products of the MHC. *Nature*, **332**:35.

Kincade, P., Lee, G., Pietrangeli, C., Hayashi, S-I. and Gimble, J. (1989). Cell and molecules that regulate B lymphopoiesis in bone marrow. *Ann. Rev. Immunol.*, **7**:111.

Kishimoto, T.K. (1993). The selectins. In *Structure, Function and Regulation of Molecules involved in Leukocyte Adhesion*. Edited by Lipsky, P., *et. al.*, pp. 107. New York, Springer-Verlag.

Knol, E., Kansas, G., Tedder, T., Schleimer, R. and Bochner, B. (1993). Human eosinophils use L-selectin to bind to endothelial cells under non-static conditions. *J. Allergy Clin. Immunol.*, **91**:334.

Labow, M., Norton, C., Rumberger, J., Lombard-Gillooly, K., Shuster, D., Hubbard, J., Bertko, R., Knaack, P., Terry, R., Harbison, M. *et al.* (1994). Characterisation of E-selectin deficient mice: demonstration of overlapping function of endothelial selectins. *Immunity*, **1**:709.

Lasky, L.A. (1992). Selectins: Interpreters of Cell-Specific carbohydrate information during inflammation. *Science*, **258**:964.

Lasky, L.A., Singer, M.S., Dowbenko, D., Imai, Y., Henzel, W.J., Grimley, G., Fennie, C., Gillet, N., Watson, S.R. and Rosen, S.D. (1992). An endothelial ligand for L-selectin is a novel mucin-like molecule. *Cell*, **69**:927.

Lasky, L.A., Singer, M.S., Yednock, T.A., Dowbenko, D., Fennie, C., Rodriguez, H., Nguyen, T., Stachel, S. and Rosen, S.D. (1989). Cloning of a lymphocyte homing receptor reveals a lectin domain. *Cell*, **56**:1045.

Lewinsohn, D.M., Bargatze, R.F. and Butcher, E.C. (1987). Leukocyte-endothelial cell recognition: Evidence of a common molecular mechanism shared by neutrophils, lymphocytes and other leukocytes. *J. Immunol.*, **138**:4313.

Ley, K., Bullard, D., Arbones, M., Bosse, R., Vestweber, D., Tedder, T. and Beaudet, A. (1995). Sequential contribution of L- and P-selectin to leukocyte rolling *in vivo*. *J. Exp. Med.*, **181**:669.

Ley, K., Gaehtgens, P., Fennie, C., Singer, M.S., Lasky, L.A. and Rosen, S.D. (1991). Lectin-like cell adhesion molecule 1 mediates leukocyte rolling in mesenteric venules *in vivo*. *Blood*, **77**:2553.

Ley, K., Tedder, T.F. and Kansas, G.S. (1993). L-selectin can mediate leukocyte rolling in untreated mesenteric venules *in vivo* independent of E- or P-selectin. *Blood*, **82**:1632.

Mackay, C. (1992). Migration pathways and immunologic memory among T lymphocytes. *Semin. Immunol.*, **4**:51.

Mackay, C.R., Marston, W. and Dudler, E. (1992). Altered patterns of T cell migration through lymph nodes and skin following antigen challenge. *Eur. J. Immunol.*, **22**:2205.

Marchesi, V.T. and Gowans, J.L. (1964). The migration of lymphocytes through the endothelium of venules in lymph nodes. *Proc. R. Soc. Lond.*, **159**:283.

Mayadas, T.N., Johnson, R.C., Rayburn, H., Hynes, R.O. and Wagner, D.D. (1993). Leukocyte rolling and extravasation are severely compromised in P-selectin deficient mice. *Cell*, **74**:541.

Mebius, R.E. and Watson, S.R. (1993). L- and E-selectin recognise the same naturally occurring ligands on HEV. *J. Immunol.*, **151**:3252.

Mebius, R.E., Dowbenko, D., Williams, A., Fennie, C., Lasky, L.A. and Watson, S.R. (1993). Expression of GlyCAM-1, an endothelial ligand for L-selectin, is affected by afferent lymphatic flow. *J. Immunol.*, **151**:6769.

Mebius, R.E., Streeter, P.R., Breve, J., Duijvestijn, A.M. and Kraal, G. (1991). The influence of afferent lymphatic vessel interruption on vascular addressin expression. *J. Cell Biol.*, **115**:85.

Michie, S.A., Streeter, P.R., Bolt, P.A., Butcher, E.C. and Picker, L.J. (1993). The human peripheral lymph node vascular addressin — An inducible endothelial antigen involved in lymphocyte homing. *Amer. J. Pathol.*, **143**:1688.

Miller, J. (1969). Studies of the phylogeny and ontogeny of the specialised lymphatic venules. *Lab. Invest.*, **21**:284.

Needham, L.K. and Schnaar, R.L. (1993). The HNK-1 reactive sulfoglucuronyl glycolipids are ligands for L-selectin and P-selectin but not E-selectin. *Proc. Natl. Acad. Sci.*, **90**:1359.

Nelson, R.M., Dolich, S., Aruffo, A., Cecconi, O. and Bevilacqua, M.P. (1993). Higher affinity oligosaccharide ligands for E-selectin. *J. Clin. Invest.*, **91**:1157.

Paavonen, T. and Renkonen, R. (1992). Selective expression of sialyl-Lewis X and Lewis A epitopes, putative ligands for L-selectin, on peripheral lymph-node high endothelial venules. *Am. J. Pathol.*, **141**:1259.

Picker, L., Treer, J., Ferguson, D., Collins, P., Bergstresser, P. and Terstappen, L. (1993). Control of lymphocyte recirculation in man II. Differential regulation of the cutaneous lymphocyte associated antigen, a tissue — selective homing receptor for skin-homing T cells. *J. Immunol.*, **150**:1122.

Picker, L.J., Treer, J.R., Ferguson-Darnell, B., Collins, P., Buck, D. and Terstappen, L.W.M. (1993). Control of Lymphocyte Recirculation in Man I. Differential regulation of the peripheral lymph node homing receptor L-selectin on T cells during the virgin to memory cell transition. *J. Immunol.*, **150**:1105.

Rosen, S., Singer, M., Yednock, T. and Stoolman, L. (1985). Involvement of sialic acid on endothelial cells in organ-specific lymphocyte recirculation. *Science*, **228**:1005.

Rosen, S.D. and Bertozzi, C.R. (1994). The selectins and their ligands. *Current Opinion in Cell Biology*, **6**:663.

Sawada, M., Takada, A., O I, Takahashi, N., Tateno, H., Sakamoto J. and Kannagi, R. (1993). Specific expression of a complex sialyl Lewis X antigen on high endothelial venules of human lymph nodes: Possible candidate for L-selectin ligand. *Biochem. Biophys. Res. Commun.*, **193**:337.

Shimizu, Y. and Shaw, S. (1993). Mucins in the mainstream. *Nature*, **366**:630.

Siegelman, M.H., de Rijn, M.V. and Weissman, I.L. (1989). Mouse lymph node homing receptor cDNA clone encodes a glycoprotein revealing tandem interaction domains. *Science*, **243**:1165.

Siegelman, M.H. and Weissman, I.L. (1989). Human homologue of mouse lymph node homing receptor: Evolutionary conservation at tandem cell interaction domains. *Proc. Natl. Acad. Sci.*, **86**:5562.

Spertini, O., Luscinskas, F., Gimbrone Jr., M. and Tedder, T. (1992). Monocyte attachment to activated human vascular endothelium *in vitro* is mediated by leukocyte adhesion molecule-1 (L-selectin) under nonstatic conditions. *J. Exp. Med.*, **175**:1789.

Spertini, O., Luscinskas, F., Kansas, G., Munro, J., Griffin, J., Gimbrone Jr., M. and Tedder, T (1991). Leukocyte adhesion molecule-1 (LAM-1, L-selectin) interacts with an inducible endothelial cell ligand to support leukocyte adhesion. *J. Immunol.*, **147**:2565.

Springer, T.A. (1994). Traffic Signals for Lymphocyte Recirculation and leukocyte emigration: the multistep paradigm. *Cell*, **76**:301.

Stamper, H.B.J. and Woodruff, J.J. (1976). Lymphocyte homing into lymph nodes; *In vitro* demonstration of the selective affinity of recirculating lymphocytes for high endothelial venules. *J. Exp. Med.*, **144**:772.

Stoolman, L. and Rosen, S. (1983). Possible role for cell-surface carbohydrate-binding molecules in lymphocyte recirculation. *J. Cell Biol.*, **96**:1535.

Stoolman, L., Tenforde, T. and Rosen, S. (1984). Phosphomannosyl receptors may participate in the adhesive interaction between lymphocytes and high endothelial venules. *J. Cell Biol.*, **99**:1535.

Stoolman, L.M. (1992). Selectins (LEC-CAMs): Lectin-like receptors involved in lymphocyte recirculation and leukocyte recruitment. In *Cell Surface Carbohydrates and Cell Development.*

Streeter, P.R., Berg, E.L., Rouse, B.T.N., Bargatze, R.F. and Butcher, E.C. (1988). A tissue-specific endothelial cell molecule involved in lymphocyte homing. *Nature*, **331**:41.

Streeter, P.R., Rouse, B.T.N. and Butcher, E.C. (1988). Immunohistologic and functional characterization of a vascular addressin involved in lymphocyte homing into peripheral lymph nodes. *J. Cell Biol.*, **107**:1853.

Tamatani, T., Kuida, K., Watanabe, T., Koike, S. and Miyasaka, M. (1993). Molecular mechanisms underlying lymphocyte recirculation. III. Characterization of the LECAM-1 (L-selectin)-dependent adhesion pathway in rats. *J. Immunol.*, **150**:1735.

Tedder, T.F., Isaacs, C.M., Ernst, T.J., Demetri, G.D., Adler, D.A. and Disteche, C.M. (1989). Isolation and chromosomal localization of cDNAs encoding a novel human lymphocyte cell surface molecule, LAM-1. Homology with the mouse lymphocyte homing receptor and other human adhesion proteins. *J. Exp. Med.*, **170**:123.

Thome, R. (1898). Endothelien als phagocyten (aus den Lymphdrussen von Macacus cynomalgus). *Arch. Microsk. Anat.*, **52**:820.

Uckun, F. (1990). Regulation of human B-cell ontogeny. *Blood*, **76**:1908.

Varki, A. (1993). Biological roles of oligosaccharides: All the theories are correct. *Glycobiology*, **3**:97.

von Andrian, U.H., Chambers, J.D., Berg, E.L., Michie, S.A., Brown, D.A., Karolak, D., Ramezani, L., Berger, E.M., Arfors, K.E. and Butcher, E.C. (1993). L-selectin mediates neutrophil rolling in inflamed venules through sialyl Lewis X-dependent and -independent recognition pathways. *Blood*, **82**:182.

von Andrian, U.H., Chambers, J.D., McEvoy, L.M., Bargatze, R.F., Arfors, K.E. and Butcher, E.C. (1991). Two-step model of leukocyte-endothelial cell interaction in inflammation: Distinct roles for LECAM-1 and the leukocyte beta2 integrins *in vivo*. *Proc. Natl. Acad. Sci.*, **88**:7538.

von Boehmer, H. (1988). The development biology of T lymphocytes. *Ann. Rev. Immunol.*, **6**:309.

Watson, S., Imai, Y., Fennie, C., Geoffroy, J.S., Rosen, S.D. and Lasky, L.A. (1990). A homing receptor-IgG chimera as a probe for adhesive ligands of lymph node high endothelial venules. *J. Cell Biol.*, **110**:2221.

Watson, S.R., Fennie, C. and Lasky, L.A. (1991). Neutrophil influx into an inflammatory site inhibited by a soluble homing receptor-IgG chimera. *Nature*, **349**:164.

Watson, S.R., Imai, Y., Fennie, C., Geoffrey, J., Singer, M., Rosen, S.D. and Lasky, L.A. (1991). The complement binding-like domains of the murine homing receptor facilitate lectin activity. *J. Cell Biol.*, **115**:235.

Yang, X-D., Karin, N., Tisch, R., Steinman, L. and McDevitt, H.O. (1993). Inhibition of insulitis and prevention of diabetes in nonobese diabetic mice by blocking L-selectin and very late antigen 4 adhesion receptors. *Proc. Natl. Acad. Sci.*, **90**:10494.

Yednock, T.A., Butcher, E.C., Stoolman, L.M. and S.D., R. (1987). Receptors involved in lymphocyte homing: Relationship between a carbohydrate-binding receptor and the MEL-14 antigen. *J. Cell Biol.*, **104**:725.

Yednock, T.A. and Rosen, S.D. (1989). Lymphocyte homing. *Adv. Immunol.*, **44**:313.

11 Glycoprotein Ligands for the Two Endothelial Selectins

Dietmar Vestweber

Institute for Cell Biology, ZMBE, University of Münster, Von-Esmarch-Str. 56, D-48149 Münster, Germany

E- and P-selectin are involved in the initial tethering and rolling of leukocytes along the blood vessel wall, as was shown by studies of gene deficient mice and by analysing the *in vivo* effects of selectin-blocking antibodies. In addition, blocking the binding function of the endothelial selectins has proven to be beneficial in various animal models of inflammation and models of ischemia/reperfusion damage. This has raised much interest in the identification of the physiological ligands of the endothelial selectins. The current knowledge about the presently identified ligands will be summarized in this chapter.

DIFFERENT CONCEPTS FOR THE IDENTIFICATION OF SELECTIN — LIGANDS

Already before the N-terminal lectin domain of the selectins had been discovered by sequencing, the lectin character of L-selectin had been described. A mannose-6-phosphate rich polysaccharide (PPME) from the yeast *Hansenula hostii* as well as the sulphated fucose polymer fucoidin had been shown to inhibit lymphocyte binding to high endothelial venules (HEV) in cryostat section assays (Yednock *et al.*, 1987). Direct evidence for the lectin activity of L-selectin was presented by ELISA assays in which immunopurified L-selectin, immobilized on plastic, was bound to the yeast polysaccharide PPME (Imai *et al.*, 1990). Endogenous carbohydrate structures which are present on human neutrophils and contain α(2,3)-sialylated, α(1,3)-fucosylated lactosaminoglycan (known as sialylated Lewis X, sLe^x) were soon shown to bind to E-selectin (Phillips *et al.*, 1990; Walz *et al.*, 1990). The same carbohydrate structure was also found to bind to P-selectin (Polley *et al.*, 1991) and to L-selectin (Foxall *et al.*, 1992). Derivatives of sLe^x like the stereoisomer sLe^a and SO_4-Le^x were found to bind as well or even better (Green *et al.*, 1992; Yuen *et al.*, 1992).

The regulating step which controls the expression of sLe^x and its derivatives is catalysed by fucosyl-transferases. It was already demonstrated in 1990 that expressing a fucosyl transferase in COS cells or in CHO cells was sufficient to render these cells adhesive for E-selectin (Lowe *et al.*, 1990). This gave rise to the hypothesis that no specific scaffold molecules exist as carriers for functional carbohydrate ligands of E-selectin and that just any or at least many molecules could serve as carriers or presenters for selectin-binding carbohydrate structures. Interestingly, the same fucosyltransferase which when transfected, made COS and CHO cells bind to E-selectin was not sufficient to make them bind to P-selectin (Sako *et al.*, 1993). Thus, it is possible that COS and CHO cells express specific scaffold molecules which can be modified to become ligands for E-selectin, but do not bind to P-selectin. Alternatively, it is possible, that the structural requirements for E-selectin binding might be less stringent than the requirements for P-selectin binding, and that an ectopically expressed fucosyl transferase in COS or CHO cells may artificially generate many fucosylated structures which could serve as E-selectin — but not as P-selectin-ligands. The final answer to these questions cannot yet be given.

It is also not yet known whether the authentic carbohydrate ligands for the selectins are commonly presented on lipids or on protein backbones. While glycolipids were shown in *in vitro* assays to be able to bind to the selectins (Aruffo *et al.*, 1990; Handa *et al.*, 1991; Needham and Schaar, 1993; Green *et al.*, 1992; Yuen *et al.*, 1992) and were even shown to be able to support rolling when coated onto plastic (Alon *et al.*, 1995) it is considered to be less likely that they would be sufficiently accessible on the cell surface to mediate the rolling process *in vivo*. Binding sites of cell adhesion molecules probably need to be located beyond a certain distance away from the lipid bilayer in order to be able to efficiently contact adhesion molecules on partner cells (Springer, 1990; Chan and Springer, 1992). However, it has been reported that protease treatment of HL60 cells does not block binding to E-selectin in static adhesion assays (Larsen *et al.*, 1992), arguing for a possible function of glycolipids in such assays. In contrast the binding of other myeloid cell lines such as the mouse neutrophil progenitor 32DCl3 is sensitive to protease treatment (Lenter *et al.*, 1994; Steininger *et al.*, 1992). In recent years the search for authentic selectin ligands on myeloid cells has mainly concentrated on glycoproteins.

The identification of glycoprotein ligands was tried by three different approaches. The first was based on the fact that sLe^x and its derivatives can bind to the selectins and that antibodies against sLe^x can even block the binding of human myeloid cells to E- and P-selectin. Thus, proteins which carried sLex were suspected to be ligands and analysed in more detail. The other approach was based on the use of selectin affinity probes. By offering basically all detergent extractable, cellular proteins of myeloid cells to a selectin affinity probe, ligands were searched which would bind with high selectivity and with an affinity sufficiently high to withstand extensive washing procedures. In a third approach, ligands were identified by expression cloning, using a recombinant form of the selectin as panning reagent and screening cDNA expression libraries of myeloid cells expressed in COS cells. All three approaches have been successfull and the identified ligands will be described below.

Among those ligands which can be selectively isolated from cell extracts with selectin affinity probes, two categories of glycoprotein ligands were found: those which carry the functional carbohydrate moieties on O-linked side chains, the so-called "sialomucins" and those which carry them on N-linked side chains. Only the latter group of ligands were found to be monospecific for one of the two endothelial selectins.

Table 1 Glycoproteinligands of P-selectin

Molecular weight	*Polypeptide name*	*Source*	*Requirements for binding*[+]	*Comments*	*References*
220 kD homodimer (reduced 110 kD)	PSGL-1	hum. neutrophils hum. activated T cells HL60 cells	– sialylation – fucosylation – Core 2 branching of O-linked oligosaccharides – tyrosine sulfation – proteolytic processing of pro-form	sialomucin; binds also to E-selectin	Sako *et al.*, 1993 Moore *et al.*, 1992; 1995 Patel *et al.*, 1995 Vachino *et al.*, 1995
230/130 kD (less 230 kd form after reduction)	mouse PSGL-1	mouse bone marrow cells (80% neutrophils) 32Dcl3 cells	– sialylation – no N-linked oligosacharides required	OSGPase sensitive; possibly the mouse homolog of PSGL-1; binds also to E-selectin	Lenter *et al.*, 1994 Borges and Vestweber, in preparation
160 kD (red. 80 kD)	polypeptide unknown	mouse bone marrow cells (80% neutrophils) 32Dcl3 cells HL60 cells	– sialylation – N-linked carbohydrates	OSGPase sensitive; does not bind to E-selectin	Lenter *et al.*, 1994
160 kD	polypeptide unknown	human neutrophils			Moore *et al.*, 1992
40–60 kD	Heat stable antigen (HSA)*	mouse bone marrow cells (80% neutrophils) mouse B lymphoblasts	– sialylation – N-linked carbohydrates	does not bind to E-selectin	Sammar *et al.*, 1994 Aigner *et al.*, 1995

Abbreviations: OSGPase: O-sialoglycoprotease;
*Affinity isolation with P-selectin not yet reported; [+] All listed ligands require Ca^{2+} for binding

LIGANDS FOR P-SELECTIN

The P-selectin Glycoprotein Ligand-1 (PSGL-1)

The first successfull identification of a P-selectin ligand was based on the use of the platelet derived, purified P-selectin membrane protein as an affinity probe (Moore *et al.*, 1992). Iodinated P-selectin bound to a 250/120kD pair of proteins on Western blots from human neutrophils. The same pair of proteins was also detected from ^{3}H-glucosamin labeled HL60 cells by affinity isolation with immobilized P-selectin. The 250 kD protein turned out to be the disulfide linked dimeric form of the 120 kD protein. In addition, a 160 kD protein was described as a possible ligand candidate in this study, which was detected with ^{125}I-P-selectin in Western blots under non-reducing conditions.

Soon after this study, a 220 kD homodimeric ligand for human P-selectin was identified by expression cloning on HL60 cells and named P-selectin glycoprotein ligand-1 (PSGL-1) (Sako *et al.*, 1993). The protein contains 15 consecutive Ser/Thr rich decameric repeats which are typical sites for the addition of clustered O-linked carbohydrate side chains. Thus, PSGL-1 belongs to the class of sialomucins. Transfection of the full length cDNA of this protein into COS cells rendered these cells adhesive for P-selectin, provided they had been co-transfected with the fucosyl transferase III (FT III). Thus, fucosylation of PSGL-1 was imperative for binding. COS-cells transfected with FT III alone or co-transfected with the cDNAs for the mucin CD43 and FT III showed no binding to P-selectin. A soluble, recombinant form of PSGL-1 made in COS cells not only supported binding of P-selectin-transfected cells but also of cells which were transfected with E-selectin. This was the first reported evidence that PSGL-1 could function as a ligand for both endothelial selectins. The identity of PSGL-1 with the affinity isolated 250/120 kD ligand identified by Moore *et al.* (1992) was demonstrated by antibodies against a PSGL-1 peptide and by microsequencing of the purified 250/120 kD ligand (Moore *et al.*, 1994).

Searching for glycoprotein ligands on mouse myeloid cells, a 230/130 kD pair of proteins was identified as a ligand for P-selectin as well as E-selectin (Lenter *et al.*, 1994). In this case, identification was based on the use of selectin-immunoglobulin fusion proteins as affinity probes. These fusion proteins contained the first four protein domains of either mouse P- or E-selectin (lectin domain, EGF-repeat and the first two short consensus repeats) fused to the Fc-part of human IgG1, resulting in a dimeric, antibody like structure. The 230/130 kD pair of proteins was isolated from detergent extracts of metabolically labelled mouse neutrophils by both selectin-Ig fusion proteins with the same efficiency. A similar pair of proteins with slightly smaller apparent molecular weight could be identified with the soluble mouse selectin fusion proteins from certain HL60 subclones, though not from others (due to heterogeneities between different populations of HL60 cells). These proteins could be precipitated with a rabbit antiserum raised against a soluble recombinant form of human PSGL-1 (Borges and Vestweber, unpublished observation). Affinity isolation of human PSGL-1 with an E-selectin-Ig affinity probe was also shown by Asa *et al.* (1995). It is likely that the 230/130 kD pair of proteins is the mouse homolog of human PSGL-1.

Whether the binding of PSGL-1 to E-selectin is of similar affinity as the binding to P-selectin is still debated. Using soluble, truncated forms of human E-selectin and P-selectin coated onto plastic, it was shown in non-equilibrium binding assays with iodinated PSGL-1 that binding occurred with 50 fold higher affinity to P-selectin than to

E-selectin (Moore *et al.*, 1994). However, binding of PSGL-1 to E-selectin seems to be strong enough to mediate, at least in part, rolling of human neutrophils on E-selectin. The adhesion blocking antibody PL-1 against human PSGL-1 which can block rolling of neutrophils on P-selectin (Moore *et al.*, 1995) partially blocks rolling of human neutrophils on plastic coated E-selectin, but not attachment in static adhesion assays (Patel *et al.*, 1995).

PSGL-1 is not restricted to myeloid cells. Flow cytometric analysis of peripheral blood T cells with a rabbit antiserum against human PSGL-1 allowed to detect the antigen on the majority of T-cells (Alon *et al.*, 1994). Furthermore, it was shown that certain T-cell clones which expressed PSGL-1 could bind to plastic coated P-selectin under shear force. This attachment could be partially blocked with the anti PSGL-1 serum, although very high concentrations of the serum were necessary. A non-functional form of PSGL-1, which does not bind to P-selectin was found widely distributed on lymphocytic cells (Vachino *et al.*, 1995). Several enzymes which posttranslationally process and modify PSGL-1 are involved in the generation of the functional P-selectin-binding form of PSGL-1. In addition to a fucosyltransferase the Core 2 $\beta(1,6)$ GlcNAc transferase is necessary, which generates branched core 2 O-linked oligosaccharide side chains, with the branch attached to the C-6 of the Ser/Thr attached GalNac (Vachino *et al.*, 1995). Both enzyme activities were found to be elevated in stimulated peripheral blood T cells, which correlated with increased binding of these cells to P-selectin. The majority of the O-linked oligosaccharide side chains on PSGL-1 purified from human myeloid cells was found to contain only very little core 1 type side chains (Moore *et al.*, 1994). In addition to glycosyltransferases the intracellular paired basic amino acid converting enzyme (PACE), which cleaves the pro-form of the protein, is involved in generating a functional form of PSGL-1. B-cells, which express PSGL-1 in the non- cleaved form on the cell surface and do not bind to P-selectin, become adhesive for P-selectin after incubation with the soluble form of PACE (Vachino *et al.*, 1995).

PSGL-1, like the well described L-selectin ligands GlyCAM-1 and CD34, is a sialomucin, thus, it carries dense clusters of O-linked carbohydrate side chains which are rich in sialic acid. As it is typical for sialomucins, PSGL-1 is readily digested by a sialomucin specific protease the O-sialoglycoprotease (OSGPase) from *Pasteurella haemolytica*. Treatment of myeloid cells with this protease also blocks attachment to P-selectin, which is in good agreement with an essential function of PSGL-1 as P-selectin ligand. Interestingly, treatment with OSGPase does not block attachment to E-selectin in several reports, questioning again the necessity of PSGL-1 for cell attachment to E-selectin (Steininger *et al.*, 1992; Lenter *et al.*, 1994; Alon *et al.*, 1994).

The remarkable specificity and affinity of the interaction between P-selectin and PSGL-1 raised the question which structural entity on PSGL-1 would be required for binding. It was clearly shown that N-linked carbohydrate side chains were not responsible for binding, since treatment of the purified protein with endoglycosidase F or Peptide:N-glycosidase F did not destroy the binding capacity of PSGL-1 (Moore *et al.*, 1994; Lenter *et al.*, 1994). Since removal of sialic acid destroyed the selectin-binding activity, obviously the O-linked carbohydrate side chains were necessary for the binding activity. As for the other sialomucin type selectin ligands it was suggested that clusters of saccharide patches would define the binding site. Since other sialomucins are known, which are also rich in sLex, yet cannot be considered as selectin ligands (Moore *et al.*,

1992), it was difficult to explain how the short O-linked side chains would generate the remarkable selectin-specificity. Norgard *et al.* (1993) have suggested that clusters of common oligosaccharides could present uncommon "clustered saccharide patches" generated by multiple oligosaccharides that are spaced closely enough to restrict their motion. This hypothesis has been described in more detail in an excellent recent review by Varki (1994). Since the carbohydrate binding site in the lectin domain is too small to accomodate large patches of carbohydrate side chains (Graves *et al.*, 1994), Varki suggests that a cluster of common carbohydrate side chains generates carbohydrate epitopes which are unique (by forcing a single side chain into an unusual conformation or by combining certain groups of different side chains into one binding epitope).

A more detailed analysis of the binding sites on PSGL-1 has revealed an additional structural element which is necessary for binding and may help to determine the remarkable ligand-specificity (Pouyanis and Seed, 1995; Sako *et al.*, 1995; Wilkins *et al.*, 1995). This element is located in the amino terminal region in front of the 15 consecutive decameric repeats which carry the clustered O-linked carbohydrate side chains. Deletion of amino acids 19 to 57 (with amino acid 1 being the start methionine of the coding sequence) resulted in the complete loss of binding activity of PSGL-1. Adding the first 100 amino acids of PSGL-1 to the amino terminus of the mucins CD34 or CD43 made these mucins capable of binding to P-selectin. The region which was responsible for this transfer of binding specificity could be further nailed down to the sequence between amino acids 38 and 57. This sequence contains three consensus sites for tyrosine sulfation and fusion proteins containing this PSGL-1 region were shown to become sulfated. Sulfation was abolished by mutating these tyrosine residues. Inhibition of sulfation in PSGL-1 transfected COS cells by treating the cells with chlorate blocked P-selectin binding but not binding to E-selectin. Thus in addition to clusters of O-linked carbohydrates, tyrosine sulfation is — directly or indirectly — necessary for the binding of PSGL-1 to P-selectin. The combination of both structural elements defines PSGL-1 as a ligand for P-selectin.

The requirement for ligand-sulfation is shared between L-selectin and P-selectin. The L-selectin ligands GlyCAM-1 and CD34 are also both sulfated (Imai *et al.*, 1991) and it was shown for GlyCAM-1 that it does not bind to L-selectin in the absence of sulfate (Imai *et al.*, 1993). However, the sulfate groups on GlyCAM-1 are located on carbohydrate side chains (Hemmerich and Rosen, 1994). The sulfo-transferases which are responsible for these modifications have not yet been characterized. While the ones which transfer sulfate to GlyCAM-1 and CD34 seem to be expressed in a very restricted fashion, the sulfotransferase which mediates tyrosine-sulfation is expressed in CHO cells as well as COS cells. The obvious importance of these enzymes for the control of the expression of functionally active L- and P-selectin ligands makes these enzymes a major target for future research activities. Interestingly, sulfation seems not to be of importance for E-selectin ligands, since the E-selectin glycoprotein ligand ESL-1 (see below) is not sulfated (Levinovitz and Vestweber, unpublished observation). Furthermore, HNK-1 reactive sulfoglucuronyl glycolipids bind well to P-selectin and L-selectin and are not bound by E-selectin.

Other Ligands for P-selectin

Using mouse P-selectin-IgG as affinity probe, a 160 kD glycoprotein could be isolated from ^{35}S-methionine/^{35}S-cysteine labeled mouse and human myeloid cell lines (Lenter *et al.*, 1994). In contrast to the 230/130 kD pair of proteins, which was identified in the same studies and most likely represents PSGL-1, the 160 kD protein did not bind to E-selectin. Binding to P-selectin required sialic acid and N-linked carbohydrates. Reducing the isolated protein by boiling in the presence of 50 mM dithiothreitol before electrophoresis lead to the disappearance of a detectable protein band. Only reduction of the protein which was cut from a polyacrylamide gel after a first electrophoresis under non-reducing conditions, allowed to detect the protein in a second electrophoresis at an apparent molecular weight of 80 kD. Despite extensive efforts, it was not yet possible to prepare sufficient amounts of this protein for further analysis. A 160 kD protein could also be detected with ^{125}I-labeled human P-selectin in Western blots of HL 60 cells (Moore *et al.* 1992). However, this protein was not found in affinity isolation experiments and was also not shown to run in polyacrylamide gel electrophoresis at an apparent molecular weight of 80 kD under reducing conditions. Whether both proteins are related is unknown.

A better characterized ligand for P-selectin is the cell surface protein heat stable antigen (HSA), also described as CD24 in human. This protein has a very small protein core of only 27 amino acids and is extensively glycosylated. It is expressed on many different cells and its apparent molecular weight varies between different cell types ranging from 28–35 kDa in embryonic brain, 35–45 kDa on erythrocytes, to 40–70 kDa in lymphoid cells (14–20). Selectin-binding was analysed in ELISA assays with different forms of HSA isolated from different cell types (Sammar *et al.*, 1994). In these assays, HSA was coated onto plastic and offered to soluble mouse P-selectin-IgG or mouse E-selectin-IgG. HSA from B lymphoblasts and from neutrophils strongly supported P-selectin-binding while HSA from red blood cells did not. Interestingly in no case was binding of E-selectin-IgG observed. In a second study (Aigner *et al.*, 1995) it could be shown that binding of monocytic cells and mouse neutrophils to endotoxin stimulated endothelioma cells could be blocked by a mAb against HSA as well as by a mAb against mouse P-selectin. Similarly binding of HSA coated latex beads to activated endothelioma cells or platelets was blocked by mAbs against HSA and P-selectin. Staining of HSA coated latex beads with P-selectin-IgG was dependent on divalent cations and was sensitive to treatment of the beads with endoglycosidase F or neuraminidase. Again no binding was observed with E-selectin-IgG. Except for the 160 kD protein, which is still not characterized, HSA is the first glycoprotein which requires N-linked carbohydrates for the binding to P-selectin and which is specific for P-selectin. Thus, different from PSGL-1 whose carbohydrates are recognized by both selectins, HSA must carry carbohydrate side chains, which can only be accomodated by the P-selectin carbohydrate binding site. Alternatively, both selectins may recognize similar carbohydrate side chains on HSA with low affinity while high affinity binding of P-selectin may be enabled by a second structural determinant, which cannot be recognized by E-selectin. The existence of monospecific ligands which only bind to one of the two endothelial selectins allows each selectin to exert different functions on different types of leukocytes.

One may argue that the binding requirements for E-selectin are just less stringent or even less specific than those for P-selectin, for the following reasons: 1. Fucosyl transferase transfected CHO or COS cells strongly bind to E-selectin but not to P-selectin; 2. Many more glycoproteins have been reported to represent ligands to E-selectin than to P-selectin (see below); 3. Both selectins bind to PSGL-1, yet only the binding of P-selectin requires sulfation of PSGL-1, while E-selectin binds to non- sulfated PSGL-1 as well as to the sulfated protein. In this respect it is remarkable that a ligand such as HSA can only be recognized by P-selectin and not by E-selectin. It will be interesting to analyse the carbohydrate moieties on HSA, which represent the binding site for P-selectin.

LIGANDS FOR E-SELECTIN

The E-selectin Ligand-1 (ESL-1)

Using a mouse E-selectin-Ig fusion protein a 150 kD glycoprotein, which was named E-selectin ligand-1 (ESL-1) was affinity isolated as the major ligand from the ^{35}S-methionine/^{35}S-cysteine labeled mouse neutrophilic progenitor cell line 32Dcl3 as well as from mouse bone marrow neutrophils (Levinovitz *et al.*, 1993). Binding of ESL-1 was Ca^{2+}-dependent and required sialic acid as posttranslational modification. In single step affinity purifications this ligand was isolated from non-labeled 32Dcl3 cells as the only protein which could be detected by silverstaining or by staining with Coomassie blue.

Subsequent studies revealed that ESL-1 did not bind to P-selectin and was not a sialomucin type of ligand (Lenter *et al.*, 1994). In contrast to the sialomucins, ESL-1 required N-linked carbohydrates for binding to E-selectin. Purifying larger quantities of this ligand by isolation with an E-selectin-Ig affinity matrix allowed to obtain several internal peptide sequences by microsequencing. Based on this information ESL-1 was cloned from a cDNA library made from poly A^+-RNA of the mouse neutrophilic cell line 32Dcl3.

Transfection experiments with the ESL-1 cDNA revealed that the ESL–1 glycoprotein could only be affinity-isolated from transfected CHO cells when co-expressed with a fucosyltransferase (Steegmaier *et al.*, 1995). Thus, like for PSGL-1, modification with fucose is a necessary requirement for binding to E-selectin. Specific antibodies against ESL-1 recognized a 150 kD protein not only on myeloid cells but also on other leukocytes and on many non-leukocytic cells from different species, even on CHO and COS cells. However, E-selectin-IgG only bound to the 150 kD protein species from myeloid cells or from cells which had been transfected with fucosyl transferase. Thus, like for other known selectin ligands, fucosylation controls the expression of the ESL-1 glycoform which is able to bind to E-selectin.

Involvement of ESL-1 in the binding of mouse myeloid cells to E-selectin could be demonstrated. An immunoglobulin fusion protein of ESL-1 containing the complete extracellular part of ESL-1 and modified by fucose was able to support attachment of E-selectin transfected CHO cells under shear force. No binding was seen to a fucose-containing L-selectin-Ig fusion protein. Furthermore, affinity-purified polyclonal antibodies against ESL-1 blocked the binding of the neutrophilic cell line 32Dcl3 to immobilized E-selectin-IgG. Similarly, the binding of these cells and of mouse PMNs to cytokine-induced mouse endothelioma cells could be blocked by the anti-ESL-1 antibodies. The cytokine-induced endothelioma cells expressed E- and P-selectin. Since the binding of PMNs was predominantly mediated by P-selectin in these assays, the inhibitory effect of the anti-ESL-1 antibodies was only detectable when P-selectin was blocked simultaneously with a mAb against P-selectin. This is additional proof that ESL-1 is a ligand specific for E-selectin and not for P-selectin.

The amino acid sequence of ESL-1 is 94% identical (over 1,078 amino acids) to a novel chicken cysteine-rich fibroblast growth factor receptor (CFR) (Burrus *et al.*, 1992), except for a unique 70 amino acid amino-terminal domain of mature ESL-1. Based on this homology it was tested, whether ESL-IgG would support the binding of ^{125}I-labeled FGF-2. Indeed specific binding, which could be competed with a 1000 fold excess of unlabeled FGF was observed, yet this binding was not saturable and its significance is questionable (Steegmaier and Vestweber, unpublished observation). The biological role of CFR is not yet known.

The existance of the unique 70 amino acid domain at the N-terminus of ESL-1 in the context of the strong structural homology of the rest of mouse ESL-1 to CFR suggested, that ESL-1 may be a splicing variant of the mouse equivalent of CFR. Interestingly the 70 amino acid domain contains a peculiar stretch of glutamines interrupted by a few prolines. Such a structure is not found in CFR. Stretches of glutamines or prolines have been demonstrated to be potential sites for protein oligomerization (Gerber *et al.*, 1994; Stott *et al.*, 1995). Recently, it was possible to identify a splicing variant of ESL-1 in brain tissue of mouse embryos (day 13), which lacks most of the 70 amino acid N-terminal domain of ESL-1 (Blanks and Vestweber, unpublished observation). This variant lacked the complete glutamine, proline rich region. The function of this splicing variant is still unknown.

The strong structural homology of ESL-1 to the chicken FGF-receptor allows one to speculate that ESL-1 may have some signalling function. Since E-selectin mediates the initial cell contact between leukocytes and endothelial cells, a step which is followed by the activation of leukocyte integrins and by chemotaxis, it is intriguing to speculate that ESL-1 may be involved in triggering signals which lead to such activation events. Indeed evidence has been reported that soluble E-selectin can induce activation of PMNs (Lo *et al.*, 1991). However, so far no evidence has been presented that ESL-1 or the chicken CFR are able to mediate signal transduction.

Another highly homologous protein to ESL-1 was found in the rat. Cloning of the rat protein MG160 revealed an overall amino acid sequence similarity of 98% (Gonatas *et al.*, 1995). This protein had originally been identified as a Golgi protein in neuronal tissue (Croul *et al.*, 1990) and in several other cell types. The function of MG160 is not yet known. Analysing the subcellular distribution of ESL-1, we could localize it by indirect immunofluorescence in the Golgi of permeabilized cells and by flow cytometry on the cell surface of neutrophilic (32Dcl3 cells and neutrophils) and lymphoid cells (Steegmaier and Vestweber, manuscript submitted). Cell surface staining was confirmed by cell surface biotinylation and by cell surface immunoprecipitations in which antibodies only had access to surface proteins of intact cells. In addition, ESL-1^{high} and ESL-1^{low} expressing cells, sorted by flow cytometry gave rise to ESL-1 immunoprecipitation signals of high and low intensity, as expected. Thus, ESL-1 is found at two different locations, in the Golgi and on the cell surface. It is likely, that ESL-1 has additional functions, besides mediating the binding of myeloid cells to E-selectin.

A similar dual expression pattern on the cell surface and in the Golgi was also recently reported for the 55 kD tumor necrosis factor-α (TNF-α) receptor. In human endothelial cells it was predominantly found in the Golgi (Bradley *et al.*, 1995), although it is well known to be the essential receptor on these cells for mediating proinflammatory and cytotoxic activities of TNF (Tartaglia *et al.*, 1992; Vandenabeele *et al.*, 1995). The

Golgi- located receptor molecules could not be chased to the cell surface (Bradley *et al.*, 1995) and it is neither known, what regulates their distribution between the Golgi and the cell surface nor is it known, for what purpose a large amount of this TNF receptor stays in the Golgi.

Immunogold scanning electron microscopy of the mouse lymphoma cell line K46 revealed that ESL-1 on the cell surface is preferentially located on microvilli (Steegmaier, Borges, Berger, Schwartz and Vestweber; manuscript submitted). A similar distribution was shown before for L-selectin (Picker *et al.*, 1991), for the P-selectin ligand PSGL-1 (Moore *et al.*, 1995) and for the integrin $\alpha_4\beta_7$ (Berlin *et al.*, 1993). Each of these adhesion molecules is known to mediate very early tethering/rolling interactions between flowing leukocytes and the endothelial cell surface. In the case of L-selectin it was shown very elegantly, that its location on the microvilli indeed improved its ability to initiate cell contacts (von Andrian *et al.*, 1995). Using CD44/L-selectin chimeric molecules it was demonstrated that the transmembrane and intracellular domains of CD44 (which is usually excluded from microvilli) targeted the extracellular part of L-selectin to the planar body. Only L-selectin located on microvilli was able to initiate cell contacts under flow efficiently, while L-selectin-CD44 chimeric molecules which were excluded from microvilli, initiated leukocyte rolling under flow only very inefficiently. In agreement with these findings, β2-integrins which are essential for leukocyte adhesion to endothelium but which are not able to initiate contacts under flow conditions, are excluded from microvillous processes. In this context, the localization of ESL-1 on microvilli is a strong indication for its possible role in very early steps of leukocyte endothelial contact formation.

Other Ligands for E-selectin

ESL-1 was not the only glycoprotein ligand which could be affinity-isolated with E-selectin-Ig from metabolically labelled mouse PMNs. In addition, a protein running as a sharp band of 250 kD apparent MW (230 kD under non-reducing conditions) in SDS-polyacrylamide gel electrophoresis (SDS-PAGE) was seen, which was not detectable on any tested myeloid cell line (Levinovitz *et al.*, 1993; Lenter *et al.*, 1994). Like ESL-1, this protein did not bind to P-selectin-Ig. As a much weaker signal a pair of proteins running at 230 and 130 kD in SDS-PAGE was isolated (Lenter *et al.*, 1994). These proteins were detected with similar efficiency in affinity isolation experiments with P-selectin-Ig and most likely represent mouse PSGL-1 (see above).

In a similar approach, a 250 kD ligand (280 kD under reducing conditions) was isolated from bovine γ/δ T lymphocytes, using as affinity probe complete human E-selectin purified from transfected L-cells and immobilized with the help of a non-adhesion blocking anti E-selectin mAb (Walcheck *et al.*, 1993). Like the selectin ligands in the mouse, this glycoprotein ligand also is not recognized by either the anti sLe^x antibody CSLEX-1 or the anti sLe^a antibody HECA 452.

Several other glycoproteins were defined as ligands for E-selectin, based either on observations that antibodies against them blocked leukocyte binding to E-selectin or that the immobilized glycoprotein could support the binding of E-selectin expressing cells. The most prominent of these ligands is L-selectin, which was suggested to serve as a carbohydrate-presenting ligand for E-selectin and P-selectin (Kishimoto *et al.*, 1991; Picker *et al.*, 1991). Only L-selectin from human neutrophils, but not from human lymphocytes, was

Table 2 Glycoproteinligands of E-selectin

Molecular weight	*Polypeptide name*	*Source*	*Requirements for binding*⁺	*Comments*	*References*
150 kD (non-red. 130 kD)	ESL-1	mouse bone marrow cells (80% neutrophils) 32Dcl3 cells	– sialylation – fucosylation – N-linked carbohydrates	no sialomucin; does not bind to P-selectin	Levinovitz *et al.*, 1993 Steegmaier *et al.*, 1995
250 kD (non-red. 230 kD)	polypeptide unknown	mouse bone marrow cells (80% neutrophils)		not found on cell lines; does not bind to P-selectin	Levinovitz *et al.*, 1993 Lenter *et al.*, 1994
250 kD (non-red. 280 kD)	polypeptide unknown	bovine γ/δ T cells	– sialylation	affinity purified with E-selectin	Walcheck *et al.*, 1993
100 kD	L-selectin*	human neutrophils	– sialylation	could be affinity-isolated with E-selectin-Ig (Lenter, Zöllner and Vestweber, manuscript submitted)	Picker *et al.*, 1991 Kishimoto *et al.*, 1991
160 kD	CD66/CD67* antigens	human neutrophils	– sialylation	only shown for a mixture of antigens	Kuijpers *et al.*, 1992
90–120 kD	lamp-1*	human colon carcinoma cells	– sialylation	Expression level correlates with cell binding efficiency in adhesion assays	Sawada *et al.*, 1993

*Affinity isolation with E-selectin not yet reported; ⁺ All listed ligands require Ca^{2+} for binding able to support binding of E-selectin transfected cells. In addition, antibodies against L-selectin blocked binding of PMNs to cells expressing E-selectin or P-selectin. In adhesion assays under flow conditions the contact formation or tethering of PMNs to an E-selectin coated surface could be blocked, while the rolling velocity was unaffected (Lawrence *et al.*, 1994, Patel *et al.*, 1995). It has not yet been published whether the interaction between L-selectin as a carbohydrate presenter and E-selectin as a lectin would be of sufficient affinity to withstand extensive washing procedures in affinity isolation experiments. However, we have found recently (Lenter, Zöllner and Vestweber, manuscript submitted) that L-selectin from human neutrophils can be affinity isolated with E-selectin-IgG. In agreement with the published *in vitro* adhesion data, L-selectin from human lymphocytes could not be isolated with E-selectin-IgG. The fact that L-selectin from mouse neutrophils did not bind to E-selectin in these experiments documents an important difference between both species.

At present it is difficult to decide, whether L-selectin, which is certainly essential for the initiation of leukocyte interactions with activated endothelium at sites of inflammation, mainly acts in these processes as a lectin or as a carbohydrate- presenting molecule for E-selectin. There is good indirect evidence that cytokine inducable ligands on endothelium exist, which present carbohydrates to L-selectin. This is based on *in vitro* adhesion assays with human endothelial cells which had been activated with cytokines for 24 h, a time period after which the endothelial selectins were not detected any more on the cell surface (Spertini *et al.*, 1991; Brady *et al.*, 1992). The identification of these ligands is still a major goal in the field.

Other suggested E-selectin-ligands include members of the NCA-family (non specific cossreactive antigens) present on human neutrophils (Kuijpers *et al.*, 1992), a subpopulation of the β2-integrins which carry sLe^x (Kotovuori *et al.*, 1993), and the heavily sLe^x-modified lysosomal membrane protein lamp-1 (Sawada *et al.*, 1993). Indeed, increasing the cell surface expression of lamp-1 on transfected cells correlated with an increase in the binding of these cells to E-selectin. It is conceivable that various colon carcinoma cells which display increased levels of lamp-1 at the cell surface may bind via lamp-1 to E-selectin expressing endothelium. However, it is not yet known for any of these ligands, whether they can bind to E-selectin with sufficient affinity to allow affinity isolation with E-selectin, as described above for the "high affinity" ligands. Whether "high affinity" or "low affinity" is necessary for a selectin-ligand to be of physiological relevance is currently being debated (van der Merwe *et al.*, 1994; Varki *et al.*, 1994).

Functional Diversity or Redundancy of the Two Endothelial Selectins and Their Ligands

Eperiments with selectin blocking monoclonal antibodies and with mice deficient for selectin genes have clearly demonstrated, that each of the two endothelial selectins is involved in leukocyte rolling along the blood vessel wall and in leukocyte extravasation (see chapter of Bullard and Beaudet). For most of the analysed *in vivo* models, the two endothelial selectins support the same process. Blocking of only one of them has a much weaker or even no effect on leukocyte extravasation than blocking both. This clearly demonstrates that they can support extravasation of the same subpopulations of leukocytes in the same vascular beds. The fact that P- and E-selectin can both bind to the same ligand, PSGL-1, and can also be expressed with similar kinetics on the endothelial cell surface, as was shown for mouse, rat and bovine endothelium (although not for human endothelial cells) suggests, that both selectins can be functionally redundant, possibly even by mediating the very same molecular step in the extravasation process. While this is certainly possible, the two endothelial selectins may also mediate different steps. They can be induced on the endothelial cell surface under conditions where the other selectin is absent (Geng *et al.*, 1990; Gotsch *et al.*, 1994).

As was reported by Alon *et al.* (1994) they support rolling of human lymphocytes via different ligands. While rolling of lymphocytes on P-selectin (in *in vitro* adhesion assays) was completely dependent on sialomucins, with PSGL-1 being the major ligand, rolling on E-selectin was independent of sialomucin. If this holds true *in vivo*, it would allow different subpopulations of lymphocytes to bind to only one of the two endothelial selectins. We found recently that under certain conditions ESL-1 can also be found as an E-selectin binding glycoform on lymphoid cells (Borges and Vestweber, manuscript in preparation). It is still open, whether PSGL-1 is as an efficient ligand for E-selectin as it is for P-selectin. While this was suggested based on affinity isolation experiments and *in vitro* adhesion assays (Lenter *et al.*, 1994; Asa *et al.*, 1995), detailed quantiative studies suggest that PSGL-1 binds more avidly to P- than to E-selectin (Moore *et al.*, 1994). Using non-flow adhesion assays it was shown that PSGL-1 needs to be tyrosine-sulfated to mediate cell adhesion to P-selectin, while cell adhesion to E-selectin was independent on sulfation (Pouyanis and Seed, 1995; Sako *et al.*, 1995). Thus, it is even conceivable, that on certain leukocyte subpopulations fucosylated, none-sulfated PSGL-1 could serve as an E-selectin-specific ligand. However, such a case has not yet been described.

Not much is known about the functions of the ligands for the endothelial selectins. PSGL-1 is certainly the best characterized ligand. It has been shown to mediate binding of myeloid cells to P- and E-selectin in non-flow adhesion assays (Sako *et al.*, 1993; Sako *et al.*, 1995, Pouyanis and Seed, 1995) as well as under flow conditions (Patel *et al.*, 1995). Involvement of ESL-1 in the binding of mouse neutrophils and the mouse neutrophilic cell line 32Dcl3 to E-selectin has also been demonstrated, although only in non-flow cell adhesion assays. PSGL-1 was located by immunogold labelling on microvillous processes which are likely to be destined to mediate the very first contacts of leukocytes with the endothelial surface. The same localization was found by immunogold scanning electron microscopy for ESL-1 on lymphoma cells (Steegmaier, Borges, Berger, Schwartz and Vestweber; manuscript submitted). Which of the ligands described today mediate the various signal transduction processes, which are induced by binding of the endothelial selectins to the leukocyte cell surface is not yet known. P-selectin is able to increase the secretion of Monocyte Chemotactic Protein-1 (MCP-1) and of TNF-α by monocytes, when these cells are stimulated by platelet activating factor (PAF) (Weyrich *et al.*, 1995). Similarly, binding of monocytes to P-selectin primed these cells for increased synthesis of PAF upon stimulation of their β-glucan receptor (Elstad *et al.*, 1995). Furthermore, P-selectin was shown to induce the expression of tissue factor on monocytes, although P-selectin presenting platelets were much more efficient in these assays than P-selectin presenting CHO cells, arguing for co-stimulatory signals on the platelet surface (Celi *et al.*, 1994). For E-selectin a chemotactic activity on human neutrophils was demonstrated (Lo *et al.*, 1991). None of the known ligands have yet been defined to be responsible for these functions.

A common theme for all glycoprotein ligands of the selectins is that the protein back bones are more broadly distributed than the selectin binding glycoforms. Thus, the modifying enzymes which generate the correct post-translational modifications necessary for selectin-binding are of central importance for the regulation of cell-type specific expression of the selectin ligands. Fucosylation is essential for the binding function of ESL-1 as well as PSGL-1. It will be interesting to analyse, which of the five known fucosyltransferases modify ESL-1 and PSGL-1 in neutrophils and whether both ligands are modified by the same transferase. It is still surprising and not yet understood, how the expression of a single fucosyltransferase only generates one or two high affinity glycoprotein ligands in a cell. One explanation could be, that the protein backbone itself may be directly involved in the formation of the binding site, possibly via a second non-carbohydrate based post translational modification. This could indeed be the correct explanation for the tyrosine-sulfated PSGL-1. No such requirement has yet been found for ESL-1. Another explanation would be that glycosyl-transferases may act in a very selective way, so that just very few proteins would acquire highly specific carbohydrate side chains which would be exclusive for these proteins. A molecular machinery inside the Golgi which would guarantee the generation of such specific and exclusive modifications is yet unknown. The third explanation is that the protein scaffold determines "the unique clustering of relatively common oligosaccharides" which form binding sites of sufficient specificity for selectin binding (Varki *et al.*, 1994). The fact, that ESL-1 just requires a few N-linked carbohydrate side chains instead of clusters of short O-linked chains, may argue against the latter hypothesis.

References

Aigner, S., Ruppert, M., Hubbe, M., Sammar, M., Sthoeger, Z., Butcher, E.C., Vestweber, D. and Altevogt, P. (1995). Heat stable antigen (mouse CD24) supports myeloid cell binding to endothelial and paltelet P-selectin. *Intern. Immunol.*, **7**:1557–1565.

Alon, R., Rossiter, H., Wang, X., Springer, T.A. and Kupper, T.S. (1994). Distinct cell surface ligands mediate T lymphocyte attachment and rolling on P and E selectin under physiological flow. *J. Cell Biol.*, **127**:1485–1495.

Alon, R., Feizi, T., Yuen, C-T., Fuhlbrigge, R.C. and Springer, T.A. (1995). Glycolipid ligands for selectins support leukocyte tethering and rolling under physiological flow conditions. *J. Immunol.*, **154**:5356–5366.

Aruffo, A., Kolanus, W., Walz, G., Freedman, P. and Seed, B. (1991). CD62/P-selectin recognition of myeloid and tumor cell sulfatides. *Cell*, **67**:35–44.

Asa, D., Raycroft, L., Ma, L., Aeed, P.A., Kaytes, P.S., Elhammer, A.P. and Geng, J-G. (1995). The P-selectin glycoprotein ligand functions as a common human leukocyte ligand for P- and E-selectins. *J. Biol. Chem.*, **270**:11662–11670.

Berlin, C., Berg, E.L., Briskin, M.J., Andrew, D., Kilshaw, P., Holzmann, B., Weissman, I.L., Hamann, A. and Butcher, E.C. (1993). $\alpha_4\beta_7$ integrin mediates lymphocyte binding to the mucosal vascular addressin MAdCAM-1. *Cell*, **74**:185–195.

Bradley, J.R., Thiru, S. and Pober, J.S. (1995). Disparate localization of 55-kd and 75-kd tumor necrosis factor receptors in human endothelial cells. *Am. J. Pathol.*, **146**:27–32.

Brady, H.R., Spertini, O., Jimenez, W., Brenner, B.M., Marsden, P.A., and Tedder, T.F. (1992). Neutrophils, monocytes, and lymphocytes bind to cytokine-activated kidney glomerular endothelial cells through L-selectin (Lam-1) *in vitro*. *J. Immunol.*, **149**:2437–2444.

Burrus, L.W., Zuber, M.E., Lueddecke, B.A. and Olwin, B.B. (1992). Identification of a cysteine-rich receptor for fibroblast growth factors. *Mol. Cell. Biol.*, **12**:5600–5609.

Celi, A., Pellegrini, G., Lorenzet, R., De Blasi, A., Ready, N., Furie, B.C. and Furie, B. (1994). P-selectin induces the expression of tissue factor on monocytes. *Proc. Natl. Acad. Sci. USA*, **91**:8767–8771.

Chan, P-Y. and Springer, T.A. (1992). Effect of lengthening lymphocyte function-associated antigen 3 on adhesion to CD2. *Mol. Biol. Cell*, **3**:157–166.

Croul, S., Mezitis, S.G.E., Stieber, A., Chen, Y., Gonatas, J.O., Goud, B. and Gonatas, N.K. (1990). Immunocytochemical visualization of the Golgi apparatus in several species, including human, and tissues with an antiserum against MG-160, a sialoglycoprotein of rat Golgi apparatus. *J. Histochem. Cytochem.*, **38**:957–963.

Elstad, M.R., La Pine, T.R., Cowley, F.S., Mcever, R.P., Mcintyre T.M., Prescott, S.M. and Zimmerman, R.P. (1995). P-selectin regulates Platelet-Activating Factor synthesis and phagocytosis by monocytes. *J. Immnunol.*, **155**:2109–2122.

Foxall, C., Watson, S.R., Dowbenko, D., Fennie, C., Lasky, L.A., Kiso, M., Hasegawa, A., Asa, D. and Brandley, B.K. (1992). The three members of the selectin receptor family recognize a common carbohydrate epitope, the sialyl Lewis^x oligosaccharide. *J. Cell Biol.*, **117**:895–902.

Geng, J-G., Bevilacqua, M.P., Moore, K.L., McIntyre, T.M., Prescott, S.M., Kim, J.M., Bliss, G.A., Zimmerman, G.A. and McEver, R.P. (1990). Rapid neutrophil adhesion to activated endothelium mediated by GMP-140. *Nature,* **343**:757–760.

Gerber, H-P., Seipel, K., Georgiev, O., Höfferer, M., Hug, M., Rusconi, S. and Schaffner, W. (1994). Transcriptional activation modulated by homopolymeric glutamine and proline stretches. *Sciences,* **263**:808–811.

Gonatas, J.O., Mourelatos, Z., Stieber, A., Lane, W.S., Brosius, J. and Gonatas, N.K. (1995). MG160, a membrane sialoglycoprotein of the medial cisternae of the rat Golgi apparatus, binds basic fibroblat growth factor and exhibits a high level of sequence identity to a chicken fibroblast growth factor receptor. *J. Cell. Sci.*, **108**:457–467.

Gotsch, U., Jäger, U., Dominis, M. and Vestweber, D. (1994). Expression of P-selectin on endothelial cells is upregulated by LPS and TNF-α *in vivo. Cell Adhesion Commun.*, **2**:7–14.

Graves, B.J., Crowther, R.L., Chandran, C., Rumberger, J.M., Li, S., Huang, K-S., Presky, D.H., Familletti, P.C., Wolitzky, B.A. and Burns, D.K. (1994). Insight into E-selectin/ligand interaction from the crystal structure and mutagenesis of the lec/EGF domains. *Nature*, **367**:532–538.

Green, P.J., Tamatani, T., Watanabe, T., Miyasaka, M., Hasegawa, A., Kiso, M., Yuen, C.T., Stoll, M.S. and Feizi, T. (1992). High affinity binding of the leukocyte adhesion molecule L-selectin to 3′-sulphated-Le^a and -Le^x oligosaccharides and the predominance of sulphate in this interaction demonstrated by binding studies with a series of lipid-linked oligosaccharides. *Biochem. Biophys. Res. Commun.*, **188**:244–251.

Handa, K., Nudelman, E.D., Stroud, M.R., Shiozawa, T. and Hakomori, S. (1991). Selectin GMP-140 (CD62; PADGEM) binds to sialosyl-Le^a and sialosyl-Le^x, and sulfated glycans modulate this binding. *Biochem. Biophys. Res. Commun.*, **181**:1223–1230.

Hemmerich, S. and Rosen, S.D. (1994). 6′-sulfated sialyl Lewis x is a major capping group of GlyCAM-1. *Biochemistry*, **33**:4830–4835.

Imai, Y., True, D.D., Singer, M.S. and Rosen, S.D. (1990). Direct demonstration of the lectin activity of gp90^{MEL}, a lymphocyte homing receptor. *J. Cell Biol.*, **111**:1225–1232.

Imai, Y., Singer, M.S., Fennie, C., Lasky, L.A. and Rosen, S.D. (1991). Identification of a carbohydrate-based endothelial ligand for a lymphocyte homing receptor. *J. Cell Biol.*, **113**:1213–1221.

Imai, Y., Lasky, L.A. and Rosen, S.D. (1993). Sulphation requirement for GlyCAM-1, an endothelial ligand for L-selectin. *Nature*, **361**:555–557.

Kishimoto, T.K., Warnock, R.A., Jutila, M.A., Butcher, E.C., Lane, C., Anderson, D.C. and Smith, C.W. (1991). Antibodies against human neutrophil LECAM-1 (LAM-1/LEU-8/DREG-56 antigen) and endothelial cell ELAM-1 inhibit a common CD18-independent adhesion pathway *in vitro. Blood*, **78**:805–811.

Kotovuori, P., Tontti, E., Pigott, R., Shepard, M., Kiso, M., Hasegawa, A., Renkonen, R., Nortano, P., Altieri, D.C. and Gahmberg, C.G. (1993). The vascular E-sectin binds to the leukocyte integrins CD11/CD18. *Glycobiol.*, **3**:131–136.

Kuijpers, T.W., Hoogerwerf, M., van der Laan, L.C.W., Nagel, G., van der Schoot, C.E., Grunert, F. and Roos, D. (1992). CD66 nonspecific cross-reacting antigens are involved in neutrophil adherence to cytokine-activated endothelial cells. *J. Cell Biol.*, **118**:457–466.

Larsen, G.R., Sako, D., Ahern, T.J., Shaffer, M., Erban, J., Sajer, S.A., Gibson, R.M., Wagner, D.D., Furie, B.C. and Furie, B. (1992). P-selectin and E-selectin: Distinct but overlapping leukocyte ligand specificities. *J. Biol. Chem.*, **267**:11104–11110.

Lawrence, M.B., Bainton, D.F. and Springer, T.A. (1994). Neutrophil tethering to and rolling on E-selectin are separable by requirement for L-selectin. *Immunity.*, **1**:137–145.

Lenter, M., Levinovitz, A., Isenmann, S. and Vestweber, D. (1994). Monospecific and common glycoprotein ligands for E- and P-selectin on myeloid cells. *J. Cell Biol.*, **125**:471–481.

Levinovitz, A., Mühlhoff, J., Isenmann, S. and Vestweber, D. (1993). Identification of a glycoprotein ligand for E-selectin on mouse myeloid cells. *J. Cell Biol.*, **121**:449–459.

Lo, S.K., Lee, S., Ramos, R.A., Lobb, R., Rosa, M., Chi-Rosso, G. and Wright, S.D. (1991). Endothelial-leukocyte adhesion molecule 1 stimulates the adhesive activity of leukocyte integrin CR3 (CD11b/CD18, Mac-1 $\alpha_M\beta_2$) on human neutrophils. *J. Exp. Med.*, **173**:1493–1500.

Lowe, J.B., Stoolman, L.M., Nair, R.P., Larsen, R.D., Berhend, T.L. and Marks, R.M. (1990). ELAM-1 dependent cell adhesion to vascular endothelium determined by a transfected human fucosyltransferase cDNA. *Cell*, **63**:475–484.

Moore, K.L., Stults, N.L., Diaz, S., Smith, D.F., Cummings, R.C., Varki, A. and McEver, R.P. (1992). Identification of a specific glycoprotein ligand for P-selectin (CD62) on myeloid cells. *J. Cell Biol.*, **118**:445–456.

Moore, K.L., Eaton, S.F., Lyons, D.E., Lichenstein, H.S., Cummings, R.D. and McEver, R.P. (1994). The P-selectin glycoprotein ligand from human neutrophils displays sialylated, fucosylated, O-linked poly-N-acetyllactosamine. *J. Biol. Chem.*, **269**:23318–23327.

Moore, K.L., Patel, K.D., Bruehl, R.E., Fugang, L., Johnson. D.A., Lichenstein, H.S., Cummings, R.D., Bainton, D.F. and McEver, R.P. (1995). P-selectin glycoprotein ligand-1 mediates rolling of human neutrophils on P-selectin. *J. Cell Biol.*, **128**:661–671.

Needham, L.K. and Schnaar, R.L. (1993). The HNK-1 reactive sulfoglucuronyl glycolipids are ligands for L-selectin and P-selectin but not E-selectin. *Proc. Natl. Acad. Sci. USA*, **90**:1359–1363.

Norgard, K.E., Moore, K.L., Diaz, S., Stults, N.L., Ushiyama, S., McEver, R.P., Cummings, R.D. and Varki, A. (1993). Characterization of a specific ligand for P-selectin on myeloid cells. *J. Biol. Chem.*, **268**:12764–12774.

Patel, K.D., Moore, K.L., Nollet, M.U. and McEver, R.P. (1995). Neutrophils use both shared and distinct mechanisms to adhere to selectins under static and flow conditions. *J. Clin. Invest.*, **96**:1887–1896.

Phillips, M.L., Nudelman, E., Gaeta, F.C.A., Perez, M., Singhal, A.K., Hakomori, A.I. and Paulson, J.C. (1990). ELAM-1 mediates cell adhesion by recognition of a carbohydrate ligand, sialyl-Le^x. *Science*, **250**:1130–1132.

Picker, L.J., Warnock, R.A., Burns, A.R., Doerschuk, C.M., Berg, E.L. and Butcher, E.C. (1991). The neutrophil selectin LECAM-1 presents carbohydrate ligands to the vascular selectins ELAM-1 and GMP-140. *Cell*, **66**:921–933.

Polley, M.J., Phillips, M.L., Wayner, E., Nudelman, E., Singhal, A.K., Hakomori, S.I. and Paulson, J.C. (1991). CD62 and endothelial cell-leukocyte adhesion molecule 1 (ELAM-1) recognize the same carbohydrate ligand, sialyl-Lewis x. *Proc. Natl. Acad. Sci. USA*, **88**:6224–6228.

Pouyani, T. and Seed, B. (1995). PSGL-1 recognition of P-selectin is controlled by a tyrosine sulfation consensus at the PSGL-1 amino terminus. *Cell*, **83**:333–343.

Sako, D., Chang, X.J., Barone, K.M., Vachino, G., White, H.M., Shaw, G., Veldman, G.M., Bean, K.M., Ahern, T.J., Furie, B., Cummings, D.A. and Larsen, G.R. (1993). Expression cloning of a functional glycoprotein ligand for P-selectin. *Cell*, **75**:1179–1186.

Sako. D., Comess, K.M., Barone, K.M., Camphausen, R.T., Cumming, D.A. and Shaw, G.D. (1995). A sulfated peptide segment at the amino terminus of PSGL-1 is critical for P-selectin-binding. *Cell*, **83**:323–331.

Sammar, M., Aigner, S., Hubbe, M., Schirrmacher, V., Schachner, M., Vestweber, D. and Altevogt, P. (1994). Heat-stable antigen (CD24) as ligand for mouse P-selectin. *Intern. Immunol.*, **6**:1027–1036.

Sawada, R., Lowe, J.B. and Fukuda, M. (1993). E-selectin-dependent adhesion efficiency of colonic carcinoma cells is increased by genetic manipulation of their cell surface lysosomal membrane glycoprotein-1 expression levels. *J. Biol. Chem.*, **268**:12675–12681.

Spertini, O., Luscinskas, F.W., Kansas, V.X., Munro, J.M., Griffin, V., Gimbrone, jr. M.A. and Tedder, T.F. (1991). Leukocyte adhesion molecule-1 (LAM-1, L-selectin) interacts with an inducible endothelial cell ligand to support leukocyte adhesion. *J. Immunol.*, **147**:2565–2573.

Springer, T.A. (1990). Adhesion receptors of the immune system. *Nature*, **346**:425–34.

Steegmaier, M., Levinovitz, A., Isenmann, S., Borges, E., Lenter, M., Kocher, H., Kleuser, B. and Vestweber, D. (1995). The E-selectin-ligand ESL-1 is a variant of a receptor for fibroblast growth factor. *Nature*, **373**:615–620.

Steininger, C.N., Christopher, A.E., Leimgruber, R.M., Mellors, A. and Welply, J.K. (1992) The glycoprotease of *Pasteurella haemolytica* A1 eliminates binding of myeloid cells to P-selectin but not to E-selectin. *Biochem. Biophys. Res. Commun.*, **188**:760–766.

Stott, K., Blackburn, J.M., Butler, P.J.G. and Perutz, M. (1995). Incorporation of glutamine repeats makes protein oligomerize: implications of neuro-degenerative diseases. *Proc. Natl. Acad. Sci. USA*, **92**:6509–6513.

Tartaglia, L.A. and Goeddel, D.V. (1992). Two TNF receptors. *Immunol. Today.*, **13**:151–153.

Vachino, G., Chang, X.J., Veldman, G.M., Kumar, R., sako, D., Fouser, L.A., Berndt, M.C. and Cumming, D.A. (1995). P-selectin glycoprotein ligand-1 is the major counter-receptor for P-selectin on stimulated T cells and is widely distributed in non-functional form on many lymphocytic cells. *J. Biol. Chem.*, **270**:21966–21974.

Vandenabeele, P., Declercq, W., Beyaert, R. and Fiers, W. (1995). Two tumor necrosis factor receptors: structure and function. *Trends Cell Biol.*, **5**:392–399.

van der Merwe, P.A. and Barclay, A.N. (1994). Transient intercellular adhesion: the importance of weak protein-protein interactions. *TIBS*, **19**:354–358.

Varki, A. (1994). Selectin ligands. *Proc. Natl. Acad. Sci. USA*, **91**:7390–7397.

von Andrian, U.H., Hasslen, S.R., Nelson, R.D., Erlandsen, S.L. and Butcher, E.C. (1995). A central role for microvillous receptor presentation in leukocyte adhesion under flow. *Cell*, **82**:989–999.

Walcheck, B., Watts, G. and Jutila, M.J. (1993). Bovine γ/δ T cells bind E-selectin via a novel glycoprotein receptor: First characterization of a lymphocyte/E-selectin interaction in an animal model. *J. Exp. Med.*, **178**:853–863.

Walz, G., Aruffo, A., Kolanus, W., Bevilacqua, M. and Seed, B. (1990). Recognition by ELAM-1 of the sialyl-Le^x determinant on myeloid and tumor cells. *Science*, **250**:1132–1135.

Weyrich, A.S., McIntyre T.M., McEver, R.P., Prescott, S.M. and Zimmerman, G.A. (1995). monocyte tehering by P-selectin regulates Monocyte Chemotactic Protein-1 and Tumor Necrosis Factor-α secretion. *J. Clin. Invest.*, **95**:2297–2303.

Wilkins, P.P., Moore, K.L., McEver, R.P. and Cummings, R.D. (1995). Tyrosine sulfation of P-selectin glycoprotein ligand-1 is required for high affnity binding to P-selectin. *J. Biol. Chem.*, **270**:22677–22680.

Yednock, T.A., Butcher, E.C., Stoolman, L.M. and Rosen, S.D. (1987). Receptors involved in lymphocyte homing: Relationship between a carbohydrate-binding receptor and the MEL-14 antigen. *J. Cell Biol.*, **104**:725–731.

Yuen, C.T., Lowson, A.M., Chai, W., Larkin, M., Stoll, M.S., Stuart, A.C., Sullivan, F.X., Ahern, T.J. and Feizi, T. (1992). Novel sulfated ligands for the cell adhesion molecule E-selectin revealed by the neoglycolipid technology among O-linked oligosaccharides on an ovarian cystadenoma glycoprotein. *Biochem.*, **31**:9126–9131.

12 P-selectin and Signal Transduction

Bruce Furie and Barbara C. Furie

Center for Hemostasis and Thrombosis Research, Division of Hematology-Oncology, New England Medical Center and the Departments of Medicine and Biochemistry, Tufts University School of Medicine, Boston MA 02111

P-selectin is a cell adhesion molecule that mediates the interaction of platelets (Larsen *et al.*, 1989) and endothelial cells (Geng *et al.*, 1990) with monocytes and neutrophils. An integral membrane protein with a molecular weight of 140,000 (Hsu-Lin *et al.*, 1984), this protein is composed of a lectin domain, an epidermal growth factor domain, a series of consensus repeat domains, a transmembrane region and a short cytoplasmic tail (Johnston *et al.*, 1989). This protein resides in the membrane of the alpha granule (Stenberg *et al.*, 1985; Berman *et al.*, 1986) and of the Weibel-Palade body of the endothelial cell (Bonfanti *et al.*, 1989; McEver *et al.*, 1989). Upon cell activation and degranulation, the protein is rapidly translocated to the plasma membrane where it functions in cell adhesion. P-selectin binds to a counterreceptor, PSGL-1, expressed on leukocytes of the myeloid lineage (Sako *et al.*, 1993; Moore *et al.*, 1994). The interaction of P-selectin with leukocytes leads to capture of leukocytes on the vascular surface, as demonstrated by *in vitro* (Larsen *et al.*, 1989; Geng *et al.*, 1990; Lawrence and Springer, 1991) and *in vivo* experiments (Mayadas *et al.*, 1993).

The role of P-selectin in cell-cell interaction is now well established. However, whether P-selectin-mediated docking of platelets and endothelial cells to leukocytes, specifically neutrophils or monocytes, might initiate signal transduction within the leukocyte has only recently come under study. Upon our discovery that P-selectin mediates platelet-leukocyte interaction (Larsen *et al.*, 1989), we were intrigued by the earlier findings of Niemetz (1972) and of Niemetz and Marcus (1974) that the incubation of mononuclear cells, including monocytes, with platelets led to the expression of procoagulant activity that was subsequently shown to be due to tissue factor. Under certain conditions that lead to *de novo* synthesis of tissue factor, monocytes have been known to express tissue factor activity on their surface (Niemetz, 1972; Colucci *et al.*, 1983). Monocytes can be activated by endotoxin (Semeraro *et al.*, 1985), immune complexes (Rothberger *et al.*, 1977), cytokines (Conkling *et al.*, 1988) and platelets (Niemetz and Marcus, 1974; Osterud and Bjorklid, 1982; Lorenzet *et al.*, 1985), leading to tissue factor expression. Tissue factor, a membrane protein that is a receptor for Factor VII and Factor

VIIa, leads to the formation of the tissue factor/Factor VIIa complex that activates Factor IX and Factor X (Furie and Furie, 1988). Activation of blood coagulation culminates in the generation of thrombin and the formation of a fibrin clot.

Since activated platelets induce tissue factor on mononuclear leukocytes, presumably monocytes, we examined directly the effect of P-selectin on the expression of tissue factor activity in monocytes. These experiments are confounded by the fact that endotoxin is a potent inducer of tissue factor activity on monocytes. Thus, all reagents had to be prepared free of endotoxin. The measurement of tissue factor activity using a biologic clotting assay proved extremely sensitive. To confirm the findings of Niemetz and Marcus (1974), platelets were incubated with mononuclear cells for 16 hours, and the procoagulant activity measured. As shown in Figure 1, neither platelets alone nor mononuclear cells alone generated procoagulant activity. However, when platelets and mononuclear cells were incubated together, procoagulant activity was induced. This procoagulant activity was due to tissue factor in that an inhibitory anti-tissue factor monoclonal antibody, HTF1, blocked procoagulant activity generated by the cells. Similarly, blocking anti-P-selectin antibodies inhibited the generation of P-selectin.

Given the potential complexity of the analysis of platelet-monocyte interaction in this system, we were interested in defining whether P-selectin on platelets was capable, in the absence of other small or large molecules, of inducing the synthesis of tissue factor (Celi *et al.*, 1994). Purified P-selectin that was rendered endotoxin-poor stimulated tissue factor expression on mononuclear leukocytes in a dose-dependent manner. Chinese hamster

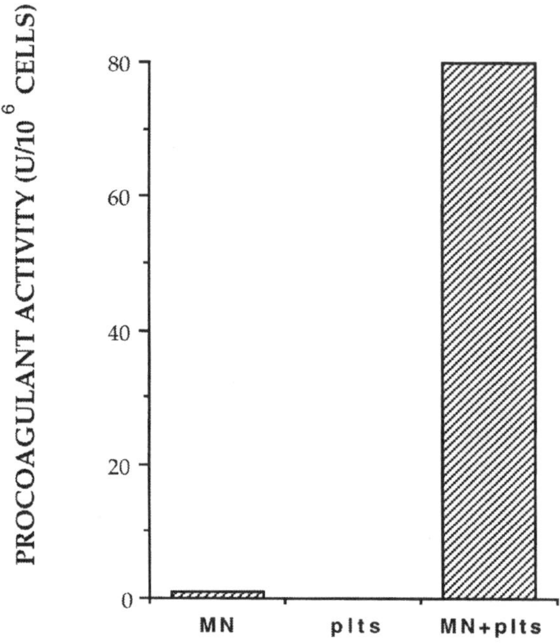

Figure 1 Interaction of platelets and monocytes leads to tissue factor activity. Monocytes (MN) alone, platelets alone (plts), or platelets with monocytes were incubated for 16 hours, and then assayed for procoagulant activity. The procoagulant activity is due to tissue factor.

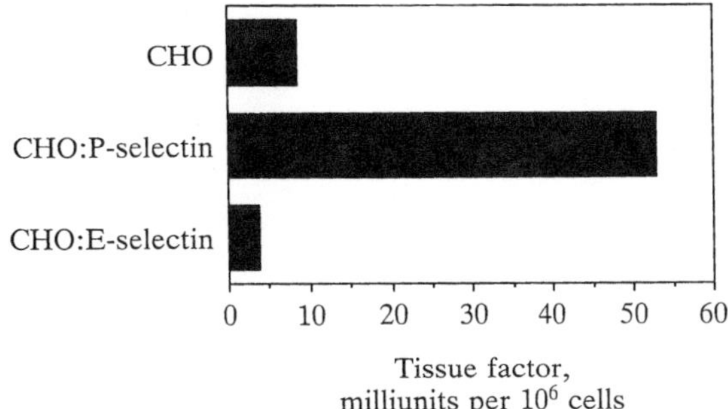

Figure 2 P-selectin expressed on CHO cells upregulates tissue factor activity on monocytes. CHO, CHO: P-selectin or CHO:E-selectin cells were cultured to confluency. Percoll gradient-purified monocytes were added and incubated for 6 hrs prior to assay for tissue factor activity. From Celi *et al.*, 1994; with permission.

ovary cells expressing P-selectin stimulated tissue factor procoagulant activity in purified monocytes whereas untransfected Chinese hamster ovary cells and CHO cells expressing E-selectin did not. The latter represents a critical control since CHO cells expressing E-selectin will bind to monocytes, but the E-selectin receptor is likely distinct from the P-selectin receptor under the conditions of these experiments (see below). Anti-P-selectin antibodies inhibited the effects of purified P-selectin and CHO:P-selectin on monocytes. Incubation of CHO:P-selectin with monocytes leads to the development of tissue factor mRNA in monocytes and to the expression of tissue factor antigen on the monocyte surface. These results indicate that P-selectin upregulates the expression of tissue factor on monocytes as well as mediating the binding of platelets and endothelial cells with monocytes and neutrophils. The binding of P-selectin to monocytes in the area of vascular injury may be a component of a mechanism that initiates thrombosis and fibrin generation for wound healing. However, this mechanism of tissue factor expression must be independent of those required for primary hemostasis.

Stimulated endothelial cells rapidly co-express P-selectin and platelet activating factor (PAF) upon exposure to agonists such as histamine or thrombin (Lorant *et al.*, 1991). It has been suggested that upon tethering of leukocytes to stimulated endothelial cells by P-selectin, the PAF released from the endothelial cell indirectly upregulates CD11/CD18 integrins on the leukocyte surface that can then further support endothelial cell-leukocyte binding. P-selectin alone does not ellicit this response, but rather requires the priming action of PAF (Lorant *et al.*, 1993). This priming action is inhibited by blocking of the PAF receptor on leukocytes. Zimmerman and colleagues concluded that cell adhesion and juxtacrine activation are both critical components of cell-cell interaction during inflammation. When monocytes bind to P-selectin in the presence of PAF, P-selectin is able to regulate the synthesis of monocyte chemotactic protein-1 and tumor necrosis factor-α (Weyrich *et al.*, 1995). However, P-selectin alone or PAF alone fail to induce a measureable effect. P-selectin binding to monocytes specifically enhances nuclear translocation of NF-κB, a transcription factor required for expression of a number of early genes involved in inflammation.

If P-selectin can upregulate the synthesis of tissue factor in monocytes in the absence of accessory molecules (Celi *et al.*, 1994) while P-selectin and PAF are required for upregulation of monocyte chemotactic protein-1 and tumor necrosis factor-α in monocytes (Weyrich *et al.*, 1995), does P-selectin induce two separate pathways or is the seeming disparity between these results due to experimental or analytical differences? Although the definitive studies have not been completed, it is most likely that the P-selectin is sufficient to stimulate low levels of expression of a number of active polypeptides, including tissue factor, monocyte chemotactic protein-1 and tumor necrosis factor-α; however, it is most likely that only tissue factor was detected because exquisitely sensitive assays for its detection were employed. Tissue factor expression by P-selectin is stimulated another 10–20 fold by 12-HETE and also by PAF (Lorenzet *et al.*, unpublished). Thus, it would appear that P-selectin is sufficient to induce certain signalling pathways in monocytes, and the magnitude of this signal is greatly enhanced by small bioactive molecules. There is likely to be a single pathway in which P-selectin binds to PSGL-1, the P-selectin receptor on monocytes; liganding of PSGL-1 leads to intracellular signalling events, including translocation of NF-κB and the upregulation of genes involved in the inflammatory response. These pathways remain to be elucidated.

Because the alpha granules of platelets undergo rapid exocytosis upon cell stimulation, with concommitant fusion of the granule membrane into the plasma membrane, we explored whether phosphorylation of P-selectin might accompany the biochemical and morphologic changes associated with platelet activation (Crovello *et al.*, 1993). Platelets undergo rapid granule secretion and upregulation of receptor function upon stimulation with specific extracellular agonists, including thrombin, ADP, epinephrine and collagen. Phosphorylation of many platelet proteins accompanies the activation of platelets and the initiation of signal transduction pathways. To determine whether phosphorylation of P-selectin accompanies platelet activation, we compared P-selectin isolated from resting and activated platelets labeled with [^{32}P]-o-phosphate (Crovello *et al.*, 1993). In the autoradiogram of the SDS gel of P-selectin immunoprecipitated from resting and activated platelets, the major band observed by immunoprecipitation of thrombin-activated platelets with AC1.2, a non-blocking anti-P-selectin antibody, corresponds to P-selectin. P-selectin isolated from thrombin-activated platelets was phosphorylated while P-selectin derived from unstimulated platelets was not phosphorylated (Figure 3). The increase in radioactivity in P-selectin upon thrombin activation ranged from 10- to 20-fold. A low level of ^{32}P in P-selectin from resting platelets was observed despite efforts to minimize platelet activation, but generally about 5% of resting platelets are activated in the course of platelet preparation. The kinetics of phosphorylation of P-selectin were studied over the initial 10 minutes following thrombin activation. The highest levels of P-selectin phosphorylation were observed at about 30 seconds after activation with thrombin, but phosphorylation could be observed as early as 5 seconds after activation and peak incorporation was observed between 15 and 30 seconds. Subsequently, the degree of total P-selectin phosphorylation decreased with a half-life of about 60 seconds. These results are consistent with a complex phenomenon of P-selectin phosphorylation and dephosphorylation following cell activation.

Phosphoamino acid analysis of phosphorylated P-selectin revealed the identity of the phosphoamino acids. Following acid hydrolysis of phosphorylated P-selectin, phosphoserine, phosphothreonine and phosphotyrosine were identified (Crovello *et al.*, 1993). In a similar series of experiments, Fujimoto and McEver (1993) only identified phosphoserine and phosphothreonine; no phosphotyrosine was reported. However, the presence of phos-

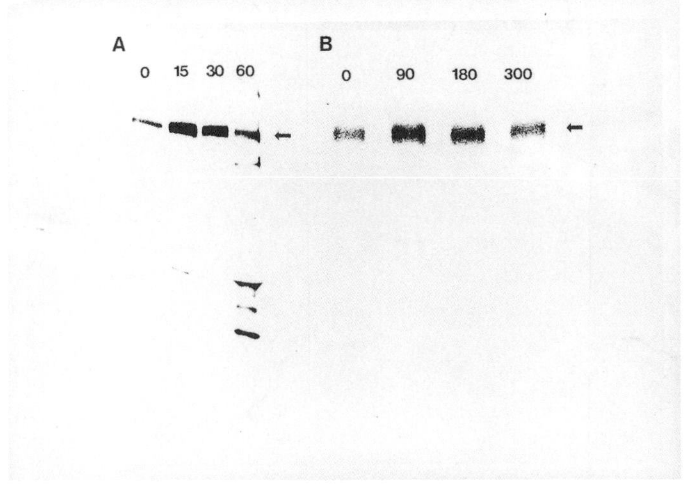

Figure 3 Phosphorylation of P-selectin in platelets during thrombin or collagen stimulation. P-selectin was immunoprecipitated and subjected to SDS gel electrophoresis. ^{32}P in the gel was visualized with a Phosphorlmager; the position of P-selectin migration (MW 140,000) is indicated by the arrow. The time (seconds) between cell activation and termination of activation is indicated above each lane. (A) Thrombin-stimulated platelets. (B) Collagen-stimulated platelets. From Crovello *et al.*, 1995; with permission.

photyrosine was confirmed independently by Modderman *et al.* (1994). When phosphorylated P-selectin was subjected to alkaline hydrolysis, both phosphohistidine and phosphotyrosine were identified (Crovello *et al.*, 1995). The tryptic digest of phosphorylated P-selectin and chromatographic separation of the tryptic peptides allowed for purification of the C-terminal tryptic peptide, representing most of the cytoplasmic tail of the protein. This peptide was the only peptide that was phosphorylated upon platelet activation. Pronase digestion and two-dimensional chromatography of the hydrolysate of this peptide allowed definitive identification of phosphohistidine, phosphotyrosine, phosphoserine, and phosphothreonine (Crovello *et al.*, 1995) (Figure 4). Edman degration of the tryptic phosphopeptide confirmed the presence and location of two phosphohistidine residues.

Thus, phosphorylation of P-selectin occurs on serine, threonine, tyrosine and histidine residues upon platelet activation. Given the complexity of these kinetics, we asked whether the kinetics of phosphorylation-dephosphorylation were different for each phosphoamino acid. By performing phosphoamino acid analysis on immunoprecipitated P-selectin as a function of time following thrombin activation of platelets, we observed changes in the apparent concentrations of phosphoserine, phosphothreonine, phosphotyrosine and phosphohistidine. Upon thrombin stimulation, maximal phosphoserine concentrations were reached at 30 seconds and remained elevated. However, phosphothreonine, phosphotyrosine and phosphohistidine reached maximal concentration at 15–30 seconds and then decreased over the next 5 minutes. These results suggest the rapid phosphorylation of histidine, serine, threonine and tyrosine residues on P-selectin and the rapid dephosphorylation of histidine, threonine and tyrosine.

Phosphorylation of P-selectin occurred with a number of known platelet agonists. Immunoprecipitation of P-selectin from platelets stimulated with ADP, epinephrine, thrombin, collagen or the synthetic peptide Ser-Phe-Leu-Leu-Arg based upon the

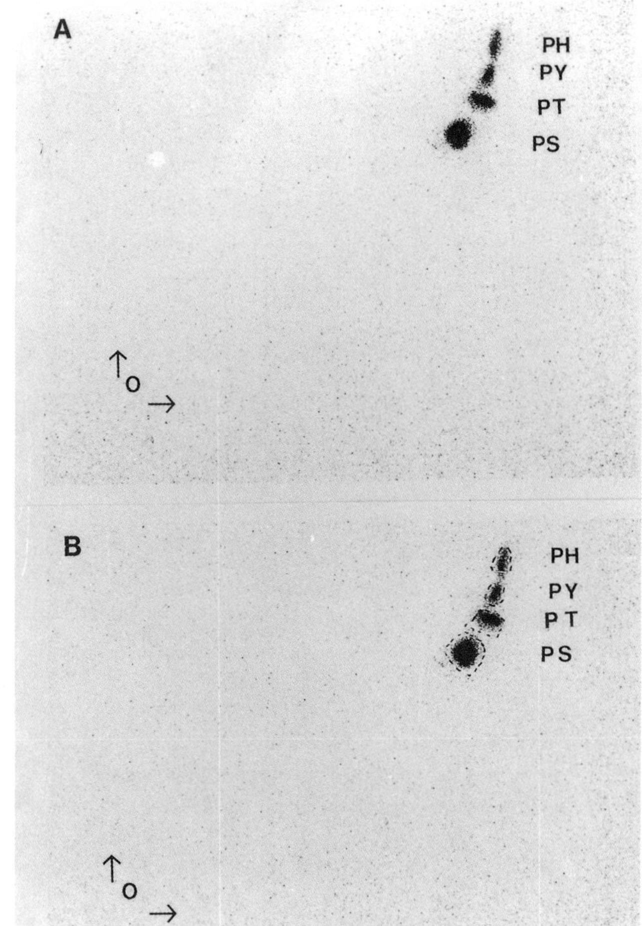

Figure 4 Phosphoamino acid analysis of the P-selectin tryptic phosphopeptide from activated platelets following enzymatic proteolysis. Two dimensional thin layer chromatography of the Pronase digest of the purified phosphopeptide was performed using an ethanol/ammonium hydroxide solvent system in the first dimension and a methanol/ammonium hydroxide solvent system in the second dimension. Radioactive phosphoamino acids (PH, PS, PY, and PT) were identified by autoradiography (upper panel). The same image is shown (lower panel) but the locations of the positions of the internal standards (phosphoserine, PS; phosphothreonine, PT; phosphotyrosine, PY; phosphohistidine, PH) identified by ninhydrin staining are overlayed (---). From Crovello *et al.*, 1995; with permission.

structure of the thrombin receptor led to phosphorylation of P-selectin during platelet activation.

Transient phosphorylation of histidine characterizes the two-component systems in bacteria that control important physiologic functions, but analogous events have not been implicated in signal transduction in mammalian cells. Although serine, threonine and tyrosine are constitutively phosphorylated on some platelet proteins, platelet activation leads to marked increase in protein phosphorylation. This increased phosphorylation appears to be closely linked to the occupancy of certain receptors inasmuch as binding of glycoprotein IIb–IIIa to fibrinogen requires prior tyrosine phosphorylation (Shattil, 1993). Until the discovery of phosphohistidine in P-selectin (Crovello *et al.*, 1995), phosphohistidine had not been observed in platelets. In this work, Crovello *et al.* (1995) demonstrated that P-selectin phosphorylation during human platelet activation includes the transient generation of phosphohistidine on the C-terminal cytoplasmic tail. Histidine phosphorylation has not been previously implicated in mammalian signal transduction pathways. However, phosphohistidine is a known component of eukaryotic proteins and the enzymatic machinery for the synthesis of phosphohistidine and for its dephosphorylation is present in lower eukaryotic and mammalian cells.

The importance of phosphohistidine as an intermediate in the two-component bacterial system and the presence of some genes encoding proteins involved in this system in early eukaryotes suggest a role for phosphohistidine in mammalian cell signalling (Swanson *et al.*, 1994). Histidine phosphorylation is involved in signalling, such as chemotaxis, porin expression and nitrogen metabolism (Stock *et al.*, 1990). Two-component systems exist in lower eukaryotic signaling pathways since analysis of *S. cerevisae* has implicated a transmembrane histidine kinase in regulation of an osmosensing MAP kinase cascade (Maeda *et al.*, 1994). Mammalian cells may, however, express independent histidine kinases and protein phosphatases that are highly regulated and responsible for transient phosphohistidine appearance.

In summary, the discovery of adhesion molecules that direct contact between specific cells has now led to the study of the signalling events that this cell contact triggers. Degranulation of platelets and endothelial cells is accompanied by the phosphorylation of P-selectin. The physiologic role of this event is not currently known. P-selectin on the cell surface interacts with a counterreceptor on leukocytes, and stimulates the *de novo* synthesis of biologic activities important for inflammation and wound healing. The details of this signalling pathway remain to be elucidated.

Acknowledgments

This work was supported by a grant (HL51926) from the National Institutes of Health. We are indebted to our laboratory colleagues who have contributed to the work described. We are especially grateful to Drs. Alessandro Celi, Coleen Crovello, and Roberto Lorenzet.

References

Berman, C.L., Yeo, E.L., Wencel-Drake, J., Furie, B.C., Ginsberg, M.L. and Furie, B. (1986). A platelet alpha granule membrane protein that is associated with the plasma membrane after activation: characterization and subcellular localization of platelet activation-dependent granule-external membrane protein. *J. Clin. Invest.*, **78**:130–137.

Bonfanti, R., Furie, B.C., Furie, B. and Wagner, D.D. (1989). PADGEM is a component of Weibel-Palade bodies in endothelial cells. *Blood*, **73**:1109–1112.

Celi, A., Pellegrini, G., Lorenzet, R., De Blasi, A., Ready, N., Furie, B.C. and Furie, B. (1994). P-selectin induces the expression of tissue factor on monocytes. *Proc. Natl. Acad. Sci. USA*, **91**:8767–8771.

Colucci, M., Balconi, G., Lorenzet, R., Pietra, A., Locati, D., Donati, M.B. and Semeraro, N. (1983). Cultured human endothelial cells generate tissue factor in response to endotoxin. *J. Clin. Invest.*, **71**:1893–1896.

Conkling, P.R., Greenberg, C.S. and Weinberg, J.B. (1988). Tumor necrosis factor induces tissue factor-like activity in human leukemia cell line U937 and peripheral blood monocytes. *Blood*, **72**:128–133.

Crovello, C.S., Furie, B.C and Furie, B. (1993). Rapid phosphorylation and selective dephosphorylation of P-selectin accompanies platelet activation. *J. Biol. Chem.*, **268**:14590–14593.

Crovello, C.S., Furie, B.C. and Furie, B. (1995). Histidine phosphorylation of P-selectin accompanies ligand-induced stimulation of human platelets: a novel pathway for cell activation-dependent signal transduction. *Cell*, **82**:279–286.

Fujimoto, T. and McEver, R.P. (1993a). The cytoplasmic domain of P-selectin is phosphorylated on serine and threonine residues. *Blood*, **82**:1758–1766.

Furie, B. and Furie, B.C. (1988). Molecular basis of blood coagulation. *Cell*, **53**:505–518.

Geng, J.G., Bevilacqua, M.P., Moore, K.L., McIntyre, T.M., Prescott, S.M., Kim, J.M., Bliss, G.A., Zimmerman, G.A. and McEver, R.P. (1990). Rapid neutrophil adhesion to activated endothelium mediated by GMP-140. *Nature*, **343**:757–760.

Hsu-Lin, S.C., Berman, C.L., Furie, B.C., August, D. and Furie, B. (1984). A platelet membrane protein expressed during platelet activation. *J. Biol. Chem.*, **259**:9121–9126.

Johnston, G.I., Cook, R.G. and McEver, R.P. (1989). Cloning of GMP-140, a granule membrane protein of platelets and endothelium: Sequence similarity to proteins involved in cell adhesion and inflammation. *Cell*, **56**:1033–1044.

Larsen, E., Celi, A, Gilbert, G.E., Furie, B.C., Erban, J.K., Bonfanti, R., Wagner, D.D. and Furie, B. (1989). PADGEM Protein: A receptor that mediates the interaction of activated platelets with neutrophils and monocytes. *Cell*, **59**:305–312.

Larsen, E., Palabrica, T., Sajer, S., Gilbert, G.E., Wagner, D.D., Furie, B.C. and Furie, B. (1990). PADGEM-dependent adhesion of platelets to monocytes and neutrophils is mediated by a lineage-specific carbohydrate, LNF III (CD15). *Cell*, **63**:467–474.

Lawrence, M.B. and Springer, T.A. (1991). Leukocytes roll on a selectin at physiologic flow rates: distinction from and prerequisite for adhesion through integrins. *Cell*, **65**:859–873.

Lorant, D.E., Topham, M.K., Whatley, R.E., McEver, R.P., McIntyre, T.M., Prescott, S.M. and Zimmerman, G.A. (1993). Inflammatory roles of P-selectin. *J. Clin. Invest.*, **92**:559–570.

Lorant, D.E., Patel, K.D., McIntyre, T.M., McEver, R.P., Prescott, S.M. and Zimmerman, G.A. (1991). Coexpression of GMP-140 and PAF by endothelium stimulated by histamine or thrombin: A juxtacrine system for adhesion and activation of neutrophils. *J. Cell Biol.*, **115**:223–234.

Lorenzet, R., Niemetz, J., Marcus, A.J. and Broekman, M.J. (1985). Enhancement of mononuclear procoagulant activity by platelet 12-hydroxyeicosatetranoic acid. *J. Clin. Invest.*, **78**:418–423.

Maeda, T., Wurgler-Murphy, S.M. and Saito, H. (1994). A two component system that regulates an osmosensing MAP kinase cascade in yeast. *Nature*, **369**:242–245.

McEver, R.P., Beackstad, J.H., Moore, K.L., Marshall-Carlson, L. and Bainton, D.F. (1989). GMP-140, a platelet alpha-granue membrane protein, is also synthesized in vascular endothelial cells and is localized in Weibel-Palade bodies. *J. Clin. Invest.*, **84**:92–.

Modderman, P.W., von dem Borne, A.E.G.K. and Sonnenberg, A. (1994). Tyrosine phosphorylation of P-selectin in intact platelets and in a disulphide linked complex with immunoprecipitated pp60 c-src. *Biochem. J.*, **299**:613–621.

Moore, K.L., Eaton, S.F., Lyons, D.E., Lichenstein, H.S., Cummings, R.D. and McEver, R.P. (1994). The P-selectin glycoprotein ligand from human neutrophils displays sialylated fucosylated, O-linked poly-N-acetyllactosamine. *J. Biol. Chem.*, **269**:23318–23327.

Mayadas, T.N., Rayburn, H., Johnson, R., Hynes, R.O. and Wagner, D.D. (1993). Leukocyte rolling and extravasation are severly compromised in P-selectin-deficient mice. *Cell*, **74**:541–554.

Niemetz, J. (1972). Coagulant activity of leukocytes. Tissue factor activity. *J. Clin. Invest.*, **51**:307–313.

Niemetz, J. and Marcus, A.J. (1974). The stimulatory effect of platelets and platelet membranes on the procoagulant activity of leukocytes. *J. Clin. Invest.*, **54**:1437–1443.

Osterud, B. and Bjorklid, E. (1982). The production and availability of tissue thromboplastin in cellular populations of whole blood exposed to various concentrations of endotoxins. *Scand. J. Hematol.*, **29**:175–187.

Rothberger, H., Zimmermann, T.S., Spielberg, H.L. and Vaughan, J.H. (1977). Leukocyte procoagulant activity: enhancement of production *in vitro* by IgG and antigen-antibody complexes. *J. Clin. Invest.*, **59**:549–.

Sako, D., Chang, X-J., Barone, K.M., Vachino, G., White, H.M., Shaw, G., Veldman, T., Bean, K.M., Ahern, T.J., Furie, B., Cumming, D.A. and Larsen, G.R. (1993). Expression cloning of a functional glycoprotein ligand for P-selectin. *Cell*, **75**:1179–1186.

Semeraro, N., Biondi, A., Lorenzet, R., Locati, D., Mantovani, A. and Donati, M.B. (1983). Direct induction of tissue factor synthesis by endotoxin in human macrophages from diverse anatomical sites. *Immunol.*, **50**:529–535.

Shattil, S.J. (1993). Regulation of platelet anchorage and signaling by integrin αIIb βIII. *Thromb. Haemostas.*, **70**:224–228.

Stenberg, P.E., McEver, R.P., Shuman, M.A., Jacques, Y.V. and Bainton, D.F. (1985). A platelet alpha-granule membrane protein (GMP140) is expressed on the plasma membrane after activation. *J. Cell Biol.*, **101**:880–886.

Stock, J.B., Stock, A.M. and Mottonen, J.M. (1990). Signal transduction in bacteria. *Nature*, **344**:395–400.

Swanson, R.V., Alex, S.A. and Simon M.I. (1994). Histidine and aspartate phosphorylation: two-component systems and the limits of homology. *Trends Biochem. Sci.*, **19**:485–490.

Weyrich, W.S., McIntyre, T.M., McEver, R.P., Prescott, S.M. and Zimmerman, G.A. (1995). Monocyte tethering by P-selectin regulates monocyte chemotactic protein-a and tumor necrosis factor-α secretion. *J. Clin. Invest.*, **95**:2297–2303.

INDEX